METHODS IN MOLECULAR BIOLOGY™

Series Editor
John M. Walker
School of Life Sciences
University of Hertfordshire
Hatfield, Hertfordshire, AL10 9AB, UK

For further volumes:
http://www.springer.com/series/7651

Difference Gel Electrophoresis (DIGE)

Methods and Protocols

Edited by

Rainer Cramer

Department of Chemistry, University of Reading, Reading, UK

Reiner Westermeier

SERVA Electrophoresis GmbH, Heidelberg, Germany

💥 Humana Press

Editors
Rainer Cramer
Department of Chemistry
University of Reading
Reading, UK

Reiner Westermeier
SERVA Electrophoresis GmbH
Heidelberg, Germany

ISSN 1064-3745   e-ISSN 1940-6029
ISBN 978-1-61779-572-5   e-ISBN 978-1-61779-573-2
DOI 10.1007/978-1-61779-573-2
Springer New York Dordrecht Heidelberg London

Library of Congress Control Number: 2011945848

© Springer Science+Business Media, LLC 2012
All rights reserved. This work may not be translated or copied in whole or in part without the written permission of the publisher (Humana Press, c/o Springer Science+Business Media, LLC, 233 Spring Street, New York, NY 10013, USA), except for brief excerpts in connection with reviews or scholarly analysis. Use in connection with any form of information storage and retrieval, electronic adaptation, computer software, or by similar or dissimilar methodology now known or hereafter developed is forbidden.
The use in this publication of trade names, trademarks, service marks, and similar terms, even if they are not identified as such, is not to be taken as an expression of opinion as to whether or not they are subject to proprietary rights.

Printed on acid-free paper

Humana Press is part of Springer Science+Business Media (www.springer.com)

Preface

Protein analysis is increasingly becoming a cornerstone in deciphering the molecular mechanisms of life. Proteomics, the large-scale and high-sensitivity analysis of proteins, is already pivotal to the new life sciences such as Systems Biology and Systems Medicine. Proteomics, however, relies heavily on the past and future advances of protein purification and analysis methods. In-depth protein and protein interaction analysis would be impossible without these advances. It is this progress being made available through techniques such as DIGE that is enormously important for pushing the boundaries in the analysis of the cellular machinery and larger biological systems even further.

Apart from mere "stamp collections", most analytical measurements have to be somewhat quantitative. This is particularly true for the analysis of proteins and protein contents in order to find out at what level these small molecular engines are involved in the fundamental processes of life. Thus, the proteomic research community has become increasingly aware of the fact that most proteomic analyses have to fulfill strict requirements with regard to their quantitative aspects. Protein or peptide identification and characterization without quantification might just be sufficient in areas such as Proteogenomics where this information can be used to curate genomic data or delineate function. However, once a protein is characterized and known to be present, it is essential to obtain quantitative information. Only this quantitative data obtained from different time points and conditions will enable a comprehensive understanding of the dynamics of protein contents in biological systems and, thus, the elucidation of how these systems function and react.

Consequently, many quantitative proteomic analysis methods were devised shortly after Proteomics was born. Most of these are based on quantification using the mass spectral readout from methods of metabolic isotope labeling (e.g., SILAC), chemical isotope labeling (e.g., iTRAQ or TMT), and increasingly from simple non-labeling methods such as MS ion signal comparison in so-called label-free approaches. However, virtually all MS-based quantitative proteomic analyses favor the analysis at the peptide level which inherently excludes quantification at the protein level, i.e., of protein isoforms. Thus, one often-cited strength of proteomics, the analysis of posttranscriptional and -translational changes, is fundamentally hampered through the inadequacy of these methods in protein isoform quantification. DIGE, being able to quantify proteins in their intact form, is one of a few methods that can facilitate this type of analysis and still provides the protein isoforms in an MS-compatible state for further identification and characterization with high analytical sensitivity.

This volume introduces the concept of DIGE and its advantages in quantitative measurements with the specific focus on proteomic analyses. It provides detailed protocols and important notes on the practical aspects of DIGE with both generic and specific applications in the various areas of quantitative protein analysis. As such this volume can be used by novices with some background in biochemistry or molecular biology, who want to widen their portfolio of quantitative techniques in protein analysis, as well as by experts in proteomics, who would like to deepen their understanding of DIGE and its employment in many hyphenations and application areas. With its many protocols, applications, and methodological variants, it is also a unique reference for all who seek fundamental details on the working principle of DIGE and ideas for a possible future use of DIGE in novel analytical approaches.

The chapters in this volume have been divided into four categories. Starting with the basics of DIGE the reader acquires a sound background in the technique and its practical details with a focus on the planning of a DIGE experiment and its data analysis. The next chapters introduce various DIGE methods. While most of these have been employed by scientists world-wide, some are more novel and provide a glance at what is at the horizon in the DIGE world. The final sets of chapters provide a good overview of the wide range of DIGE applications from Clinical Proteomics to Animal, Plant and Microbial Proteomics applications.

Reading, UK *Rainer Cramer*
Heidelberg, Germany *Reiner Westermeier*

Contents

Preface.. v
Contributors... xi

PART I FUNDAMENTALS

1 DIGE: Past and Future... 3
   Jonathan S. Minden

2 The Basics of 2D DIGE... 9
   Phil Beckett

3 Multifluorescence 2D Gel Imaging and Image Analysis................. 21
   Ingo Vormbrock, Sonja Hartwig, and Stefan Lehr

4 Assessing Signal-to-Noise in Quantitative Proteomics:
   Multivariate Statistical Analysis in DIGE Experiments............... 31
   David B. Friedman

5 Analysis of Proteins Using DIGE and MALDI Mass Spectrometry......... 47
   *Witold M. Winnik, Robert M. DeKroon, Joseph S.Y. Jeong,
   Mihaela Mocanu, Jennifer B. Robinette, Cristina Osorio,
   Nedyalka N. Dicheva, Eric Hamlett, and Oscar Alzate*

6 Synthesis and Validation of Cyanine-Based Dyes for DIGE............. 67
   *Michael E. Jung, Wan-Joong Kim, Nuraly K. Avliyakulov,
   Merve Oztug, and Michael J. Haykinson*

PART II METHODS

7 2D DIGE Saturation Labeling for Minute Sample Amounts............... 89
   Georg J. Arnold and Thomas Fröhlich

8 Proteomic Analysis of Redox-Dependent Changes Using
   Cysteine-Labeling 2D DIGE... 113
   Hong-Lin Chan, John Sinclair, and John F. Timms

9 Analysis of Protein Posttranslational Modifications Using
   DIGE-Based Proteomics... 129
   *Robert M. DeKroon, Jennifer B. Robinette, Cristina Osorio,
   Joseph S.Y. Jeong, Eric Hamlett, Mihaela Mocanu,
   and Oscar Alzate*

10 Comparative Analyses of Protein Complexes by Blue Native DIGE....... 145
   Katrin Peters and Hans-Peter Braun

11 2D DIGE Analysis of Protein Extracts from Muscle Tissue............. 155
   Cecilia Gelfi and Sara De Palma

12 Combination of Highly Efficient Hexapeptide Ligand Library-Based Sample Preparation with 2D DIGE for the Analysis of the Hidden Human Serum/Plasma Proteome 169
Sonja Hartwig and Stefan Lehr

13 2D DIGE Analysis of Serum After Fractionation by ProteoMiner™ Beads 181
Cynthia Liang, Gek San Tan, and Maxey C.M. Chung

14 Study Design in DIGE-Based Biomarker Discovery 195
Alexandra Graf and Rudolf Oehler

15 Comparative 2D DIGE Analysis of the Depleted Serum Proteome for Biomarker Discovery 207
Megan Penno, Matthias Ernst, and Peter Hoffmann

PART III APPLICATIONS IN CLINICAL PROTEOMICS

16 Differential Gel-Based Proteomic Approach for Cancer Biomarker Discovery Using Human Plasma 223
Keun Na, Min-Jung Lee, Hye-Jin Jeong, Hoguen Kim, and Young-Ki Paik

17 2D DIGE for the Analysis of RAMOS Cells Subproteomes 239
Marisol Fernández and Juan Pablo Albar

18 Application of Saturation Labeling in Lung Cancer Proteomics 253
Gereon Poschmann, Barbara Sitek, Bence Sipos, and Kai Stühler

19 Proteomic Profiling of the Epithelial-Mesenchymal Transition Using 2D DIGE 269
Rommel A. Mathias, Hong Ji, and Richard J. Simpson

20 Method for Protein Subfractionation of Cardiovascular Tissues Before DIGE Analysis 287
Athanasios Didangelos, Xiaoke Yin, and Manuel Mayr

21 Application of DIGE and Mass Spectrometry in the Study of Type 2 Diabetes Mellitus Mouse Models 299
Celia Smith, Davinia Mills, and Rainer Cramer

22 Evaluating the Efficacy of Subcellular Fractionation of Blast Cells Using Live Cell Labeling and 2D DIGE 319
Yin Ying Ho, Megan Penno, Michelle Perugini, Ian Lewis, and Peter Hoffmann

PART IV APPLICATIONS IN ANIMAL, PLANT, AND MICROBIAL PROTEOMICS

23 DIGE Analysis of Plant Tissue Proteomes Using a Phenolic Protein Extraction Method 335
Christina Rode, Traud Winkelmann, Hans-Peter Braun, and Frank Colditz

24 Native DIGE of Fluorescent Plant Protein Complexes 343
Veronika Reisinger and Lutz Andreas Eichacker

25 An Overview of 2D DIGE Analysis of Marine (Environmental) Bacteria 355
Ralf Rabus

26 Application of 2D DIGE in Animal Proteomics 373
Ingrid Miller

Index *397*

Contributors

JUAN PABLO ALBAR • *Laboratorio de Proteómica, Centro Nacional de Biotecnología, Consejo Superior de Investigaciones Científicas (CSIC), Madrid, Spain*
OSCAR ALZATE • *Department of Cell and Developmental Biology, University of North Carolina, Chapel Hill, NC, USA*
GEORG J. ARNOLD • *Laboratory for Functional Genome Analysis LAFUGA, Gene Center, Ludwig-Maximilians-University, Munich, Germany*
NURALY K. AVLIYAKULOV • *Department of Biological Chemistry, David Geffen School of Medicine at UCLA, Los Angeles, CA, USA*
PHIL BECKETT • *GE Healthcare Bio-Sciences, Piscataway, NJ, USA*
HANS-PETER BRAUN • *Institute for Plant Genetics, Leibniz Universität Hannover, Hannover, Germany*
HONG-LIN CHAN • *Institute of Bioinformatics and Structural Biology, National Tsing Hua University, Hsinchu, Taiwan*
MAXEY C.M. CHUNG • *Department of Biochemistry and Biological Sciences, Yong Loo Lin School of Medicine and Faculty of Science, National University of Singapore, Singapore*
FRANK COLDITZ • *Institute for Plant Genetics, Leibniz Universität Hannover, Hannover, Germany*
RAINER CRAMER • *Department of Chemistry, University of Reading, Reading, UK*
ROBERT M. DEKROON • *Department of Cell and Developmental Biology, University of North Carolina, Chapel Hill, NC, USA*
SARA DE PALMA • *Dipartimento di Scienze e Tecnologie Biomediche, Università degli Studi di Milano, Segrate, Italy*
NEDYALKA N. DICHEVA • *Program of Molecular Biology and Biotechnology, School of Medicine, University of North Carolina, Chapel Hill, NC, USA*
ATHANASIOS DIDANGELOS • *Cardiovascular Division, King's College London, London, UK*
LUTZ ANDREAS EICHACKER • *Center of Organelle Research (CORE), University of Stavanger, Stavanger, Norway*
MATTHIAS ERNST • *Ludwig Centre for Cancer Research, Parkville, VIC, Australia*
MARISOL FERNÁNDEZ • *Laboratorio de Proteómica, Centro Nacional de Biotecnología, Consejo Superior de Investigaciones Científicas (CSIC), Madrid, Spain*
DAVID B. FRIEDMAN • *Proteomics Laboratory, Mass Spectrometry Research Center, Vanderbilt University School of Medicine, Nashville, TN, USA*
THOMAS FRÖHLICH • *Laboratory for Functional Genome Analysis LAFUGA, Gene Center, Ludwig-Maximilians-University, Munich, Germany*
CECILIA GELFI • *Dipartimento di Scienze e Tecnologie Biomediche, Università degli Studi di Milano, Segrate, Italy*
ALEXANDRA GRAF • *Section of Medical Statistics, Center for Medical Statistics, Informatics and Intelligent Systems, Medical University of Vienna, Vienna, Austria*
ERIC HAMLETT • *Department of Cell and Developmental Biology, University of North Carolina, Chapel Hill, NC, USA*

SONJA HARTWIG • *Institute of Clinical Biochemistry and Pathobiochemistry, German Diabetes Center, Düsseldorf, Germany*
MICHAEL J. HAYKINSON • *Department of Biological Chemistry, David Geffen School of Medicine at UCLA, Los Angeles, CA, USA*
YIN YING HO • *Adelaide Proteomics Centre, University of Adelaide, Adelaide, Australia*
PETER HOFFMANN • *Adelaide Proteomics Centre, University of Adelaide, Adelaide, Australia*
HYE-JIN JEONG • *Yonsei Proteome Research Center, Yonsei University, Seoul, South Korea*
JOSEPH S.Y. JEONG • *Department of Cell and Developmental Biology, University of North Carolina, Chapel Hill, NC, USA*
HONG JI • *Ludwig Institute for Cancer Research, Parkville, VIC, Australia*
MICHAEL E. JUNG • *Department of Chemistry and Biochemistry, University of California Los Angeles, Los Angeles, CA, USA*
HOGUEN KIM • *Department of Pathology, Yonsei University College of Medicine, Seoul, South Korea*
WAN-JOONG KIM • *Department of Chemistry and Biochemistry, University of California Los Angeles, Los Angeles, CA, USA*
MIN-JUNG LEE • *Integrated OMICS for Biomedical Science, Yonsei University, Seoul, South Korea*
STEFAN LEHR • *Institute of Clinical Biochemistry and Pathobiochemistry, German Diabetes Center, Düsseldorf, Germany*
IAN LEWIS • *Division of Haematology, SA Pathology, Adelaide, Australia*
CYNTHIA LIANG • *Department of Biological Science, Faculty of Science, National University of Singapore, Singapore*
ROMMEL A. MATHIAS • *Ludwig Institute for Cancer Research, Parkville, VIC, Australia*
MANUEL MAYR • *Cardiovascular Division, King's College London, London, UK*
INGRID MILLER • *Department for Biomedical Sciences, University of Veterinary Medicine Vienna, Vienna, Austria*
DAVINIA MILLS • *The BioCentre, University of Reading, Reading, UK*
JONATHAN S. MINDEN • *Department of Biological Sciences, Carnegie Mellon University, Pittsburgh, PA, USA*
MIHAELA MOCANU • *UNC Systems-Proteomics Center, Program of Molecular Biology and Biotechnology, School of Medicine, University of North Carolina, Chapel Hill, NC, USA*
KEUN NA • *Graduate Program in Functional Genomics, Yonsei University, Seoul, South Korea*
RUDOLF OEHLER • *Department of Surgery, Medical University of Vienna, Vienna, Austria*
CRISTINA OSORIO • *UNC Systems-Proteomics Center, Program of Molecular Biology and Biotechnology, School of Medicine, University of North Carolina, Chapel Hill, NC, USA*
MERVE OZTUG • *Department of Biological Chemistry, David Geffen School of Medicine at UCLA, Los Angeles, CA, USA*
YOUNG-KI PAIK • *Department of Biochemistry, Yonsei University, Seoul, South Korea*

MEGAN PENNO • *Adelaide Proteomics Centre, University of Adelaide, Adelaide, Australia*
MICHELLE PERUGINI • *Division of Haematology, SA Pathology, Adelaide, Australia*
KATRIN PETERS • *Institute for Plant Genetics, Leibniz Universität Hannover, Hannover, Germany*
GEREON POSCHMANN • *Molecular Proteomics Laboratory, Heinrich-Heine-Universität, Düsseldorf, Germany*
RALF RABUS • *Institute of Biology and Chemistry of the Marine Environment (ICBM), University Oldenburg, Oldenburg, Germany; Max Planck Institute for Marine Microbiology, Bremen, Germany*
VERONIKA REISINGER • *Center of Organelle Research (CORE), University of Stavanger, Stavanger, Norway*
JENNIFER B. ROBINETTE • *UNC Systems-Proteomics Center, Program of Molecular Biology and Biotechnology, School of Medicine, University of North Carolina, Chapel Hill, NC, USA*
CHRISTINA RODE • *Institute for Plant Genetics, Leibniz Universität Hannover, Hannover, Germany*
RICHARD J. SIMPSON • *Ludwig Institute for Cancer Research, Parkville, VIC, Australia*
JOHN SINCLAIR • *Cancer Proteomics Laboratory, EGA Institute for Women's Health, University College London, London, UK*
BENCE SIPOS • *Institute of Pathology, Eberhard-Karls-Universität Tübingen, Tübingen, Germany*
BARBARA SITEK • *Medizinisches Proteom-Center, Ruhr-Universität Bochum, Bochum, Germany*
CELIA SMITH • *Department of Chemistry, University of Reading, Reading, UK*
KAI STÜHLER • *Molecular Proteomics Laboratory, Heinrich-Heine-Universität, Düsseldorf, Germany*
GEK SAN TAN • *Department of Biological Science, Faculty of Science, National University of Singapore, Singapore*
JOHN F. TIMMS • *Cancer Proteomics Laboratory, EGA Institute for Women's Health, University College London, London, UK*
INGO VORMBROCK • *Institute of Clinical Biochemistry and Pathobiochemistry, German Diabetes Center, Düsseldorf, Germany*
REINER WESTERMEIER • *SERVA Electrophoresis GmbH, Heidelberg, Germany*
TRAUD WINKELMANN • *Institute for Floriculture and Woody Plant Science, Leibniz Universität Hannover, Hannover, Germany*
WITOLD M. WINNIK • *NHEERL Proteomics Research Core, U.S. Environmental Protection Agency, Research Triangle Park, NC, USA*
XIAOKE YIN • *Cardiovascular Division, King's College London, London, UK*

Part I

Fundamentals

Chapter 1

DIGE: Past and Future

Jonathan S. Minden

Abstract

This chapter provides a brief historical perspective of the development of difference gel electrophoresis, from its inception to commercialization and beyond.

Key words: Difference gel electrophoresis, Historical perspective

1. Introduction

As the inventor of difference gel electrophoresis (DIGE), I was asked to contribute a historical perspective of this method. This story is similar to other inventions where something was created to solve an unmet need. In the case of DIGE, I wanted to understand protein changes in different mutant backgrounds. The unusual element of this story is that many of the ancillary tools needed to make the invention work did not exist when the idea was first hatched. It took more than a dozen years before these other technologies progressed far enough to make the entire process feasible. In this chapter, I will recount the history of the invention of DIGE—a humbling exercise that underscores my naiveté. As with any method, there are always ways to enhance and improve it. While the chemistry of DIGE is fairly well established and stable, the ancillary tools are continually being improved. However, there are persistent technological issues that will need to be resolved. In addition to the historical past of DIGE, I will outline some of the future challenges I see for DIGE.

2. The Past

2.1. The Conception of DIGE

DIGE (which my lab pronounces as dee-gay, owing to the Turkish heritage of Mustafa Ünlü, the graduate student who synthesized the first pair of DIGE dyes) was initially conceived when I was a first year graduate student at Albert Einstein College of Medicine. I was tasked with analyzing a set of *Dictyostelium discoideum* temperature-sensitive, motility mutants. In 1981, the Dictyostelium genome was not known, and there were very few genetics tools for identifying the affected genes. Two-dimensional gel electrophoresis (2DE), which was only about 6 years old at the time, seemed like the best route for discovering the biochemical changes associated with these mutations. I wanted to compare these cells at the permissive and restrictive temperatures. After a couple of abortive tries at running parallel 2DE gels, I realized that an internal control was needed. This inspiration probably came from one of the graduate classes I was taking at the time. The idea was to differentially label the two samples in separate reactions and then combine the two reactions so that they could be run on the same 2DE gel. Getting this idea to work presented several problems: what sort of tags should be used, how will the tags be attached to the proteins, will the tags effect how the proteins run on 2DE gels, how would the tags be detected…?

2.2. What Sort of Tags Should Be Used?

Proteins could be labeled metabolically in vivo or chemically in vitro. I wanted to make a versatile, user-friendly method that could be used on any source of protein. Since only a small number of model organisms are amenable to metabolic labeling schemes, chemical labeling after cell lysis seemed to be the most universal approach. Should the tag be radioactive, colored, or fluorescent? Radioactivity was rejected because autoradiography often requires very long exposures and discrimination between radionuclides can be difficult. Colored tags would only work for very abundant proteins since color detection requires considerable amounts of absorbing material, particular in gels where the path length is 1–2 mm. Fluorescence seemed like the best choice since one can synthesize compounds with very different fluorescent spectra for easy discrimination. Also, one can detect minute amounts of fluorescent material, which I learned from the fluorescence microscopy we were doing in the cell biology lab in which I was working.

2.3. The Selection of Fluorescent Labels

I had no idea what fluorescent molecules to use for DIGE. I poured through chemical catalogs and books on fluorescent dyes, but none of the existing fluorescent compounds really fit the bill. A key point in selecting the right pair of dyes was that the dyes should have the same charge as each other and they should have similar masses. Another issue was that the dyes should be pH insensitive since they will experience a broad pH range during isoelectric focusing. This latter consideration eliminated fluorescein, the most popular fluorescent dye of the day. The dyes should not change the charge of the protein to which they were attached, which raised the issue of how to attach the dyes to the proteins. The two most reactive amino acid residues are lysine and cysteine. Lysine's primary amine provided an easy route to coupling the dye to the protein, but the coupling process eliminates lysine's positive charge. This meant that the DIGE dyes had to have an amino group like lysine or an intrinsic positive charge. Given the pKa of the ε-amino group of lysine and the 3–10 pH range of isoelectric focusing, a quaternary amine was a reasonable substitute for a primary amine. Alternatively, one could couple the DIGE dyes via cysteine residues. At that point in time, it was not known what fraction of proteins had at least one cysteine, and I was frankly unfamiliar with the coupling chemistry, so I put cysteine labeling on the backburner.

2.4. The Synthesis of the First DIGE Dyes

Since the literature search did not yield suitable DIGE dyes, I tried to design and synthesize my own dyes, but my chemistry knowledge was too limited. So I sent a letter to Kodak asking for help synthesizing the dyes. I got a nice letter back from their lawyers saying that they would help as long as I signed a form releasing my invention rights to Kodak. This was before the biotechnology industry took hold, and academic institutions rarely had technology transfer offices. I was in uncharted territory. This also coincided with my changing research advisors and a major shift in my research direction. Consequently, I left the idea behind for 10 years, during this period I completed my PhD in DNA replication and my postdoctoral training in cell and developmental biology. In retrospect, this hiatus was a good thing. In 1981, fluorescence imaging systems were too insensitive, and protein sequencing was only done by Edman sequencing—mass spectrometric protein identification was only in its infancy.

In 1991, I accepted a faculty position at Carnegie Mellon University because of its reputation for interdisciplinarity and entrepreneurship. My lab primarily studies Drosophila embryo development, particularly how cells change shape during development. We were studying a specific cell shape change that had been extensively analyzed by genetic dissection. I was puzzled by the fact that none of the genetically identified genes required for this cell shape altering process involved the cytoskeleton or its regulators. I reasoned that cytoskeleton was required for so many cellular functions that mutations in these genes would prevent the embryo from developing far enough. But I was certain that changes in the cytoskeleton and other proteins must occur during this process. This led me to dust off my old notebooks and reinvestigate DIGE. An equally important impetus for revisiting DIGE was that my lab was around the corner from Alan Waggoner's, the father of cyanine

(Cy) dyes. Alan and I went to lunch one day and I recounted the four criteria for designing DIGE dyes. He said "oh, that's easy" and immediately drew two structures on a napkin. True to CMU's open-door, interdisciplinary policy, I recruited a first year graduate student, Mustafa Ünlü, from the Chemistry Department. The first pair of DIGE dyes, Cy3-NHS and Cy5-NHS, were synthesized in a few months. These are lysine-reactive dyes that have a net positive charge and differ in molecular weight by 2 Da.

2.5. The First DIGE Experiments

Since there were no fluorescent-gel imagers that detected Cy3 and Cy5 at that time, we had to build our own imager. The very first imager was constructed from a 12-bit, cooled CCD camera mounted on a darkroom enlarger stand. The light source was a 35-mm slide projector mounted on a repurposed rail from another darkroom enlarger stand, and the fluorescent filters were mounted in a manual filter turret. This primitive imager was housed in a darkened room. The initial images were encouraging, but needless to say this was not a very light-tight arrangement. The next-generation imager was made more light-tight by building the imager around an IKEA cabinet with holes cut in the top and sides for the CCD camera and illuminator, respectively. This imager was sufficient to begin optimizing the protein labeling reaction. Our initial plan was to saturation label all lysine residues of all proteins in a cellular extract. This gave very poor results since the proteins tended to precipitate. In retrospect, it was not a good idea to saturation label lysine as this addition would increase the protein mass by about 25%. Instead, substoichiometric labeling turned out to be the best way to maintain protein solubility and avoid size heterogeneity due to multiple dyes molecules bound. In substoichiometric (or minimal) labeling, about 2–3% of all lysine residues are labeled. This translates to about 5% of all proteins having a single fluorescent dye bound, while the rest are unlabeled. This set of innovations led to the first published report on DIGE (1).

2.6. Taking DIGE to Commercialization

A great deal of my lab's efforts was assisted by Lans Taylor and Alan Waggoner's Center for Light Microscopy and Biotechnology, an NSF Science Technology Center (STC). Lans and Alan had spun off several of their inventions into companies that were eventually purchased by larger companies. Amersham, plc (now part of GE Healthcare) was interested in licensing the rights to the STC's suite of cyanine dyes, including the DIGE dyes, which Lans and Alan negotiated masterfully. Working with Amersham to commercialize DIGE was an important learning experience. They were meticulous in establishing a robust, reliable protocol. Since the initial development of Cy3-NHS and Cy5-NHS, Amersham/GE has introduced Cy2-NHS for three-color minimal labeling and Cy3-mal and Cy5-mal for saturation labeling of cysteines. Amersham/GE also developed fluorescent-gel imaging systems and image analysis software dedicated to DIGE proteomics. These reagents and tools now make DIGE one of the most versatile and sensitive comparative proteomics methods currently available. This volume is a testament to the broad applicability and robustness of DIGE.

3. The Future

The goal of comparative proteomics is to discover protein changes between cells and tissues under a variety of conditions and circumstances. Implicit in this goal is the desire to detect all protein species within the proteome. Given the chemical complexity of the proteome and its large number of different protein species, separating the proteome into discrete entities is a virtually impossible task, but we are obligated to do as best we can.

Historically, conventional 2DE has had difficulty in resolving proteins that are very large (>250,000 Da) or very basic (>9.5 pI, many of which are ribosomal proteins). Mass spectrometry-centric methods also have difficulties with proteins outside these ranges. Fortunately, these proteins represent a very minor component of the proteome. 2DE also has difficulty resolving integral membrane proteins. This may be due to their hydrophobicity, which causes protein aggregation, and glycosylation, which causes heterogeneity in mass and charge so that membrane proteins do not appear as discrete spots. Important strides in improving 2DE's capacity to resolve membrane proteins have been made by methods such as introducing novel detergents, lipid removal, and deglycosylation (2–4). Finally, concern has been raised about protein overlap or comigration on 2DE gels. The development of narrow pH-range isoelectric focusing gels and very large-format 2DE gels has greatly improved the resolution of 2DE gels where over 10,000 protein species are now detectable (5, 6). These improvements have significantly increased the resolving power of 2DE. One can expect even further advances since many laboratories continue to study new ways to increase the resolution of 2DE. These efforts have improved resolution relative to protein mass and pI, but comparatively little has been done to improve resolution relative to protein abundance.

Proteins exist in cells over an approximately 10^5-fold concentration range, while in serum the concentration range is on the order of tens of millions fold. Currently, DIGE imagers are capable of detecting proteins over a 20,000-fold concentration range. In contrast, conventional mass spectrometers have a dynamic range of about 1,000-fold. There is a clear need to improve the sensitivity and dynamic range of DIGE fluorescent-gel imagers. I am heartened by the fact that technology for detecting single-molecule fluorescence already exists. Hence, it should be possible to build the next-generation fluorescence imager that has the requisite dynamic range to detect proteins over a million-fold concentration range. Once this dynamic range goal is achieved, the next goal will be to isolate sufficient quantities of protein from these very rare protein spots to be identified by mass spectrometry. I am confident that these methods are within our reach. Thus, I feel that the future for DIGE is very bright (please forgive the pun).

Acknowledgments

The development of DIGE would not have been possible without the efforts and dedication of my students and staff (Mustafa Ünlü, Liz Morgan, Chris Lacenere, Surya Viswanathan, Lei Gong, Mamta Puri, Anupam Goyal, and Susan Down).

References

1. Ünlü, M., Morgan, M. E., and Minden, J. S. (1997) Difference gel electrophoresis: a single gel method for detecting changes in protein extracts, *Electrophoresis* 18, 2071–2077.

2. Helling, S., Schmitt, E., Joppich, C., Schulenborg, T., Mullner, S., Felske-Muller, S., Wiebringhaus, T., Becker, G., Linsenmann, G., Sitek, B., Lutter, P., Meyer, H. E., and Marcus, K. (2006) 2-D differential membrane proteome analysis of scarce protein samples, *Proteomics* 6, 4506–4513.

3. Comunale, M. A., Mattu, T. S., Lowman, M. A., Evans, A. A., London, W. T., Semmes, O. J., Ward, M., Drake, R., Romano, P. R., Steel, L. F., Block, T. M., and Mehta, A. (2004) Comparative proteomic analysis of de-N-glycosylated serum from hepatitis B carriers reveals polypeptides that correlate with disease status, *Proteomics* 4, 826–838.

4. Ruan, Y., and Wan, M. (2007) An optimized procedure for solubilization, reduction, and transfer of human breast cancer membrane-enriched fraction by 2-DE, *Electrophoresis* 28, 3333–3340.

5. Han, M. J., Herlyn, M., Fisher, A. B., and Speicher, D. W. (2008) Microscale solution IEF combined with 2-D DIGE substantially enhances analysis depth of complex proteomes such as mammalian cell and tissue extracts, *Electrophoresis* 29, 695–705.

6. Sitek, B., Sipos, B., Pfeiffer, K., Grzendowski, M., Poschmann, G., Hawranke, E., Koper, K., Kloppel, G., Meyer, H. E., and Stuhler, K. (2008) Establishment of "one-piece" large-gel 2-DE for high-resolution analysis of small amounts of sample using difference gel electrophoresis saturation labelling, *Anal Bioanal Chem* 391, 361–365.

Chapter 2

The Basics of 2D DIGE

Phil Beckett

Abstract

The technique of two-dimensional (2D) gel electrophoresis is a powerful tool for separating complex mixtures of proteins, but since its inception in the mid 1970s, it acquired the stigma of being a very difficult application to master and was generally used to its best effect by experts. The introduction of commercially available immobilized pH gradients in the early 1990s provided enhanced reproducibility and easier protocols, leading to a pronounced increase in popularity of the technique. However gel-to-gel variation was still difficult to control without the use of technical replicates. In the mid 1990s (at the same time as the birth of "proteomics"), the concept of multiplexing fluorescently labeled proteins for 2D gel separation was realized by Jon Minden's group and has led to the ability to design experiments to virtually eliminate gel-to-gel variation, resulting in biological replicates being used for statistical analysis with the ability to detect very small changes in relative protein abundance. This technology is referred to as 2D difference gel electrophoresis (2D DIGE).

Key words: Two-dimensional gel electrophoresis, 2D DIGE, CyDye, Multiplexing, Difference gel electrophoresis

1. Introduction

Two-dimensional (2D) gel electrophoresis allows for the simultaneous separation of thousands of proteins and was pioneered by O'Farrell (1), utilizing a denaturing environment. The technique separates the proteins based on their charge in the first dimension using isoelectric focusing (IEF), where the proteins will migrate to their isoelectric point (pI). In the second dimension, the proteins are separated based on molecular weight by the use of classical SDS-PAGE (2). The traditional use of carrier ampholytes to establish the pH gradient in the first dimension led to a number of issues, most notably the lack of reproducibility. This resulted in the development of immobilized pH gradients (3), leading to commercially available precast gels for the first dimension that allowed for enhanced reproducibility (since they do not rely on carrier ampholytes to establish the pH gradient). Despite this increase in reproducibility, there were still problems with gel-to-gel reproducibility requiring the use of technical replicates to remove artifacts of experimental variation. This can result in a prohibitive number of gels, especially for complex experimental designs.

A breakthrough in 2D gel electrophoresis arrived with the introduction of the ability to multiplex fluorescently labeled proteins on the same gel (4). This technique is referred to as 2D difference gel electrophoresis (2D DIGE) (see reviews (refs. 5–7)). The fluorescent dyes used are specially modified cyanine dyes (CyDye™ DIGE fluors) which are matched for molecular weight and charge and provide a useable dynamic range of up to 4 orders of magnitude. There are two approaches to the labeling; the most common approach is termed minimal labeling, where the dye binds to a restricted number of lysine residues. For certain samples an alternative approach termed saturation labeling is used, where the dyes bind to all of the accessible cysteine residues.

The minimal dyes (Cy™2, Cy3, and Cy5) all have an approximate molecular weight of 450 Da and carry a +1 charge (this replaces the +1 charge of the lysine resulting in no overall change to the pI). The dye-to-protein ratio is controlled such that only a small percentage of the total available lysine population is labeled with the CyDye to avoid multiple labels per protein. By utilizing size- and charge-matched dyes, the labeled proteins will comigrate on the 2D gel and allow precise image overlay from each sample.

For the saturation dyes (Cy3 and Cy5) the opposite strategy is used for labeling. The dye and reductant concentrations are optimized to ensure that all the reduced cysteine residues are labeled with the CyDye, resulting in an increase in signal. These dyes have a molecular weight of 680 Da and are neutrally charged. Samples labeled with saturation dyes will exhibit altered spot migrations due to the number of cysteines present and will thus display a different spot pattern compared to the minimal labeling approach. However, the samples within the same gel will comigrate such that differential analysis can still be performed.

Running differently labeled samples in a single gel and analyzing the resulting images can provide possible proteins of interest. However, to allow for biological variation the use of biological replicates for statistical confidence is necessary, so multiple gels still need to be run. To overcome the problems of gel-to-gel variation, one of the dyes is used to label a pooled internal standard (sometimes referred to as a pooled internal reference) (see Fig. 1). The pooled internal standard is comprised of all the potential detectable proteins in the experiment such that it is a combination of equal aliquots of each of the samples to be analyzed (see Table 1). The virtual elimination of gel-to-gel variation (coupled with multiplexing) now allows for the running of biological replicates such that the number of gels to be run is dramatically reduced

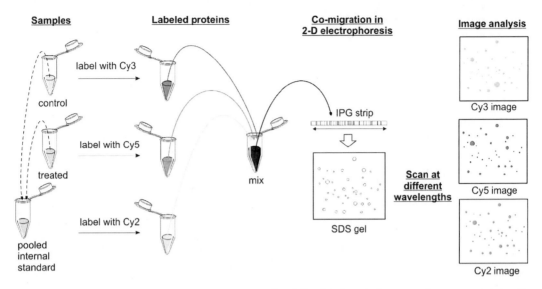

Fig. 1. Workflow for a minimal CyDye labeling experiment. After CyDye labeling, the three samples are separated on the same 2D gel and then imaged for the resulting fluorescence associated with each CyDye. Permission to reproduce from Westermeier and Scheibe (8).

Table 1
Experimental design for a minimal labeling experiment, incorporating a dye swap and including a pooled internal standard (standard). This scenario allows for looking at two different conditions: 1 and 2

	Cy2	Cy3	Cy5
Gel 1	Standard	Sample 1a	Sample 2d
Gel 2	Standard	Sample 2c	Sample 1b
Gel 3	Standard	Sample 1c	Sample 2b
Gel 4	Standard	Sample 2a	Sample 1d

Standard = s1a + s1b + s1c + s1d + s2a + s2b + s2c + s2d

The letters *a*, *b*, *c*, and *d* denote biological replicates

compared to classical detection techniques (9) (see Fig. 2). The 2D DIGE system benefits from the pooled internal standard in several ways; it is used to help normalize the signal between and within each gel by comparing the ratio of each labeled protein spot to the internal standard and then to the same protein spot in the other gels. In addition, the pooled internal standard is used as a standard map to match protein spots across multiple gels since all of the spots in the internal standard should be present across all of the gels.

Traditional 2D gel electrophores is (Silver, Coomassie, SYPRO Ruby, Deep Purple)

Need gel replicates and biological replicates.
Minimum needed for statistics is 3 of each.
No. of gels (N) = no. of samples (y) x 3 (biological reps.) x 3 (gel reps.)
$N = y \times 3 \times 3$
Example 1: control versus treated (y = 2);
$N = 2 \times 3 \times 3 = 18$ gels.
Example 2: control versus treated versus treated+drug (y=3);
$N = 3 \times 3 \times 3 = 27$ gels.

2D DIGE (Minimal Labeling)

Only need biological replicates (since gel-to-gel variation is virtually eliminated with this technique).
Minimum needed for statistics is 3 - but to properly enable reverse labeling 4 is recommended. The multiplexing capability allows three samples per gel (one of these is the internal standard so two real samples are run per gel).
No. of gels (N) = $\frac{\text{no. of samples (y)} \times 4 \text{ (biological reps.)}}{2 \text{ (since two samples on each gel)}}$ $N = \frac{y \times 4}{2}$
Example 1: control versus treated (y = 2);
$N = \frac{2 \times 4}{2} = 4$ gels.
Example 2: control versus treated versus treated+drug (y=3);
$N = \frac{3 \times 4}{2} = 6$ gels.

Fig. 2. Calculation for the minimum number of gels to be run for a traditional 2D experiment compared to a 2D DIGE experiment.

It is important with this technique that the three fluorescent dyes (Cy2, Cy3, and Cy5) are imaged with an appropriate device that can not only independently excite these fluors but is also able to distinguish between the three resulting spectra and avoid any cross talk issues, which would interfere with quantification. A laser scanner capable of blue, green, and red excitation and equipped with the appropriate band pass filters for the corresponding emission is highly recommended. Also, a suitably designed image analysis software package should be used to perform the required calculations (7). In particular, the software's ability to properly handle the codetection of the images within each gel, and the normalization against the internal standard, will influence the accuracy and reliability of the quantification.

Since the inception of 2D DIGE in 1997, there are now over 2,500 papers (as of May 2011, Ishida Y, GE Healthcare, personal communication). Many types of samples have been investigated using this technique, including a wide range of plant and animal species (7). Recently, advances have been made in furthering the utility of the technique by exploring niche applications. Such examples include cell surface labeling (10–12), reduced vs. nonreduced states (13), host cell protein monitoring (14), and samples from laser microdissection (15), to name but a few.

2. Materials

The use of high-quality electrophoresis/proteomic-grade chemicals is paramount to achieving successful experiments with resulting identifications—this is especially true of the quality of the water used in all buffers and solutions and should be of 18 MΩ or less.

2.1. First Dimension of 2D Electrophoresis

1. Rehydration buffer: 7 M urea, 2 M thiourea, 4% (w/v) CHAPS, 40 mM dithiothreitol (DTT), 0.5% (v/v) carrier ampholyte.
2. Immobilized pH gradient (IPG) strips (see Note 1).

2.2. Second Dimension of 2D Electrophoresis

1. SDS equilibration buffer: 6 M urea, 30% glycerol, 2% sodium dodecyl sulfate (SDS), 0.002% bromophenol blue, 75 mM Tris–HCl, pH 8.8 (step 1: DTT followed by step 2: iodoacetamide).
2. SDS gel: acrylamide (10%), bisacrylamide (3%), SDS (0.1%), ammonium persulfate, TEMED, 0.37 M Tris–HCl, pH 8.8 (see Note 1).
3. SDS running buffer: 25 mM Tris-base, 192 mM glycine, 0.2% SDS.
4. Bind-Silane (see Note 2).

2.3. Labeling (Minimal Dye Approach)

1. CyDye DIGE fluors (Cy2, Cy3, and Cy5).
2. Dimethyl formamide (DMF) (see Note 3).
3. Labeling buffer: 7 M urea, 2 M thiourea, 4% CHAPS, 30 mM Tris–HCl, pH 8.5.
4. 10 mM lysine.
5. pH test paper 7.5–9.5 (see Note 4).
6. 50 mM NaOH.

2.4. General Reagents

1. Ethanol.
2. Glacial acetic acid.
3. Deionized water.
4. Fluorescent stain (e.g., Deep Purple™, SYPRO® Ruby).

2.5. General Apparatus

1. First dimension electrophoresis unit.
2. Second dimension electrophoresis unit.
3. Power supply.
4. Temperature controlled recirculating water bath.
5. Ice bucket/ice.
6. Imaging device (laser scanner).
7. Analysis software.

3. Methods

This protocol describes minimal labeling. Full details for performing a saturation labeling experiment can be found in the associated product booklet (16).

3.1. Sample Preparation

The sample is prepared as for classical 2D gel electrophoresis (17), except that primary amines, carrier ampholytes, and thiols are omitted from the buffers. It is then usual to concentrate the resulting sample (e.g., by precipitation) and resuspend it in labeling buffer (7 M urea, 2 M thiourea, 4% (w/v) CHAPS, 30 mM Tris–HCl, pH 8.5) to a concentration of between 5 mg/mL and 10 mg/mL (though 1 mg/mL to 20 mg/mL have been successfully used) (see Note 5). The pH of this resulting sample is then checked with pH test paper such that the pH is between 8 and 9 (and adjusted with 50 mM NaOH if necessary). If the pH is below 8.0, then the dye will not bind, and if the pH is over 9.0, then multiple dyes can bind to the protein or to different amino acids.

The internal standard is prepared by pooling together equal aliquots of all the biological replicates in the experiment (see Table 1).

3.2. Sample Labeling

The labeling protocol (18) involves the resuspension of each lyophilized CyDye in DMF to create a stock solution of 1 mM. To limit any effects of photobleaching on the fluors, all subsequent steps are performed in the dark.

The dye-to-protein ratio is controlled at 400 pmol of dye to 50 μg of protein (though 100–1,000 pmol have been successfully used)—bulk labeling can also be performed by keeping this ratio constant. It is recommended to label the pooled internal standard with the Cy2 dye and then to perform a dye swap with each of the sample types in the experiment such that an equal number are labeled with Cy3 as with Cy5 (see Table 1; Notes 6 and 7). The labeling reaction is performed on ice for 30 min, and then, the labeling reaction is terminated by the addition of lysine to quench any unreacted dye (for 10 min on ice).

3.3. 2D Gel Electrophoresis

The labeled samples are mixed appropriately for loading onto the first dimension IPG strips, either by in-gel rehydration, cup loading, or paper-bridge loading (17). The samples are made up to the correct volume for sample loading ensuring that the final buffer concentrations are 7 M urea, 2 M thiourea, 4% CHAPS, 0.5% carrier ampholyte, and 40 mM DTT.

Standard IEF separation protocols are subsequently followed (17). After the first dimension a two-step equilibration procedure is performed. This procedure saturates the IPG strip with the SDS buffer system required for the second dimension separation.

The equilibration solution contains urea, glycerol, reductant, SDS, and tracking dye. The second equilibration step replaces the reductant with iodoacetamide to alkylate the reduced cysteine residues. The strips are then sealed with agarose on to the top of the second dimension gel and then separated for molecular weight by classical SDS-PAGE.

3.4. Gel Imaging

If the gel is to be imaged while still between the glass plates or attached to a plastic backing, then the glass or plastic must have low fluorescent properties to minimize any background issues (autofluorescence) that could compromise quantification.

The gels are imaged with a suitable fluorescent imager that is capable of exciting the three dyes independently and has the necessary band pass filters to avoid cross talk (see Note 8).

The image capture is then performed as described in the instrument manual, with the following guidelines:

(a) The final image is scanned at 100-μm resolution.
(b) The file format should be a 16 bit .tif (or similar).
(c) Steps should be taken to avoid introducing any fluorescent particles (dust, lint, etc.).
(d) The gel images should first be prescanned using a short-exposure or low-resolution setting so that the final image capture settings can be optimized to avoid saturation while taking advantage of the full dynamic range.

3.5. Image Analysis

Each gel set (three images) should be cropped to remove any areas that are redundant from the analysis (such as the dye front and IPG strip) that may interfere with spot detection and normalization. The cropping should be performed to keep similar spot patterns the same rather than using similar sized crop areas. The analysis software should allow for the use of the pooled internal standard to facilitate the normalization and spot matching procedures. It is usual for the software to incorporate some statistical tools to allow for the assignment of spots of interest that can then be exported as a pick list. These protein spots can then be excised for further analysis, such as by mass spectrometry to identify the protein.

3.6. Gel Processing for Spot Picking

For spot picking, it is necessary to poststain a designated gel with a total protein detection system (e.g., silver, Coomassie®, or ideally a fluorescent stain such as Deep Purple or SYPRO Ruby) (see Note 9).

The reason for this procedure is that if the original CyDye spot coordinates were used, then there is the possibility that only the protein with the dye attached will be picked (a small percentage of the total protein) as this has an approximate 450-Da molecular weight shift to a higher position in the gel (this is the same for all three dyes). The bulk of the protein lies at a slightly lower molecular weight and will be more of an issue for the lower molecular weight proteins, but this procedure should be performed as standard practice. Spots of interest can now be matched to the pick gel image by using the analysis software. The pick gel can be run as a separate preparative gel on its own, or extra unlabeled protein (made up as for the internal standard) can be added equally to all of the analytical gels such that each gel is then a potential pick gel.

3.7. Spot Processing and Identification

The excised protein spot can be enzymatically digested (usually with trypsin), and the resulting peptides can be analyzed with a mass spectrometer. The most commonly used techniques include matrix-assisted laser desorption/ionization (MALDI) or electrospray ionization (ESI) mass spectrometry (MS). Once a protein has been identified by MS, it can be very useful to verify its identity. This can be achieved by Western blotting if an antibody is available against the target of interest. Western blotting combined with a 2D gel can be a very powerful approach since an SDS-PAGE gel alone will not detect the posttranslational modifications that result in different charge forms of the same protein being present. If blotting an actual 2D DIGE gel from an experiment, it is important that the reporter molecule should not interfere with the signal from the CyDyes such that the antibody can be linked to an enzyme (such as horse radish peroxidase, HRP) for chemiluminescent detection or the antibody can be linked to an infrared reporter molecule. It must be remembered that the total protein from control, treated, and internal standard will now be detected, so this approach is more useful for confirmation of location and identification. However, if using a two-dye system (see Note 6) for the DIGE experiment (e.g., Cy 3 and Cy5), then the third dye (Cy2) could be used as a reporter molecule on the primary or secondary antibody.

4. Notes

1. Reproducibility of spot patterns can be facilitated by the use of precast gels for both the first and second dimension.
2. To facilitate accurate spot picking, it is strongly recommended to immobilize the gel to prevent swelling or shrinking during the staining procedure. This can be achieved by treating one of the two low fluorescent glass plates (one pair of plates is used per gel) with Bind-Silane. Another approach is to use a low fluorescent plastic-backed gel.

    Reference markers can be attached to the support surface prior to gel casting or imaging to enable more accurate spot picking with robotic instrumentation—these markers will serve as "anchor points" such that the pixel coordinates from the software can be accurately converted to picking coordinates.

3. DMF is used to reconstitute the CyDye and should be anhydrous. Poor-quality DMF will result in reduced labeling efficiency and reduced shelf life for the dyes. Water accumulation and amine-containing byproducts can be avoided by the addition of a 4 Å molecular sieve (cat. no. M2635, Sigma-Aldrich®) to absorb these impurities and the water.

4. Wider pH range test papers are not accurate enough.

5. The initial determination of protein concentration should be verified using an assay that is compatible with the reagents that are used in classical 2D gel electrophoresis. Chemicals such as urea and DTT can interfere with standard protein assays. Labeling should be performed at the same protein concentration across all the samples in the experiment.

6. The incorporation of a dye swap in a 3-dye approach negates any chance for dye bias. Utilizing a 2-dye approach will also negate this dye bias (19).

7. Please see the paper by Karp et al. (19) for a discussion on how many replicates should be run in an experiment. Another paper by Karp et al. discusses when pooling or subpooling of samples can be employed (20).

8. For the minimal labeling CyDyes, see also Table 2.

    Cy2 has an excitation maximum at 491 nm and emission maximum at 509 nm.

    Cy3 has an excitation maximum at 553 nm and emission maximum at 569 nm.

    Cy5 has an excitation maximum at 645 nm and emission maximum at 664 nm.

9. The gel should only be fixed (usually in a combination of acid and alcohol) after the gel has been imaged since the use of ethanol can interfere with the fluorescent properties of the CyDyes. The gel should not be fixed if Western blotting will be

Table 2
The colors related to the different CyDyes used in a minimal labeling experiment

CyDye	Reagent color	Laser excitation	Emission fluorescence
Cy2	Yellow	Blue	Green
Cy3	Red	Green	Orange
Cy5	Blue	Red	Red

performed. If the gels need to be stored prior to scanning, then they can be kept under SDS running buffer in a light-tight container at 4°C. It is recommended to allow the gels to warm up to room temperature before imaging, as fluorescent intensity is temperature dependent. The ethanol used should be free of hexanes or other nonalcohol organic solvent impurities that can contribute to background fluorescence.

Acknowledgments

Many thanks to Rita Marouga, Viola Ruddat, and Chris Rozanas (GE Healthcare) for their critical review and feedback.

Trademarks
CyDye™, Cy™, and Deep Purple™ are trademarks of GE Healthcare Ltd., a General Electric company.
Coomassie® is a registered trademark of ICI plc.
SYPRO® is registered trademark of Molecular Probes, Inc.
Sigma-Aldrich® is a registered trademark of Sigma Chemical Co.
2D Fluorescence Difference Gel Electrophoresis (Ettan DIGE) technology is covered by US Patent Numbers 6,043,025, 6,127,134, 6,426,190 and foreign equivalents and exclusively licensed from Carnegie Mellon University by GE Healthcare, Ltd., a General Electric company.

References

1. O'Farrell PH. (1975) High resolution two-dimensional electrophoresis of proteins. J. Biol. Chem. 250: 4007–4021.
2. Friedman, D, Hoving, S, Westermeier, R. (2009) Isoelectric focusing and two-dimensional gel electrophoresis. Methods Enymol. 463: 515–540.
3. Gorg A, Drews O, Luck C, Weiland F, Weiss W (2009) 2DE with IPGs. Electrophoresis 30: 122–132.
4. Ünlü, M, Morgan, ME, Minden JS. (1997) Difference gel electrophoresis: A single gel method for detecting changes in protein extracts. Electrophoresis. 18: 2071–2077.
5. Lilley KS, Friedman DB (2004) All about DIGE: quantification technology for differential-display 2D-gel proteomics. Expert Rev. Proteomics 1: 401–409.
6. Marouga, R, David, S., Hawkins, E (2005) The development of the DIGE system: 2D fluorescence difference gel analysis technology. Anal. Bioanal. Chem. 382: 669–678.
7. Loyland, SM, Rozanas, CR (2008) Capabilities using 2-D DIGE in Proteomics Research: The New Gold Standard for Two-dimensional Gel Electrophoresis. Methods Mol. Biol., 441: 1–18.
8. Westermeier, R., and Scheibe, B. (2008) Difference Gel Electrophoresis Based on Lys/Cys Tagging In: Posch, A. Ed. 2D PAGE: Sample preparation and Fractionation. Methods in Molecular Biology 424, Humana Press, Totowa, NJ., p. 75.
9. Alban, A., David, S., Bjorkesten, L., Andersson, C., Sloge, E., Lewis, S., Currie, I. (2003) A novel experimental design for comparative two-dimensional gel analysis: Two-dimensional difference gel electrophoresis incorporating a pooled internal standard. Proteomics 3: 36–44.
10. Mayrhofer, C., Krieger, S., Allmaier, G., Kerjaschki, D. (2006) DIGE compatible labeling of surface proteins on vital cells in vitro and in vivo. Proteomics 6: 579–585.

11. Sidibe, A., Yin, X., Tarelli, E., Xiao, Q., Zampetaki, A., Xu, Q., Mayr, M. (2007) Integrated Membrane Protein Analysis of Mature and Embryonic Stem Cell-derived Smooth Muscle Cells using a Novel Combination of Cy-dye/Biotin Labeling. Mol. Cell Proteomics 6: 1788–1797.

12. Selective labeling of cell-surface proteins using CyDye DIGE Fluor minimal dyes. Application Note 11-0033–92 AB, GE Healthcare.

13. Hurd, TR, Prime, T, Harbour, ME, Lilley, KS, Murphy, MP (2007) Detection of Reactive Oxygen Species-sensitive Thiol Proteins by Redox Difference Gel Electrophoresis. Implications for mitochondrial redox signaling. Biol. Chem. 282: 22040–22051.

14. Jin, M., Szapiel, N., Zhang, J., Hickey, J., Ghose, S. (2010) Profiling of host cell proteins by two-dimensional difference gel electrophoresis (2D-DIGE): Implications for downstream process development. Biotechnol. Bioeng. 105: 306–316.

15. Kondo, T. and Hirohashi, S. (2006) Application of highly sensitive fluorescent dyes (CyDye DIGE Fluor saturation dyes) to laser microdissection and two-dimensional difference gel electrophoresis (2D-DIGE) for cancer proteomics. Nature Protocols 1(6): 2940–2956.

16. Amersham CyDye DIGE Fluor Labeling Kit for Scarce Samples 25-8009-83 AD. Instruction manual, GE Healthcare.

17. 2D Electrophoresis, Principles and Methods. 80-6429-60 AD. Handbook, GE Healthcare.

18. Amersham CyDye DIGE Fluors (minimal dyes) for Ettan DIGE 28-9531-63 AF. Instruction manual, GE Healthcare.

19. Karp, NA, McCormick, PS, Russell, MR, Lilley KS (2007) Experimental and statistical considerations to avoid false conclusions in proteomics studies using differential in-gel electrophoresis. Mol. cell. Proteomics 6: 1354–1364.

20. Karp, NA, Lilley KS (2009) Investigating sample pooling strategies for DIGE experiments to address biological variability. Proteomics 9: 388–397.

Chapter 3

Multifluorescence 2D Gel Imaging and Image Analysis

Ingo Vormbrock, Sonja Hartwig, and Stefan Lehr

Abstract

Although image acquisition and analysis are crucial steps within the multifluorescence two-dimensional gel electrophoresis workflow, some basics are frequently not carried out with the necessary diligence. This chapter should help to prevent easily avoidable failures during imaging and image preparation for comparative protein analysis.

Key words: 2DE, Two-dimensional gel electrophoresis, Cyanine fluorescence dyes, Gel imaging, Image analysis, 2DE Multifluorescence analysis, Multifluorescence analysis, Protein visualization, Spot detection, DIGE

1. Introduction

Multifluorescence two-dimensional gel electrophoresis (2DE) is theoretically capable of detecting and quantifying thousands of protein spots in a single gel with a detection limit below 0.5 fmol of protein and a dynamic range of up to five orders of magnitude. Enormous volumes of high-quality data can be derived from multifluorescence 2DE gels provided that image acquisition, processing and analysis are performed according to optimized and standardized operating protocols. Here we describe the instrumentation and the most critical parameters for highly sensitive and reproducible fluorescence imaging, spot detection and quantification.

2. Materials

For the imaging methods described below, the preparation of a 2D DIGE gel is needed. Most of the examples are based on employing consumables and equipment from GE Healthcare (Freiburg, Germany), including CyDyes™, Typhoon™ laser scanner and DeCyder™ software.

3. Methods

3.1. Preliminary Considerations

The prerequisite for successful gel imaging and for all subsequent spot detection and quantification approaches is the use of high-quality 2D polyacrylamide gels.

The easiest but more expensive way to obtain suitable gels is to purchase commercially available pre-cast gels that comply with the standards of good manufacturing practice. There are two concepts of gels available: pre-cast gels in gel cassettes and film-backed. Before deciding on one of these concepts, its compatibility with the existing image acquisition hardware has to be evaluated. For example, in case of deciding on a film-backed gel in combination with a laser fluorescence scanner, the gel backing foil has to show low fluorescence properties and should be translucent for light in the near UV spectral range. Although pre-cast gels are commercially available, many users still prefer lab-cast gels for economic reasons. In case of deciding on casting gels in one's own lab, the whole casting procedure should be part of a very precise and well-established multifluorescence 2D gel standard operating protocol (SOP). The acrylamide monomer solution has to be made of very high-quality chemicals and should be degassed prior to gel casting. In order to obtain sets of lab-cast gels with similar physical and chemical properties, it is necessary to use a multi-gel caster. In case of using a fluorescence scanner equipped with confocal laser optics for image acquisition, the gel casting cassettes should be made of low-fluorescent glass, allowing scanning directly between the glass plates. During gel casting, special attention has to be paid to gel homogeneity and a straight gel surface: It is recommended to pre-cool the acrylamide monomeric solution with a relatively low concentration of ammonium persulfate and to overlay the gels by spraying 0.1% sodium dodecyl sulphate into the gel casting cassettes before the monomer solution starts to polymerize. After polymerization, the gels should be overlayed with 2D running buffer and stored for a reasonable period of time (about 4 days) at 4°C to ensure complete polymerization. The storage time should be defined and controlled by an SOP.

3.2. Image Acquisition

3.2.1. Imaging Hardware

2DE multifluorescence analysis (MFA) profits from two major advantages of fluorescence-based protein quantification techniques: high sensitivity and broad dynamic range. Because protein spot detection and quantification is carried out by image analysis software, it is important to generate image files with high dynamic range. Modern imaging systems are capable to generate gel images with up to 100,000 levels in measuring signal intensity. These images are usually saved as greyscales using the 16-bit tagged image file format (TIFF) which allows distinguishing only 65,536 levels. To overcome these limitations, modern imaging systems like the Typhoon™ scanner use a square root compression algorithm to convert the 100,000 possible levels of signal intensity into a proprietary 16-bit ".gel" file. If working with gel imaging hardware that creates artificial files like ".gel" files, it is important, that the image processing and analysis software is capable to handle this compressed file format.

At the moment two different imaging systems for multifluorescence 2D gel imaging are available: scanning CCD cameras and laser fluorescence scanners. There are several benefits of working with scanning CCD camera systems. They are fast, less expensive than fluorescence scanners, and they usually feature a broad range of imaging capabilities beside multifluorescence 2D gel imaging (chemiluminescence detection, transmission imaging of gels stained with visible dyes, etc.). Scanning CCD systems work with a white light source in combination with several excitation filters that generate monochromatic light to excite only one definite fluorophore. The gel to be imaged is deposited on a transillumination tray and scanned from above by a high-resolution CCD camera equipped with an appropriate emission filter. By utilizing CCD-based imaging techniques, a maximum dynamic range of about 3.5 orders of magnitude can be realized. Dynamic ranges up to five orders of magnitude can be achieved by using laser fluorescence scanners for gel imaging. In these devices, a set of lasers with different excitation wavelengths is combined with a corresponding set of emission filters. Signal detection and amplification is carried out by a photomultiplier tube (PMT). The signal amplification factor correlates with the voltage applied at the multiplier tube, so scanning sensitivity can be enhanced by increasing the PMT voltage. Modern laser scanners are equipped with confocal laser optics. This feature allows to protect the structural stability of the gels by scanning them directly inside their cassettes and to improve the image quality by reducing disturbing signals from scattered excitation light and from background fluorescence. Further aspects of fluorescence scanning are described in detail in the fluorescence imaging handbook provided by GE Healthcare (1).

3.2.2. Preparing Gels for Scanning

The imaging of 2D gels has to be performed immediately after terminating the last electrophoretic step. The scanning of an associated gel set (i.e. the gels covering one MFA experiment) has to be performed in one pass without interruptions and without changing scan parameters. In case of using a laser fluorescence scanner for gel imaging, the imaging procedure for a set of twelve large 2D gels may take several hours, so electrophoresis should run over night and has to be terminated early in the morning. After removing the gel cassettes from the electrophoresis chamber, a first washing step is carried out with deionized tap water to remove remains of running buffer. To prevent the gels from drying and shrinkage, they should be left in their cassettes until scanning. The washed cassettes can be stored as stacks in shaded plastic boxes together with wetted lint-free paper. Because fluorescence intensity is temperature dependent, all gels should be scanned at a definite temperature (2) (see Note 1).

In case of using a laser scanner with confocal optics, the gels will remain in gel cassettes even during the scanning procedure. To prevent the formation of Newton's rings, gel cassettes should not be placed directly on the scanners' glass platen, but on a holding tray that determines a small gap between the bottom of the cassette and the top of the platen. Immediately before scanning, gel cassettes have to be cleaned meticulously for a second time with highly pure water and should be dried subsequently with lint-free paper. The cleanliness of the scanner platen surface should be checked prior to each scanning procedure. Cleaning should be carried out using lint-free tissue soaked in highly pure water. In the long run, remains of fluorescence dyes may cause streaky scanning artefacts which may render the image useless for later analysis. To prevent such streaks, the gel cassettes and glass platen have to be cleaned at regular intervals with 10% hydrogen peroxide followed by a washing step with highly pure water (see Note 2).

3.2.3. Gel Scanning and Scan Parameters

To ensure the operational readiness of the imaging system, it should be turned on about half an hour before the first gel scan is performed. As mentioned above, all gels of one MFA experiment should be scanned in one single "scanning session" without any intermediate change of the scan parameters (see Note 3).

After the completion of each scan, the technical quality of the generated image should be verified before the gel is removed from the scanning device. In case of poor image quality or scan errors, it would be possible to readjust the scanner and to image the gel for a second time. Otherwise, the gel can be transferred to the fixing solution.

Besides the practical aspects of gel handling, special attention should be paid to the appropriate setting of the scan parameters. Some crucial parameters for multifluorescence 2D gel imaging will be discussed in detail below.

Gel-Specific Settings

Depending on gel size and whether the gels are scanned in gel cassettes or not, some gel-specific settings have to be adjusted.

To ensure that always one defined section of all gels within an MFA experiment is scanned, it is useful to pre-define a scan area using the scanner control software. It might be helpful to use special hardware extensions of the scanning devices, e.g. gel holding trays, that may help to adjust gels or gel cassettes. In addition, the orientation of the gel might be redefined by software set-up and should be well documented. When scanning gels inside their cassettes, the confocal optics of the laser scanning device has to be adjusted focussing 3 mm above the scanner glass platen.

Laser, Filter and Photomultiplier Settings

For multifluorescence 2D gel imaging, the imaging device has to be equipped with a defined combination of emission filters and excitation filters (CCD scanning systems) or lasers (laser fluorescence scanners). For the three CyDye™ fluorescence labels, the ideal combinations of excitation and emission wavelengths for multifluorescence 2D gel imaging are given below:

Cy2: Blue (480 nm) excitation filter/laser, 520 nm emission filter.

Cy3: Green (532 nm) excitation filter/laser, 580 nm emission filter.

Cy5: Red (633 nm) excitation filter/laser, 670 nm emission filter.

The specifications and the quality of the applied filters and lasers are crucial for a reliable separation of the three signal channels. Best possible results can be accomplished by a combination of excitation lasers with high-quality band-pass emission filters. The scanner control software provides the opportunity to select (only) the appropriate combinations of excitation and emission wavelengths.

Each multifluorescence 2D gel has to be scanned in all fluorescence channels. The scans have to be carried out as individual scans resulting in a set of individual gel images.

The intensity of the fluorescence signals emitted by labelled proteins in a gel can vary considerably depending on the total amount of protein applied to the gel and the amount of fluorescence dyes used for protein labelling. To maximize the dynamic range of the imaging device, the signal amplification of the scanning CCD camera or the PMT has to be adapted to the strength of the fluorescence signals in order to prevent signal saturation. In case of using fast CCD cameras, the adaptation can be performed "on the fly". In case of using a laser fluorescence scanner, the adaptation routine is a little bit more time consuming. As mentioned above, the control of the signal amplification is carried out by varying the PMT voltage. For voltage estimation, a pre-scan of the gel has to be performed and the generated corresponding image files have to be analysed, whether the pixel intensity values cover the full range between 1 and 100,000. In particular, the maximum pixel intensity value should be within this range, exceeding it would mean signal saturation and would prevent an accurate quantification of spot volumes by image analysis software. An example for signal

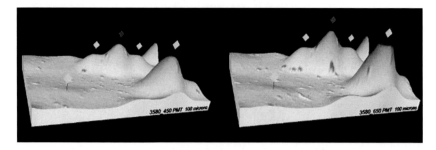

Fig. 1. Impact of PMT voltage on spot volumes. The two images show a small gel section scanned with a PMT potential of 450 V (*left*) and 650 V (*right*). At the lower PMT voltage, the volumes of all five spots within the section are quantifiable, whereas at higher voltage, the signal of one spot runs into saturation. In 3D view, signal saturation manifests itself in a characteristic peak truncation.

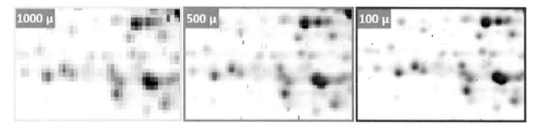

Fig. 2. Impact of pixel size on image resolution. The three images show a small 2D gel section scanned with pixel size settings between 1,000 and 100 μm. A pixel size above 500 μm is not sufficient to resolve distinct protein spots properly.

saturation is given in Fig. 1. In case that protein spots are recognizably affected by signal saturation, the PMT voltage has to be reduced to enhance the dynamic range of the gel image. The PMT voltage has to be checked for all three fluorescence channels to prevent signal saturation effects.

Pixel Size

The pixel size defines the resolution of a gel image. A small pixel size (25–100 μm) results in a high-resolution image and is associated with an extended scanning time and an increased size of the image file generated. A pixel size of 1,000 μm is typically used for fast pre-scans to estimate whether the PMT voltage has to be adjusted. 2D imaging is carried out using resolutions around 100 μm. As demonstrated in Fig. 2, this resolution is fully sufficient to perform image analysis. A further increase in resolution would only increase the size of the generated image but not its information content.

File Naming and Documentation

Apart from all technical considerations, it is of crucial importance that the generated image files can be assigned to the appropriate laser fluorescence channel. It is highly recommended to use an unambiguous file-naming scheme (i.e. extending each image file name by a short abbreviation of the associated channel) and to save image files derived from identical gels in one folder (see Note 4).

3.3. Software-Assisted Image Analysis

After completion of the image acquisition process, the entire data that constitutes an MFA experiment is exclusively contained in the generated set of gel image files. It has to be emphasized that software-assisted analysis of the image data can only provide meaningful results, when it is conducted on the base of high-quality gel images. Image analysis software is neither intended nor capable to attain satisfactory (or even any) results from low-grade gel images.

The overall workflow for multifluorescence 2D gel image analysis is similar for all software packages available on the market:

- Image cropping.
- Joining gel images in groups of a MFA experiment.
- Spot detection and matching.
- Difference analysis and extended statistical analyses.

Some important aspects of the analysis procedure are outlined below.

3.3.1. Cropping

Even when working with pre-defined scan areas, raw gel images generated by the scan device contain non-essential information that should be deleted prior to performing image analysis. This can be realized by "cropping" these images, i.e. cutting the section of interest from the raw image file and pasting it as a new image into a new file. Within one MFA experiment, all cropped image sections have to be exactly of the same size and should cover the same gel section. Cropping can be performed using specialized software that is capable of cropping all three corresponding images of one gel in parallel without losing the calibrated image raw data. After cropping, associated gel images can easily be linked to MFA experiments of the established 2D image analysis software packages.

3.3.2. Spot Matching

A typical MFA experiment usually consists of a set of several gels. After gel image acquisition, protein spots have to be detected and all spots representing identical proteins within different gel images have to be matched. For the images derived from the same gel, spots do not need to be matched because of the co-migration of differentially labelled proteins. However, considering all images derived from different gels of one extensive MFA experiment, spot matching becomes a challenging business and can justifiably be seen as the crucial step within the whole image analysis process.

Two different approaches for spot detection and spot matching are currently realized: Most image analysis software packages such as DeCyder™ (GE Healthcare) and PDQuest™ (Bio-Rad) use the more "traditional" approach, whereas the Delta2D™ (Decodon) software features an alternative spot matching concept.

The traditional approach starts with a spot detection routine to detect protein spots in every gel image combined in the MFA experiment. By using special spot detection algorithms in combination

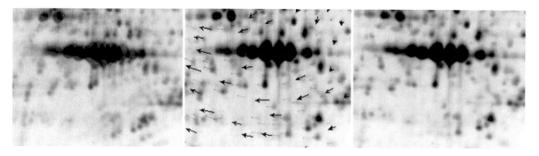

Fig. 3. The "traditional" matching approach is based on computer-aided matching between pairs of gel images. An overlay view of the gel pairs allows estimating the success of the matching procedure. In case of matching ambiguities, the overlay does not fit properly (*left image*), and further matching has to be performed manually by the software operator. Several tools like a "vector mode" may assist the operator's work (*centre image*, some vectors are highlighted by *black arrows*). Ideally, all matching ambiguities can be solved manually (*right image*).

with defined spot parameters like spot radius, spot volume, spot contrast intensity, etc. it is possible to distinguish real protein spots from dirt or scan artefacts. For spots in multifluorescence 2D gel images, several different detection algorithms are employed depending on the analysis software package used. As an example, the analysis software DeCyder™ uses a so called "co-detection" algorithm. Co-detection means that—for the definition of a possible spot—information of all three gel images is combined, whereas other spot detection algorithms only use the image information of the internal protein standard image. After completion of the spot detection routine, spot matching has to be performed within all gels of the MFA experiment. The inter-gel matching procedure is carried out using the gel images of the pooled internal protein standard. Although 2D analysis software uses advanced matching algorithms (see Fig. 3), the software operator has to check the matching results very closely for spot matching ambiguities. In particular, gel sections with considerable separation inhomogeneities might cause mismatches that have to be corrected manually if possible. Remaining spot matching ambiguities may cause incorrect or missing values (see Note 5).

To overcome spot matching conflicts and to accelerate the matching procedure, an alternative matching concept has been developed (3) and was implemented in the Delta2D™ image analysis software. Instead of performing spot detection on every individual gel image, an image warping routine compensates for running differences between different gels. After warping, corresponding spots have equal positions in all images that were engaged in the warping process. The warped images were fused in an artificial image. Subsequently, a spot detection algorithm is applied on the "fused" gel to create a consensus spot pattern. Based on this spot pattern and on the warping information, spot boundaries can be transferred from the fused image to the corresponding regions of the

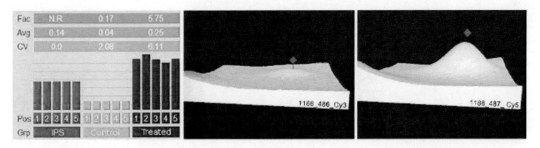

Fig. 4. Example for difference analysis on the level of a single protein spot. Image data was obtained from five individual gels ("Pos 1–5", *left image*) and arranged into three groups ("IPS", "control" and "treated") within an MFA experiment. The spot volumes were normalized on the IPS, and the average volumes ("Avg") of "IPS", "control" and "treated" were determined. Referring to the IPS average volume, the abundance variation factors ("Fac") was calculated. The individual volume information for all spots included in a MFA experiment can be displayed as peaks in 3D view.

original images. As all spot patterns originate from the consensus pattern on the fused image, spot matching without matching ambiguities can be achieved.

3.3.3. Difference Analysis

The major challenge of a multifluorescence analysis is to find biologically significant changes between different samples. One big advantage of this technique is the opportunity to link image data derived from a huge number of gels of a single MFA experiment via the internal standard images. The benefit of using internal protein standards (IPS) in complex multifluorescence 2DE-based studies has been emphasized in several publications (4).

For software-based detection and quantification of differences between two protein samples (e.g. "untreated" vs. "treated"), all gel images of these samples have to be assigned to two specific groups. A third group contains the images of the IPS. The analysis software can automatically normalize and quantify the sample spot volumes in relation to the IPS spot volume. An example for analysing the spot values obtained from a set of multifluorescence 2D gels is depicted in Fig. 4.

To enhance the statistical capabilities of the basic 2D image analysis software packages, several tools for an extended data analysis are available. These tools provide additional analysis concepts like principle component analysis or discriminant analysis as well as enhanced opportunities for a comprehensive visualization and presentation of the statistical results.

4. Notes

1. In case of using "traditional" separation gel with 12.5% T and 3% C, gels can be stored in their cassettes for several hours at RT and overnight at 4°C without losing spot quality.

2. All chemicals for the maintenance/cleaning of the gel cassettes and scanner glass platen should be stored in containments composed of chemically inert materials to prevent contamination by plasticizers with fluorescing properties.
3. If an interruption or a delay of the scanning procedure is unavoidable, e.g. for technical reasons, it is possible to store the gels in their cassettes over night at 4°C and to resume the scanning procedure the next day. To prevent gel shrinkage, the gel cassette stacks should be wrapped in plastic film. Before resuming the scan, the gel cassettes have to be brought up to room temperature and cleaned.
4. It is essential to assure correct sample assignment and gel orientation throughout the whole imaging process. In case of using lab-cast gels, it is highly recommended to number the gels with little scraps of paper placed in one corner of each cassette prior to gel casting.
5. In case of doubt whether a spot of dirt was wrongly detected as protein signal, it might be helpful to use the 3D visualization tool of the analysis software as decision-making aid.

References

1. GE Healthcare Handbook: Fluorescence imaging: Principles and Methods, GE Healthcare Life Science (2002) 63-0035–28.
2. Rozanas CR, Loyland SM (2008) Capabilities using 2-D DIGE in proteomics research: the new gold standard for 2-D gel electrophoresis. in: Liu BCS, Ehrlich JR, Eds. Tissue Proteomics: Approaches for Pathways, Biomarkers, and Drug Discovery. Humana Press, Totowa NJ. 441:1–18
3. Luhn S, Berth M, Hecker M, Bernhardt J (2003) Using standard positions and image fusion to create proteome maps from collections of two-dimensional gel electrophoresis images. Proteomics. 3(7):1117–27.
4. Friedman DB, Hill S, Keller JW, Merchant NB, Levy SE, Coffey RJ, Caprioli RM (2004) Proteome analysis of human colon cancer by two-dimensional difference gel electrophoresis and mass spectrometry. Proteomics. 4(3):793–811.

Chapter 4

Assessing Signal-to-Noise in Quantitative Proteomics: Multivariate Statistical Analysis in DIGE Experiments

David B. Friedman

Abstract

All quantitative proteomics experiments measure variation between samples. When performing large-scale experiments that involve multiple conditions or treatments, the experimental design should include the appropriate number of individual biological replicates from each condition to enable the distinction between a relevant biological signal from technical noise. Multivariate statistical analyses, such as principal component analysis (PCA), provide a global perspective on experimental variation, thereby enabling the assessment of whether the variation describes the expected biological signal or the unanticipated technical/biological noise inherent in the system. Examples will be shown from high-resolution multivariable DIGE experiments where PCA was instrumental in demonstrating biologically significant variation as well as sample outliers, fouled samples, and overriding technical variation that would not be readily observed using standard univariate tests.

Key words: DIGE, Principal component analysis, Multivariate statistics, Variation, Technical replicates, Biological replicates

1. Introduction

The complex methodologies used in many quantitative proteomics studies involving multiple experimental conditions often are comprised of small sample sets and underpowered experiments. Ultimately what is being measured in these experiments is *variation*. Utilizing multiple, independently derived (biological) replicate samples is the only way to determine if an observed change is due to variation in the signal (the biology) or the noise (technical/analytical variation or normal biological variation that is not associated with the experiment). Technical replicates (repeat analyses on the same samples) are necessary to control for analytical variation, but when this technical noise is low, the biological replicates can control for both technical and biological noise. Due to the low amount of technical noise demonstrated for DIGE (1), this technology platform enables statistically powered experiments (see Note 1) using a relatively low number of biological replicates without the need for additional technical replicates of each sample (see Note 2).

In the simplest type of quantitative proteomics experiment, there are two conditions being measured, typically some experimental condition and a control. Even here the need for independent biological repetition remains paramount; it is insufficient to simply multiplex a single experimental and control sample into the same analytical run (e.g., DIGE gel) and be able to determine if the observed change represents a relevant biological signal or technical/biological noise. Even pooling independent samples does not alleviate this issue, as once samples are pooled the ability to distinguish signal from noise is lost (see Note 3).

Using the requisite number of biological replicates, the most commonly used statistical test for quantitative proteomic changes is the univariate Student's t-test, whereby the distribution about two means is compared with the magnitude of difference between these means, and the resulting p-value reflects the likelihood that the measurements are derived from the same distribution (the null hypothesis). When experimental conditions become greater than two, then the univariate analysis of variance (ANOVA) test is commonly invoked (the t-test is a special case of the ANOVA test). Despite their commonplace usage in quantitative proteomics, these univariate tests only assess changes on a feature-by-feature basis; a single species relative to itself across all samples and conditions. Although multiple testing correction algorithms are available to compensate for univariate tests performed within large datasets (e.g., Bonferroni and Benjamini-Hochberg methods), univariate tests do not take into account the variation present in the global experimental system. Especially in the small-sample-size regime, the likelihood of measuring a change by chance increases regardless of what the univariate p-value suggests. Thus, knowledge of the global variation is essential and in some cases can influence the likelihood that univariate changes are biologically significant. The univariate p-value is only a guide, a means to an end but not an end unto itself.

In contrast, multivariate tests enable the visualization of the experimental variation on a global scale, analyzing all of the variables simultaneously. Technical noise (poor sample prep, run-to-run variation) and biological noise (normal differences between samples, especially present in clinical samples) are almost always associated with analytical datasets in quantitative proteomics and may well override any variation that arises due to actual differences related to the biological questions being tested. Multivariate tests can highlight major sources of variation within a dataset, and when performed in an unsupervised fashion, can test if this variation is consistent with the anticipated biological differences between samples as well as identify sample outliers, fouled samples, and even potentially poor experimental design.

Multivariate analyses such as principal component analysis (PCA) and hierarchical clustering are most commonly used in this realm of quantitative proteomics, and both are easily accessible using commonly employed software suites designed for DIGE analysis, such as DeCyder (GE Healthcare) and Progenesis SameSpots (Nonlinear Dynamics) (see Note 4). These multivariate analyses work essentially by comparing the expression patterns of all (or a subset of) proteins across all samples, using the variation of expression patterns to group or cluster individual samples. At the very least, unsupervised clustering of related samples adds additional confidence that a "list of proteins" changing in a DIGE experiment are not arising stochastically. At the very most, these tools offer additional insights to the underlying structure of the variation within a complex multivariable experiment and enable the discovery of protein expression changes by patterns and within subgroups that are beyond the scope of simple pairwise tests.

Using PCA as an example, this chapter will cover the steps taken in typical DIGE analyses to perform multivariate statistical tests. It will use as examples experiments with varying levels of both biological signal and technical noise to illustrate measures taken to evaluate the experimental variation and to derive meaningful information for typical DIGE analyses.

2. Materials

DeCyder software (GE Healthcare, Uppsala, Sweden) and SameSpots software (Nonlinear Dynamics, Newcastle-upon-Tyne, UK)—the versions used for the examples were 6.5 and 4.0.3779.13732, respectively—and a suitable computer for running these software packages.

3. Methods

3.1. Sources of Variation

As stated in the Introduction, ultimately what is being measured in quantitative proteomics experiments is *variation*. In the best cases, the measured variation is produced by the biology being manipulated, and insight can be drawn from an understanding of the proteins and their modified forms that give rise to this variation. This is the *signal*. However, there is also significant *noise* that can contribute to this variation, and this can result both from the technology employed to measure the variation as well as from the normal biological variation that exists between samples but is unrelated to the biology being manipulated.

Sources of technical noise in DIGE include gel-to-gel variation as well as sample handling and labeling variation; however, the use of the Cy2-labeled internal standard methodology effectively normalizes this variation (1–5). Additional sources of technical noise can be related to sample preparation (e.g., subcellular fractionation), laboratory conditions (e.g., medium composition, ambient temperature, incubation conditions), and sample procurement (e.g., tissue dissection, protein extraction). Testing independent biological replicate samples from each condition is necessary to distinguish between these different sources of variation, and PCA is an effective way to visualize the experimental variation on a global scale and enable the assessment of the major sources of variation with respect to the experimental conditions.

3.2. Principal Component Analysis

PCA is a commonly used statistical tool that is capable of reducing the complexity of multivariable space of an experimental dataset into the major sources of variation, the principal components. In the case of DIGE, the variation arises from the expression values of each resolved protein feature, and these features are registered across all samples and gels (i.e., the spot patterns are matched and thereby so are the expression values across the dataset).

One need not to understand all of the underlying mathematics of PCA, which involve covariance matrices and their resulting eigenvectors and eigenvalues, to be able to utilize the strong diagnostic and discovery capabilities that this statistical test provides. But an understanding of what PCA is doing to the dataset is necessary to properly interpret the results. Considering a plot of all of the variables in a multidimensional space (a typical DIGE analysis could contain close to 1,000 defined features each of which is represented in multiple samples), PCA distills out the major sources of variation as defined by the eigenvectors and eigenvalues associated with each square covariance matrix calculated. The longest axis through this cloud of multivariable values is defined by an eigenvector (and associated eigenvalue) as the first principal component, with the second principal component by definition being orthogonal (perpendicular eigenvector) to the first. Additional principal components (each accounting for smaller amounts of variation) can also be defined in complex datasets.

3.3. Applying PCA to DIGE Datasets

Several software tools are available for DIGE analysis, and two of the more commonly utilized tools found in published experiments, DeCyder (GE Healthcare) and Progenesis SameSpots (Nonlinear Dynamics), shall be used as examples here. Both enable the straightforward assembly of complex DIGE experimental designs using the Cy2-labeled internal standard and independent replicates from multiple experimental conditions, and both enable the visualization of global variation using PCA and other multivariate statistics. Where they differ the most is in the approach to feature detection and missing values in the dataset (see Note 5). Both approaches have their unique strengths and limitations, but ultimately they deliver essentially the same information.

Regardless of software approach, care should always be taken to yield reproducible, high-resolution two-dimensional gel electrophoretic separations, and this should include optimization steps if necessary. In addition, this chapter assumes that the CyDye labeling, electrophoresis, and image acquisition procedures documented elsewhere in this volume have been followed using proper experimental design with a mixed-sample Cy2-labeled internal standard to coordinate the appropriate number of biological replicates that have been randomized with respect to dye labeling and gel position to compensate for unanticipated dye or gel biases.

3.3.1. Quick Guide: DeCyder

1. Inspect and import all gel images.
2. Detect feature boundaries and generate ratios on each gel independently in the differential in-gel analysis (DIA) software module, and then define groups and set up experimental design and matching between gels in the biological variation analysis (BVA) software module (this can all be set up in batch mode if desired). Normalized ratios are generated in BVA at this stage.
3. Assess matching on all images in BVA and manually adjust matching in all images following prescribed methods associated with the software. Ratios are recalculated as matching is adjusted within the dataset.
4. Ensure that the picking references are not included in the analysis by breaking matches for the picking references on at least one gel in the matched set (no more than two should be necessary, depending on how many mismatches you allow for in the EDA analysis). This will avoid skewing of the PCA results due to preferential fluorescence of the reference markers.
5. Open the extended data analysis (EDA) software module and begin a new project (the project must be closed in BVA for this but can be reopened once the data have been imported into the new EDA project).
6. Create in EDA a manual base set, with 100% of spot maps where protein is present (no missing values, see Note 6) and remove the unassigned spot maps.
7. Perform the following calculations:
    (a) Differential Expression Analysis, ANOVA.
    (b) Principal Component Analysis, select option to place spot maps into the score plot and proteins into the loading plot.
    (c) For hierarchical clustering (if desired), under Pattern Analysis select spot maps vs. proteins for both Proteins and Spot maps and exp groups.

8. Inspect PCA results. The score plots (samples) and loading plots (features) are displayed on separate graphs. Default settings place PC1 on *x*-axis and PC2 on *y*-axis. These can be changed up to PC5, and up to three PCs can be viewed at once (in a 3D view that can rotate). The values assigned to each PC are analogous to the percent of the overall variation that accounts for each component (see Note 7).

9. If necessary, construct a filtered set to perform additional data reduction (see Note 8). If performed, the analysis is no longer considered to be unsupervised but can be valuable in visualizing weak signals by removing background noise (see Subheadings 3.5 and 3.6).

3.3.2. Quick Guide: SameSpots

1. Import and inspect all gel images.
2. Assign groups and experimental design.
3. Ensure that the picking references are not included in the analysis by appropriately using the "Setting Mask of Disinterest" functionality. This will avoid skewing of the PCA results due to preferential fluorescence of the reference markers.
4. Assess alignment on all images and manually adjust matching in all images following prescribed methods associated with the software.
5. Generate normalized ratios on prealigned images (see Note 9).
6. Inspect the PCA results in Progenesis Stats. The score plots (samples) and loading plots (features) are displayed on a single overlapping graph (biplot). PCA and hierarchical clustering (if desired) can be selected under the "Ask another question" tab. Default settings place PC1 on *x*-axis and PC2 on *y*-axis. The values assigned to each PC are analogous to the percent of the overall variation that accounts for each component (see Note 10).
7. If necessary, construct a filtered dataset by setting tags and clusters and reevaluate in PCA.

Since PCA organizes independent samples based on the variation between them, *post hoc* identification of the samples enables the investigator to establish the relative signal-to-noise ratio (S/N) present in a dataset. When S/N is high, one can expect proper organization of the individual samples based on the first two or three principal components. When S/N is low, additional data reduction may be necessary to determine if biological variation is present in the experiment. And when technical noise is extremely high, the principal components can reveal organization of the samples that is not based on the biology but rather on some technical aspect of the sample preparation. What follows are examples of each of these scenarios, using the PCA functionality of DeCyder for illustration of the experimental variation.

3.4. Example 1: High Signal, Low Noise

Helicobacter pylori is the strongest known risk factor for gastric adenocarcinoma, yet only a fraction of infected persons ever develop cancer. In a recent study, Franco et al. used DIGE to quantify differences between two related strains of *H. pylori*, one of which caused only gastritis in rodents whereas the other also induced adenocarcinoma (6). The virulent strain was directly derived from the nonvirulent strain after passage through an animal, leading to the expectation that only a few key changes gave rise to the more aggressive nature of the virulent strain. Membrane and cytoplasmic fractions were analyzed from both genotypes, and each was produced as four independent (biological) replicates to control for unanticipated technical variation. The resulting 16 samples were coresolved on 8 coordinated DIGE gels along with a Cy2-labeled internal standard using standard methods as described elsewhere in this volume.

The technical noise of the DIGE platform has been demonstrated to be low (1), and in this case the additional variation (noise) derived from sample handling and normal biological variation was also expected to be low because the replicates involved clonal bacterial colonies, although the fractionation into cytoplasmic and membrane preparations could produce unanticipated technical variation. PCA performed on 842 features matched across all 8 gels (no missing values) demonstrated that the majority of variation (PC1 = 80%) was consistent with differences between cytoplasmic and membrane fractions, as expected. However, the second principal component (PC2), which described the second greatest source of variation between the samples, accounted for only an additional 5% variation, but this variation organized the samples derived from the virulent vs. nonvirulent strain (see Fig. 1). Thus, PCA clearly indicated low technical noise among the independently derived samples and clearly indicated that a low but significant level of biological variation was correlated with virulence.

That these two largest sources of variation correlated with expectations about the biology, rather than with technical issues such as dye-labeling bias, sample prep number (each set was prepared on separate days using different reagents), or other unanticipated, nonbiological factors, enabled these investigators to focus on the small number of virulence-related changes with high confidence and low expectations for false discovery, and led to a number of significant findings including an amino acid substitution (cysteine-to-arginine, causing a pI shift) in a flagellar protein that affected motility. This result was readily detected via DIGE coupled with mass spectrometry-based protein identification because the mutation altered the pI of the resolved, intact protein forms but otherwise did not affect the relative expression level of the flaA protein. As such, this alteration would most likely have gone undetected in a peptide-base, bottom-up LC-MS/MS strategy.

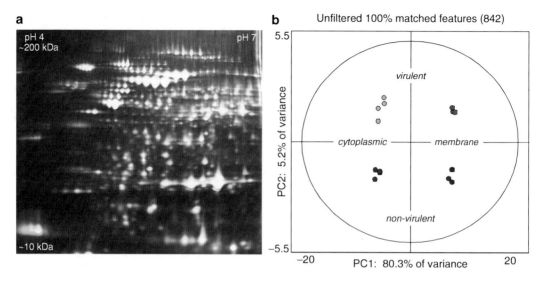

Fig. 1. Principal component analysis (PCA) showing a high signal-to-noise ratio. The cytoplasmic and membrane fractions from two related strains of *Helicobacter pylori* that differ in carcinogenic potential were analyzed by DIGE. Protein lysates were extracted and fractionated from the B128 (nonvirulent) and 7.13 (virulent) strains independently in quadruplicate, and 842 features were matched across all eight gels. (**a**) Representative DIGE gel from an 8-gel set used to coresolve the resulting 16 individual samples (labeled with either Cy3 or Cy5 using a dye-swapping strategy) along with the Cy2-labeled mixed-sample internal standard. (**b**) PCA was performed on the unfiltered dataset. The protein expression characteristics from 842 features from each individual sample are represented by each of the 16 data points in the PCA score plot. This analysis demonstrated that 80.3% of the variance (PC1) separated cytoplasmic from membrane samples as expected. An additional 5.2% of variance (PC2) separated the virulent strain 7.13 from the nonvirulent strain B128. Adapted from ref. (6).

3.5. Example 2: Low Signal, Low Noise

In another *H. pylori* study, Loh et al. investigated differential protein expression between wild-type and mutant strains deleted for the ArsS component of the ArsRS signal transduction system in response to growth in different pH media (7). The experimental design was similar to that described for Example 1, with $N=4$ independent (biological) replicates from two strains grown at two pH conditions, resulting in 16 samples coresolved across 8 DIGE gels, each of which contained an aliquot of a Cy2-labeled mixed-sample internal standard.

Six hundred and thirty-nine features were matched across all eight DIGE gels, all of which was evaluated by PCA. Using no missing values in the data (100% matching), PC1 accounted for 56.7% of the variation among these features and organized the samples by genotype, with none of the other principal components organizing the samples based on pH treatment (see Fig. 2). Thus, the biological signal with respect to genotype is high, but any signal consistent with pH-specific growth was too low to be visualized over the genotypic signal or any other technical noise present in the overall variation.

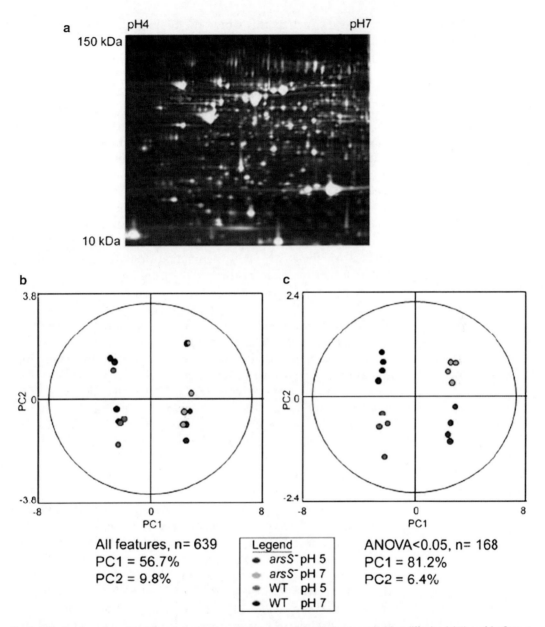

Fig. 2. PCA showing a low signal-to-noise ratio. Two related strains of Helicobacter pylori that differ by deletion of ArsS were grown in pH 5 vs. pH 7 medium, and protein expression was analyzed by DIGE. The experiment was independently repeated in quadruplicate, resulting in 16 samples coresolved across 8 DIGE gels along with a Cy2-labeled internal standard. (a) Representative DIGE gel from the 8-gel set. (b) PCA performed on the unfiltered 639 features matched across all eight gels indicated that the majority of variation among these features was consistent with strain-specific differences, but no other correlation with phenotype was revealed in the other principal components. (c) Refining the PCA to those features that exhibited a significant change between any group relative to the others (ANOVA $p < 0.05$) revealed that the second principal component was now consistent with pH-specific changes which were the subject of the study. Adapted from ref. (7).

In cases such as this where the biological signal of interest is not readily evident but the technical signal is low, additional data reduction may be helpful. In this case, low signals of interest may be revealed if the background noise (whether it be technical or biological in nature) can be removed from the PCA, and this can effectively be done by applying a statistical filter. Analysis of variants (ANOVA) is an appropriate univariate statistical test to use for this purpose when multiple variables are under consideration because it identifies features that are changing in one group relative to any of the others without specifying any pairwise comparisons (see Note 11).

In the example from Loh et al. (8), applying an ANOVA filter ($p<0.05$) to the dataset defined a subset of 168 features that were of biological significance for the experiment. Although this imposes the bias of the biological experimental design onto the PCA, it only selects those features that were changing significantly in one of the four classification groups relative to the other three, and for each feature this could be an independent classification. By removing those variations that contributed to the background noise in the original set of 639 features, PC1 (genotype) increased to 81% of the remaining variation, and now PC2 was found to organize the samples based on pH treatment, accounting for 6.4% of the variation or about ten features (see Fig. 2c). Thus, the experimental conditions appeared to induce changes of interest, but they were too subtle in magnitude and/or number to be visualized in the context of greater sources of variation in the experiment.

3.6. Example 3: Very Low Signal, Low Noise

In this third example (also from *H. pylori*, now cultured under different medium conditions), the signal is so low from any of the biological sources of variation that now dye-labeling bias appears to be ordered by the first principal component in an unfiltered analysis of 977 matched features across a six-gel DIGE experiment (see Fig. 3). Dye-labeling bias is known to exist and is typically controlled for using a dye-swap labeling scheme in the experimental design (see Note 12). That Cy3/Cy5 labeling bias appears here as the driving force behind PC1 (see Fig. 3b) indicates not only very low biological signal but also very low noise because this anticipated bias typically contributes very little to the overall variation. Thus, the finding of organization by dye labeling in PC1 is expected in the absence of significant signal or other sources of technical noise.

In this example, using a univariate ANOVA filter (imposing bias from the experimental design) prevents these changes from influencing the PCA when a dye swap is used in the experimental design (see Fig. 3c, Note 13). As was found in Example 2, applying the ANOVA filter now enables PCA to organize the samples based on the biological signal of different culture conditions. The weak biological signal represented in these 68 features is qualitatively evident by the loose nature of the sample clustering but nevertheless

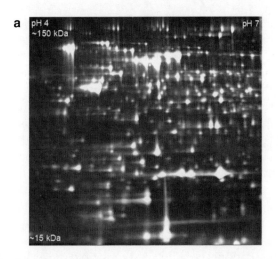

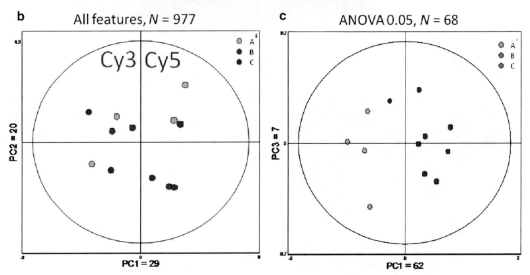

Fig. 3. PCA showing an extremely low signal-to-noise ratio. *Helicobacter pylori* cultured under three different experimental conditions: A, B, and C. The experiment was repeated four independent times, resulting in 12 samples coresolved across six DIGE gels along with a Cy2-labeled internal standard. (**a**) Representative DIGE gel from the 6-gel set. (**b**) PCA performed on the unfiltered 977 features matched across all six gels failed to organize the samples based on the anticipated biological manipulation but indicated a low level of variance (PC1 = 29%) that organized the samples based on Cy3/Cy5 dye-labeling bias. Such dye bias is known to exist; that it appears here indicates not only that the biological signal is not detected but also that the technical noise is similarly low because only dye bias, and not something else, is influencing PC1. (**c**) Refining the PCA to 68 features that exhibited a significant change between any group relative to the others (ANOVA $p < 0.05$) now begins to reveal some biologically relevant ordering of the samples based on the first two principal components.

demonstrates that the weak signal is present once the background noise (dye bias) is removed by imposing a biologically influenced filter over the data.

3.7. Example 4: Low Signal, High Noise

This last example depicts a worst-case scenario where, without inspecting the global variation with PCA, the experimental results might be misinterpreted. In this case, a primary mammalian cell line was cultured in the presence or absence of a drug at five time points, and this experiment was repeated four independent times resulting in a 20-sample, 10-gel DIGE experiment. By all conventional methods, the experiment appears to have worked very nicely as illustrated by high-resolution separations (see Fig. 4a) and the presence of some statistically significant univariate changes (see Fig. 4b). But PCA on the unfiltered dataset of 1,012 features matched across all ten gels clearly indicates that the principal components are organizing the samples based on cell passage number and not by treatment (see Fig. 4c).

Thus, a simple PCA on the unfiltered dataset can quickly determine if something went drastically wrong with the experiment and in this case saved the investigator considerable time and effort to follow up potential proteins of interest when clearly the major sources of variation were unrelated to the biology being manipulated. These findings raised the likelihood that the few changes "found" by univariate ANOVA arose stochastically and not from the drug treatment.

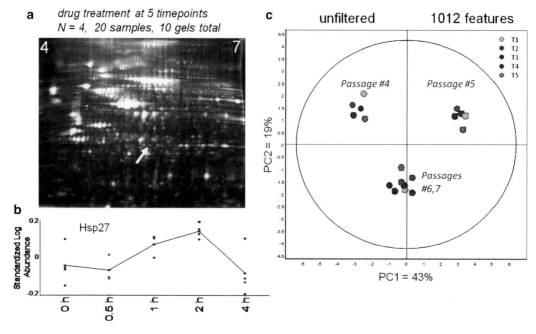

Fig. 4. PCA showing unanticipated high background noise. Primary mammalian cells were subjected to a drug treatment during four successive passages in tissue culture. Five time points were taken in each independent experiment, resulting in 20 samples that were coresolved across ten DIGE gels along with a Cy2-labeled internal standard. (a) Representative DIGE gel from the 10-gel set. (b) A protein feature (*white arrow* in *panel A*, identified by mass spectrometry as Hsp27) was identified as significantly changing (ANOVA $p<0.05$). (c) PCA performed on the unfiltered 1,012 features matched across all ten gels clearly organized the samples into three clusters defined by PC1 and PC2, with each cluster containing each of the 5 time points rather than the anticipated clustering by time point. Closer inspection of the samples revealed that the samples in each cluster came from the same cell passage number in the experiment. These results indicated that far more variation was derived from changes occurring in the primary cells during tissue culture, and this experiment was abandoned due to the tremendous risk that any observed univariate change (such as was found for Hsp27) simply arose stochastically rather than due to biologically significant reasons.

But without the utility of visualizing this very complex dataset with PCA, this high degree of potentially confounding background variation would likely have remained undetected.

4. Notes

1. Statistical power is the ability to visualize an X-fold magnitude change (effect size) at a Y-% confidence interval (e.g., $p < 0.05$) and be "correct" (usually expressed as power of 0.8, or 80% of the time it is "correct"). Statistical power depends heavily on the analytical variation of the instrumentation as well as the number of independent (biological) replicates. For example, the experimental noise of DIGE is extremely low owing to the internal standard experimental design, enabling statistically powered experiments with very few biological replicates (1).

2. Technical and biological replicates are important for ensuring the accuracy and biological significance of quantitative measurements. Technical replicates are necessary to control for variation in the analytical measurement. However, biological replicates are vital to assess whether or not changes in protein abundance/modification are descriptive of the biology rather than arising from unanticipated sources of experimental variation (e.g., inter-subject variation, sample preparation variation, analytical variation of the instrument). Technical replicates provide confidence in the result from the tested samples but do not provide any confidence to the biological relevance. Determining whether any observed changes come from the biology rather than technical variation can only be assessed with independent biological replicates. In cases where technical noise has been demonstrated to be sufficiently low, then biological replicates can also serve as technical replicates, which is what is typically done with the DIGE platform.

3. Pooling independent (biological) replicates should be done with extreme caution with respect to the statistical power of the resulting data and is not advised. If it is known *a priori* that technical variation is low between samples, then pooling can be effective, but if it is high, then a pooling strategy can be disastrous. Even with the low analytical noise of DIGE, the pooled $N=1$ comparison on a single gel assumes that the averaging of populations is reflective of biological signal. In some cases it may be valid to create subpools from a larger experiment, to either produce sufficient material or to minimize costs (the "economics of proteomics"), but in these cases it is still essential to maintain some degree of individualization of samples to retain statistical power (8).

4. These programs differ mostly in the algorithms used to detect protein features (boundaries) and gel-to-gel alignment/registration (e.g., vector-based, image warping). In general, they all provide powerful analytical tools coupled with univariate (Student's *t*-test, ANOVA) and multivariate statistical analyses (e.g., PCA and hierarchical clustering) that can be extremely beneficial in evaluating abundance changes of individual protein features as well as global expression patterns that can help discern changes that describe the biological phenotype from those that arise from unanticipated variation in the experiment.

5. SameSpots employs a strategy of image prealignment followed by the same spot boundaries applied to all features across a dataset. Whereas this leads to no missing values in the multivariable space, it can also condense information from multiple features into the same spot boundary in regions of feature crowding. The DeCyder approach defines spot boundaries on the different gels independently, and then employs a matching algorithm to register features across gels. While this yields a higher resolution for features in crowded regions, it typically involves more manual feature editing and can also lead to missing values in the dataset which must then be imputed during multivariable statistics (although there is the option to analyze only data that have been registered in every gel). Both platforms enable the editing of spot boundaries and feature registration across the gel set to abrogate these issues, and ultimately both are capable of delivering high-caliber quantitative DIGE results using both univariate and multivariate statistical analyses.

6. It is common to allow for mismatches in the dataset with DeCyder, thereby including some features that may not have a proper registration in every gel. However, this leads to missing values that then must be imputed into the multivariate analysis, which then imposes assumptions about data that do not exist. Focusing on 100% matches in DeCyder alleviates this issue at the expense of loosing real information from an uncorrected mismatch. Another approach is to construct the base set allowing for one or even two mismatches and then apply a 100% match filter to the data in step 9 in Subheading 3.3.1.

7. Up to five principal components can be selected in DeCyder, and these can be displayed in any 2- or 3-way combinations. Three-way comparisons are displayed on a three-dimensional projection that can be rotated with the computer mouse (right clicking and dragging).

8. Perform an ANOVA or Student's *t*-test in the unfiltered dataset along with the PCA and other calculations. Then, select "filter dataset" using this criterion with a *p*-value set to some meaningful threshold, such as $p < 0.05$ or 0.01. Then, repeat the PCA calculation with this new filtered dataset selected.

9. By performing the prealignment, the SameSpots approach enforces no missing values by having all features analyzed with the same spot boundaries. But this is at the expense of loosing information from features that are not well resolved on all gels (feature crowding).

10. SameSpots provides for the visualization of as many principle components as can be detected as contributing to the variance. Any of these principal components can be visualized in pairwise 2-dimensional displays.

11. When only two conditions are present (e.g., mutant vs. wild type), applying an ANOVA or Student's t-test to filter the data should necessarily designate PC1 as organizing the samples by condition, although in these cases this test may still be useful to assess the clustering within each classification to determine if the biological signals are uniform or influenced by other dimensions of principal components.

12. Dye bias is controlled for by labeling half of the samples with Cy3 and the other half with Cy5 from a given experimental condition. The groups separated by PC1 in Fig. 3b each have two members from each experimental condition, thereby negating any changes due to dye bias coming through a statistical test between biological groups.

13. Using the two-dye system, dye bias does not contribute to the experimental variation if the internal standard is always labeled with one dye (Cy3) and the experimental samples are always labeled with the other (Cy5).

References

1. Karp, N. A., and Lilley, K. S. (2005) Maximising sensitivity for detecting changes in protein expression: experimental design using minimal CyDyes, *Proteomics. 5*, 3105–3115.

2. Alban, A., David, S. O., Bjorkesten, L., Andersson, C., Sloge, E., Lewis, S., and Currie, I. (2003) A novel experimental design for comparative two-dimensional gel analysis: Two-dimensional difference gel electrophoresis incorporating a pooled internal standard, *Proteomics 3*, 36–44.

3. Friedman, D. B., Stauff, D. L., Pishchany, G., Whitwell, C. W., Torres, V. J., and Skaar, E. P. (2006) *Staphylococcus aureus* Redirects Central Metabolism to Increase Iron Availability, *PLoS Pathogens 2*, e87.

4. Friedman, D. B., Wang, S. E., Whitwell, C. W., Caprioli, R. M., and Arteaga, C. L. (2007) Multi-variable Difference Gel Electrophoresis and Mass Spectrometry: A Case Study on TGF-beta and ErbB2 signaling, *Mol Cell Proteomics 6*, 150–169.

5. Lilley, K. S., and Friedman, D. B. (2004) All about DIGE: quantification technology for differential-display 2D-gel proteomics, *Expert Rev Proteomics. 1*, 401–409.

6. Franco, A. T., Friedman, D. B., Nagy, T. A., Romero-Gallo, J., Krishna, U., Kendall, A., Israel, D. A., Tegtmeyer, N., Washington, M. K., and Peek, R. M., Jr. (2009) Delineation of a carcinogenic Helicobacter pylori proteome, *Mol Cell Proteomics 25*, 25.

7. Loh, J. T., Gupta, S. S., Friedman, D. B., Krezel, A. M., and Cover, T. L. (2010) Analysis of protein expression regulated by the Helicobacter pylori ArsRS two-component signal transduction system, *J Bacteriol 192*, 2034–2043.

8. Karp, N. A., and Lilley, K. S. (2009) Investigating sample pooling strategies for DIGE experiments to address biological variability, *Proteomics. 9*, 388–397.

Chapter 5

Analysis of Proteins Using DIGE and MALDI Mass Spectrometry

Witold M. Winnik, Robert M. DeKroon, Joseph S.Y. Jeong, Mihaela Mocanu, Jennifer B. Robinette, Cristina Osorio, Nedyalka N. Dicheva, Eric Hamlett, and Oscar Alzate

Abstract

Difference gel electrophoresis (DIGE) is a common technique for characterizing differential protein expression in quantitative proteomics. Usually a combination of enzymatic digestion and peptide analysis by mass spectrometry is used to identify differentially expressed proteins following separation and statistical analysis by DIGE. In this chapter, methods for gel spot picking, enzymatic digestion, and matrix-assisted laser desorption/ionization (MALDI) mass spectrometry (MS) for protein identification of DIGE-analyzed proteins are discussed. Two examples are given: first, a specific protein is used to test the sensitivity of the 2D DIGE/MALDI MS combination for protein quantification and identification, and second, several proteins with and without the labels typically used in DIGE are identified to demonstrate that these labels do not alter MS-based protein identification. Technical variations of protein gel spot preparation, in-gel digestion, and mass spectral protein identification are discussed.

Key words: MALDI-TOF/TOF mass spectrometry, 2D DIGE, Gel spot picking, Destaining, Dehydration, Enzymatic digestion, Peptide extraction, Data analysis, Proteomics

1. Introduction

Sensitivity of the 2D DIGE-MS (two-dimensional difference gel electrophoresis mass spectrometry) (1) technique is vital given the limited amount of tissue available for most proteomics experiments. Hence a proof-of-principle study was undertaken to determine the ability of 2D DIGE-MS to detect a protein standard at physiological levels. The sensitivity of the approach was tested by detecting decreasing amounts of the brain-derived neurotrophic factor (BDNF) protein at the low picomole and subpicomole range. Also, it is important to demonstrate that the Cy dyes, which are usually utilized to label the proteins under analysis, do not affect the mass spectrometric identification process of labeled proteins. This was demonstrated by identifying a group of Cy-dye-labeled and unlabeled proteins.

Although MS-based protein identification from 2D DIGE gels is a well-established technique (2), it has multiple procedural variations used in different laboratories. However, most of these technical variations share the following common steps: spot picking from polyacrylamide gels, protein destaining, gel dehydration, enzymatic digestion of the proteins and extraction of resulting peptides, mass spectral analysis, and MS data analysis and verification of the results. The following paragraphs present a brief overview of these steps, and more in-depth discussions are included in the protocols section and in the reference notes.

1.1. Spot Picking from Polyacrylamide Gels

Spot picking from polyacrylamide gels is performed from 2D gels either manually, using a variety of tools such as pipette tips cut to the size of the protein spot, plastic straws, metal punches, or automatically using a software-controlled robotic spot picker.

1.2. Destaining Procedures

Destaining procedures vary depending on the type of dye used to visualize and quantitate proteins in the gel. The choice of appropriate destaining procedures is important for successful MS protein analysis. In general, gel stains can be grouped into two major categories as MS-friendly and nonfriendly. For example, a traditional silver stain technique is sensitive, but it often requires a cumbersome destaining procedure that may result in poor protein digestion and poor extraction efficiency, and hence poor MS identification. On the opposite side of the staining technique spectrum is a classic Coomassie blue stain. It is less sensitive than silver stain but can be readily removed from a gel spot prior to the protein digestion and MS analysis. Colloidal Coomassie is a more sensitive version of the Coomassie blue dye. There are also several ultrasensitive fluorescent protein dyes compatible with downstream MS applications. For example, Sypro Ruby and Krypton dyes allow visualization of nanogram amounts of proteins and can be removed for subsequent analysis. A major disadvantage in the use of fluorescent dyes is in our inability to visualize these dyes with the naked eye, hence requiring fluorescence detectors, which may be extremely expensive.

1.3. Gel Dehydration

Following the destaining step, a protein-containing gel piece is usually dehydrated with acetonitrile and then rehydrated with a protease enzyme solution.

1.4. Enzymatic Digestion and Extraction of the Peptides

The most commonly used protease is trypsin because it produces predictably cleaved peptides, which are within the optimal mass range of most mass spectrometers and are, on average, long enough to produce informative tandem MS (MS/MS) spectra for peptide sequencing. MS-grade trypsin is chemically modified to minimize autolysis and the resulting MS "background" peaks. It is important for efficient digestion that proteins are denatured and fully available to the enzyme. This is ensured by S–S bond reduction and cysteine alkylation before, and often also after, electrophoresis ((1) and see Chapter 10). Different techniques have been reported to improve the efficiency of the digestion process, including sonication (3), microwave irradiation (4), high pressure (5), chemical additives, and gel shredding (6). After enzymatic digestion, peptides are extracted out of the gel into the solution.

Extraction protocols vary depending on the MS system and the method of analysis. Solid-phase extraction using C18 ZipTips is often employed to clean samples before plating them on the target plate for MALDI MS. Only low concentrations of organic solvents are allowed if an extract is to be directly loaded onto reverse-phase chromatographic material such as in C18 ZipTips, peptide microtraps, and LC columns. Although multiple extraction steps increase the amount of peptides extracted from a gel plug, significant sample loss may result from the subsequent sample transfer to a different tube and from solvent evaporation in vacuum evaporators. Trifluoroacetic acid (TFA) is commonly used to acidify peptide extracts for better ionization efficiency during MALDI-MS analysis but should not be used for electrospray ionization (ESI) MS analysis.

1.5. Mass Spectral Analysis

Mass spectral analysis can be performed with a variety of mass spectrometers with varying ionization mechanisms and mass analyzers (2). Several different ionization sources are utilized in mass spectrometry, with ESI and MALDI being the most popular types. MALDI and ESI ionization modes create different ionization conditions that affect the peptide's charge state, ionization efficiency, and observed mass spectrum. In general, MALDI produces singly charged peptide ions under vacuum, while ESI produces multiply charged ions at atmospheric pressure (AP). However, AP-MALDI is a relatively new MALDI ionization procedure that is increasingly employed with many types of mass analyzers. Since peptides respond differently to these ionization conditions, MALDI and ESI should be considered as complementary techniques.

Mass analyzers also differ in design and utility, often exploiting fragmentation of peptide ions in MS/MS applications. Fragmentation is typically collisionally induced after ionization and is often employed because it generally provides the most sensitive approach to peptide sequencing and offers one of the most reliable methods for protein identifications. Several mass analyzers dominate the field: linear ion traps, triple stage quadrupole, and TOF mass analyzers. Linear ion traps and various modern hybrid trap instruments produce high-quality MS/MS fragmentation spectra due to high ion storage capacity and fast scan times. Triple stage quadrupole mass spectrometers are ideally suited for targeted proteomics applications aimed at simultaneous monitoring from tens to hundreds of MS/MS single-ion peptide transitions in a quantitative manner. Ideally, these instruments are equipped with a high-performance liquid chromatograph delivering LC eluent to the ESI MS ion source with high accuracy at the micro- and nanoflow rates. The high accuracy of LC instruments allows the use of LC retention time as an important predictive and confirmatory parameter during the analysis of complex peptide mixtures. Triple stage quadrupole and hybrid quadrupole-ion trap instruments provide unique surveillance scans such as precursor and product ion scans, aiding in search for the fragmentation features that are common to a family of compounds. In proteomics, this feature is often used to find posttranslationally modified peptides. Commercial time-of-flight (TOF) and hybrid (TOF/TOF) instruments predominate MALDI-MS analyzers in the field due to their high speed and relatively lower costs. Many modern MS instruments provide high mass resolution, extended mass range, high mass accuracy, good calibration stability, and powerful "task-oriented" software.

1.6. Software-Based MS Data Analysis and Verification of the Results

There are numerous integrated MS data acquisition (operating systems) and data processing software packages. Data acquisition software operates an MS system and can make real-time choices during the MS data acquisition. This level of automation is designed to optimize data acquisition and results in minimum expenditure of time and sample consumption. In contrast to the data acquisition software, data processing software is usually installed on a computer or a computer cluster separate from the one that controls the MS analytical system. Modern proteomics software (ProteinPilot, GPS Explorer, Xcalibur, Scaffold, ProteoIQ, MassLynx, etc.) provides appropriate tools for protein identification from the MS and MS/MS data using one or more search engines (Mascot, Sequest, Peaks, Phenyx, X! Tandem, etc.). DIGE images can be created and annotated with the MS protein identification and DIGE quantitation results. Systems Biology software tools can then provide principal component and biological pathway network analyses, to name but a few.

In this chapter, we discuss MALDI MS-based protein identification from 2D DIGE-analyzed proteins using a 4800 MALDI-TOF/TOF mass spectrometer. The techniques described in the Subheadings 3 and 4 are routinely applied in our laboratory.

2. Materials

2.1. General Consumables

1. Pipettors: 0.1–10, 2–200, and 1,000 µL.
2. Pipette tips: 0.1–10, 2–200, and 1,000 µL.
3. 0.6-mL microcentrifuge tubes (Axygen, MCT-060-L-C).
4. 1.5-mL microcentrifuge tubes (Axygen, MCT-150-L-C).

2.2. Spot Picking

1. 70% Denatured ethanol.
2. 7% Acetic acid.
3. Ettan spot picker (GE Healthcare, Piscataway, NJ, USA).

2.3. Manual In-Gel Digestion

1. 25, 40, and 50 mM ammonium bicarbonate (ABC).
2. Acetonitrile (ACN).
3. 20 µg/mL sequencing- or MS-grade trypsin in 25 mM ABC—prepare immediately before use.
4. 0.1 and 3% trifluoroacetic acid (TFA).

2.4. MALDI-MS Sample Preparation

1. 10% TFA.
2. Dissolving solution: 50% MeOH/0.1% TFA.
3. Matrix solution: 50% ACN/0.1% TFA/5 mM ammonium citrate—prepare fresh before use.

2.5. MALDI-MS Analysis

1. Peptide calibration standard.
2. 4800 MALDI-TOF/TOF mass spectrometer (AB Sciex, Foster City, CA, USA).
3. GPS Explorer™ software (AB Sciex).
4. Mascot software (Matrix Science, London, UK).

3. Methods

3.1. Spot Picking Using a Commercial Spot Picker

1. Clean the Ettan spot picker thoroughly with 70% denatured ethanol and rinse with pure water.
2. Use a 1.4-mm pick head (included with the instrument).
3. Prime the syringe at least 5 times to remove air bubbles and set the syringe volume to 200 µL.
4. Set up the gel in the gel tray and cover it slightly with pure water (see Note 1).
5. Upload a pick list generated from the spot coordinates (see Note 2).
6. Manually detect reference markers (see Note 1).
7. Set up a 96-well plate on the plate platform included with the instrument.
8. Press "start" and let the picking robot perform spot picking automatically.
9. Upon completion of the picking, remove the water from each well, being careful not to remove the gel piece, and replace it with 5 µL of 7% acetic acid.
10. Store the plate at 4°C until further sample preparation (see Note 3).

3.2. In-Gel Protein Digestion with Trypsin

3.2.1. Destaining (see Note 4)

1. Cover the gel pieces with sufficient ACN:50 mM ABC in a 1:1 (v/v) ratio.
2. Shake the plate in a thermomixer for 10 min at 25°C.
3. Remove the destaining solution.
4. Repeat steps 1–3 until the pieces are clear (see Note 5).

3.2.2. Dehydration

1. Add 100 µL of ACN to the gel plugs.
2. Shake the plate in a thermomixer for 10 min at 25°C.
3. Discard the ACN.
4. Repeat steps 1–3.
5. Centrifuge the samples briefly to ensure that all gel spots are at the bottom (see Note 6).

3.2.3. Digestion (see Note 7)

1. Add cold trypsin solution to the samples.
2. Incubate on ice for 20 min and remove the excess enzyme solution.
3. Add 25 mM ABC to completely cover the gel plugs and incubate the plate in a thermomixer over night at 36°C.

3.2.4. Extraction of Peptides (see Note 8)

First extraction:

1. Add 50 µL of ACN on top of the buffer and shake the plate in a thermomixer for 10 min at 25°C.
2. Remove the solutions from each well and transfer individually to clean 0.6-mL Axygen tubes.

Second extraction:

3. Add 30 µL of pure water and 50 µL of ACN to each well and shake the plate in a thermomixer for 10 min at 25°C.
4. Remove the solutions and combine them with the first extraction solutions.

Third extraction:

5. Add 50–100 µL of ACN and shake the plate in a thermomixer for 10 min 25°C.
6. Remove the peptide solutions and add them to the combined extraction solutions.
7. Freeze the combined solutions for 2 h at −80°C and then lyophilize.

3.3. Preparation of Peptides for MALDI Mass Spectrometry

3.3.1. Peptide Reconstitution

1. Add 5 µL of dissolving solution into each digest-containing sample well (or tube).
2. Centrifuge for 1 min at 2,000 rpm.
3. Vortex the plate for 10 min at shaking position (or vortex tubes for 1 min).
4. Centrifuge again for 1 min at 2,000 rpm.

3.3.2. Spotting MALDI Samples on a Plate (see Note 9)

Dried-droplet technique

1. Spot 0.5–0.8 µL of sample solution on the center of a MALDI target spot. Discard the pipette tip.
2. Spot immediately 0.3–0.4 µL of matrix solution on the sample drop and mix both solutions pipetting 3–4 times if needed. Discard the pipette tip.
3. Repeat steps 1 and 2 for all samples.
4. Dry the spotted MALDI plate at room temperature.

3.4. MALDI Mass Spectrometry

3.4.1. Mass Spectrometry with the 4800 MALDI-TOF/TOF Proteomics Analyzer

MS spectra of protein digests are typically collected in the reflector positive ion mode with a mass range of 700–4,000 Da and 50 laser shots per desorption spot (1,250 shots per spectrum). The peptide MS peaks with a signal-to-noise ratio above 20 are selected for MS/MS analysis. A maximum of 45 MS/MS spectra are collected per MALDI sample spot. The precursor mass window is 200 relative resolution (fwhm). Calibration is performed internally using common trypsin autolysis peaks or externally using a standard peptide calibration mixture before each series of MS or MS/MS experiments (see Note 10).

3.4.2. MS Data Analysis

MS data search of all the spectra against the taxonomy of interest is performed using the NCBInr database, with the GPS Explorer™ Software v3.6 and Mascot search engine or with the Protein Pilot (vs. 3.0) software. Mass tolerance is usually set at 50–100 ppm for precursor ions and 0.5 Da for fragment ions. In Mascot usually up to two missed cleavages are declared, and oxidation (M) and acetylation (K) are declared as variable modifications. Cysteine alkylation is routinely performed for 2D gel analysis and hence should be declared as fixed modification (see Note 11).

3.5. Examples

3.5.1. Sensitivity of Detection

2D DIGE experiments are often based on the "minimal" labeling method, in which around 5% of total protein is covalently labeled ((1) and Chapter 10). For the experiment described below, the method was slightly changed by not adding lysine to the reaction

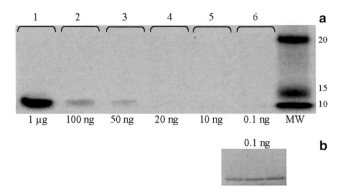

Fig. 1. 1D gel of decreasing amounts of NHS-Cy5-labeled BDNF. Amounts loaded on the gel range from 1 μg to 0.1 ng, as indicated at the bottom of the gel (**a**). The molecular weights of the protein standards are as indicated. Only the first three amounts are visible when the imager parameters are set below saturation values; (**b**) protein amounts in the 0.1 ng range can be visualized by increasing the imager parameters (PMT Voltage) above saturation values (>600 V). See text for details.

and letting the NHS dyes label the proteins further than 5%. However, the experiment is presented based on the total protein amount loaded and not on the extension of labeling.

Brain-derived neurotrophic factor (BDNF; accession number CAA62632; calculated molecular weight of 27,741 g/mol) was labeled with Cy dyes following labeling and cleanup procedures as described elsewhere ((1) and Chapter 10). A dilution series of this protein was run on a one-dimensional (1D) SDS PAGE gel (see Fig. 1). The amounts of protein loaded on the gel ranged from 1 μg (36.0 pmol) to 0.1 ng (3.6 fmol) as indicated in Fig. 1a. The gel was visualized with the Typhoon TRIO+ (GE Healthcare) scanner using the photomultiplier tube (PMT) setting below saturation levels. Protein amounts 1 μg and 100 and 50 ng were visible using these visualization parameters; by increasing PMT voltages above saturation levels (usually >600 V), it is possible to visualize lower amounts (down to 0.1 ng) as indicated in Fig. 1b. All protein bands indicated in Fig. 1a were digested and subjected to MALDI-MS analysis as described above. A variable number of BDNF peptides were identified (see Table 1), with a greater number of peptides identified in the more concentrated samples. Protein amounts 1 μg and 100, 50, and 20 ng were identified correctly as BDNF; however, lower amounts of protein were not identified. These results suggest that although it is possible to visualize and analyze protein changes in the low femtomole range (3.6 fmol in this example) with 2D DIGE, MALDI MS, as described here, is reliable for protein identification only in the 1–40 pmol range.

3.5.2. MS Identification of Cy-Labeled vs. Nonlabeled Proteins

To determine whether or not the presence of the Cy dyes affected the MS-based protein identification process, we run two 2D gels, both gels containing a total of 20 μg of a protein standard mixture.

Table 1
Protein identification results for decreasing amounts of BDNF by MALDI MS

Band	Protein name	Species	Database accession ID[a]	MW (Da)	Peptide count[b]	MS & MS/MS score[c]	Peptide sequenced ion score[d]	Scoring threshold[e]
1	BDNF	*Homo sapiens*	CAA62632	27740.9	10	240	197	64
2	BDNF	*Homo sapiens*	CAA62632	27740.9	8	227	197	64
3	BDNF	*Homo sapiens*	CAA62632	27740.9	8	249	219	64
4	BDNF	*Homo sapiens*	CAA62632	27740.9	4	100	93	64

[a]The protein sequence was searched at "http://www.ncbi.nlm.nih.gov/."
[b]Number of observed peptides matching the theoretical digest of the identified protein
[c]Combined score of the quality of the peptide mass fingerprint match and MS/MS peptide fragment ion matches
[d]Score of the quality of MS/MS peptide fragment ion matches only
[e]Significant score threshold. A hit with an "MS and MS/MS score" or an "ion score" above this value is considered a significant identification ($p < 0.05$)

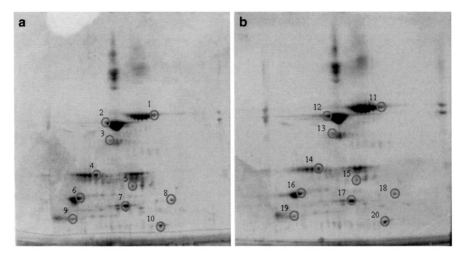

Fig. 2. Two 2D gels were run with unlabeled (**a**) and NHS-Cy5-labeled (**b**) proteins. Gels were Coomassie-stained, and equivalent protein spots (1–10 from A, and 11–20 from B) were excised from the gels, digested with trypsin, and identified by MALDI MS.

The protein standard mixture run in one gel was previously labeled with Cy5 minimal label, while the other was left unlabeled. After running, both gels were stained with Coomassie blue, and twenty equivalent protein spots were "picked" from both gels (see Fig. 2), digested with trypsin, and identified following the procedures described above. The results are shown in Table 2. All equivalent proteins, labeled and unlabeled, were identified as identical proteins; however, two pairs of protein spots (7 and 17, 10 and 20) were identified by MALDI MS as proteins not expected to be in the protein standard mixture. However, all identifications were consistent between the two gels. Despite the unexpected identifications, it is clear that the Cy5 label did not affect the protein identification process, as indicated by the MS and MS/MS scores (column 7 in Table 2) and the peptide sequenced ion scores (column 8 in Table 2), even the peptide counts were very similar (column 6 in Table 2).

4. Notes

1. When covering the gel with pure water, be careful not to add too much water, but rather add just enough to cover the gel surface, leaving the picking references visible. Otherwise, it will be difficult for the camera to detect the picking references. In some cases, e.g., after Coomassie blue staining, it is necessary to remove the acrylamide that is covering the picking references and replace the original picking tags with unstained tags.

Table 2
Results of protein identification by MALDI MS from labeled and unlabeled protein lysates. Two pairs of proteins (_underline_) were incorrectly identified

Spots in gels A and B (Fig. 2)	Protein name	Species	Database accession ID[a]	MW (Da)	Peptide count[b]	MS & MS/MS score[c]	Peptide sequenced ion score[d]	Scoring threshold[e]
1	Ovotransferrin	_Gallus gallus_	gi|83754919	75807.4	39	2,010	1735	83
2	Serum albumin precursor (_Bos taurus_)	_Bos taurus_	gi|30794280	69278.5	41	2,170	1887	83
3	Ovotransferrin	_Gallus gallus_	gi|83754919	75807.4	26	1,020	916	83
4	Concanavalin A—jack bean	_Canavalia ensiformis_	gi|72333	25556.8	13	1,190	1121	83
5	Carbonic anhydrase II (_Bos taurus_)	_Bos taurus_	gi|30466252	29095.7	11	377	326	83
6	Chain A, soybean trypsin Inhibitor	_Glycine max_	gi|3318877	20082.1	18	1,120	977	83
7	_Superoxide dismutase (Bos taurus)_	_Bos taurus_	_gi|27807109_	_15672.8_	_11_	_872_	_791_	_83_
8	Carbonic anhydrase II (_Bos taurus_)	_Bos taurus_	gi|30466252	29095.7	10	717	674	83
9	Concanavalin A—jack bean	_Canavalia ensiformis_	gi|72333	25556.8	6	562	522	83
10	_Ubiquitin (Drosophila melanogaster)_	_Drosophila melanogaster_	_gi|158767_	8540.0	7		_378_	_54_
11	Ovotransferrin	_Gallus gallus_	gi|83754919	75807.4	44	2,300	1929	83

(continued)

Table 2 (continued)

Spots in gels A and B (Fig. 2)	Protein name	Species	Database accession ID[a]	MW (Da)	Peptide count[b]	MS & MS/MS score[c]	Peptide sequenced ion score[d]	Scoring threshold[e]
12	Serum albumin precursor (*Bos taurus*)	*Bos taurus*	gi|30794280	69278.5	41	2,510	2249	83
13	Ovotransferrin	*Gallus gallus*	gi|83754919	75807.4	28	1,120	957	83
14	Concanavalin A—jack bean	*Canavalia ensiformis*	gi|72333	25556.8	13	1,220	1143	83
15	Carbonic anhydrase II (*Bos taurus*)	*Bos taurus*	gi|30466252	29095.7	12	448	391	83
16	Chain A, soybean trypsin inhibitor	*Glycine max*	gi|3318877	20082.1	15	633	525	83
17	*Superoxide dismutase* (*Bos taurus*)	*Bos taurus*	gi|27807109	15672.8	13	952	855	83
18	Carbonic anhydrase II (*Bos taurus*)	*Bos taurus*	gi|30466252	29095.7	9	609	572	83
19	Concanavalin A—jack bean	*Canavalia ensiformis*	gi|72333	25556.8	6	669	650	83
20	Ubiquitin (*Scyliorhinus torazame*)	*Scyliorhinus torazame*	gi|10719701	19471	10		584	54

[a] The protein sequence was searched at "http://www.ncbi.nlm.nih.gov/"
[b] Number of observed peptides matching the theoretical digest of the identified protein
[c] Combined score of the quality of the peptide mass fingerprint match and MS/MS peptide fragment ion matches
[d] Score of the quality of MS/MS peptide fragment ion matches only
[e] Significant score threshold. A hit with an "MS and MS/MS score" or an "ion score" above this value is considered a significant identification ($p < 0.05$)

2. Picking lists may be generated with the corresponding gel analysis software. In our laboratory, we use DeCyder 2D 7.0 (GE Healthcare) or Melanie 2D Gel Analysis Software (http://us.expasy.org/melanie/).

3. Before punching out gel spots for MS analysis, it is advisable to make holes as coordinate markers along the sides of the wet gel. Rescanning the gels at the 1:1 scale before and after spot picking shows that the excision of gel pieces is done correctly. The gels are firmly attached to the low-fluorescence glass support to prevent them from sliding during scanning and for better spot picking (1, 7).

4. Acceptable conditions for destaining gel pieces vary widely, depending on the stain and the staining method, ranging from, for example, (a) 25 mM ABC, 50% ACN at 37°C for 30 min with shaking (8), (b) two washes with ACN/25 mM ABC (70/30) (9), (c) 200 mM ABC, 40% ACN, destaining twice at 37°C for 30 min (10), and (d) 200 mM ABC, 50% ACN at 37°C for 45 min (11). Using high concentrations of ABC might require addition of larger amounts of TFA in the acidification solution to drop the pH after the digestion step, thus producing higher salt amounts that might impact microcrystal formation on a MALDI target if a prior desalting step is not included in the procedure.

5. For small gel pieces and low molecular weight proteins, it might be important to minimize predigestion time during which a gel plug is covered in solution. A study using metabolically radiolabeled recombinant proteins showed that peptide recovery from 0.5 mm gels was low and variable during fixing and staining (11). Hence, it might be important to keep gels and gel plugs for longer-term storage in a dry form (9). A multichannel automatic pipette can be used for a higher sample throughput. In order to reduce the chance of cross-contamination between sample wells, it is convenient to slice the plate-covering matt parallel to the sample columns or rows and open the matt portion over one plate lane at a time when the solvent exchanges are performed with a multichannel pipette.

6. At this point, gel plugs should be small, hard, and white. Quick sample evaporation in a vacuum evaporator will assure that gel drying is complete. Thin gel-loading pipette tips or Hamilton syringes are convenient tools for solvent removal while keeping small colorless gel plugs in place. For dehydration of small robotically picked gel plugs, it is better to use a smaller amount of ACN (approximately 50 μL) and then remove it entirely in a vacuum evaporator. Alternatively, solvent handling and evaporation can be performed using a Hamilton syringe through the pinholes made in a sample plate matt or a tube cap to prevent

gel plug contamination and to minimize the chance of an accidental removal of small gel plugs (12).

7. In-gel digestion protocols vary. For example, a Sigma-Aldrich protocol (10) recommends initially adding 20 µL of a 20-ng/µL trypsin solution prepared in a trypsin reaction buffer containing 40 mM ABC/9% ACN, followed by another 50 µL aliquot of the trypsin reaction buffer for a 4-h to an overnight digestion at 37°C. According to this protocol, the ABC/trypsin solution may be stored as frozen aliquots for up to 4 weeks. A digestion protocol from Pierce (8) requires initial incubation of a dry gel piece with a minimum volume (10–20 µL) of a trypsin solution that allows the gel piece to fully swell and hydrate at room temperature, which is then followed by the addition of 25 µL of a digestion buffer. Digestion is performed either for 4 h at 37°C or at 30°C overnight. In a similar protocol (11), dried gel pieces with or without prior cysteine alkylation are rehydrated with 20 µL of 0.02 µg/µL of sequencing-grade modified trypsin (Promega) or with 1.5 times gel volume if it is greater than 20 µL, in 40 mM ABC/10% ACN for 1 h at room temperature. An additional 50 µL of 40 mM ABC/10% ACN is then added, and the digestion continues for 16–18 h with agitation. Additional cysteine reduction and alkylation steps have been recommended in a 1D PAGE LC-MS approach for processing of silver-stained gel spots or for de novo sequencing (13). This step might be omitted to minimize the risk of protein loss and contamination, particularly if the reduction (dithiothreitol) and alkylation (iodoacetamide) are routinely performed during IPG strip equilibration in an SDS buffer prior to running the second dimension of a 2D gel. Thorough S–S bond reduction and alkylation are recommended before the second dimension in order to block the cysteine active thiol groups and prevent gel procedure artifacts, e.g., acrylamide adducts and gel streaking due to intra- and extramolecular protein covalent bonding.

8. There are several widely used peptide extraction procedures. In this laboratory, a robust extraction procedure utilizing C18 ZipTip purification is often used. Importantly, it avoids tube-to-tube transfers of peptide solutions and associated sample losses. The procedure starts with an addition of 5 µL of 3% aqueous TFA to stop the digestion and to facilitate peptide extraction from the gel plugs. The extraction is carried out at 37°C on a thermomixer (300 rpm), and the peptide-containing solution is then loaded directly onto the primed and wet C18 ZipTips (either regular or microsize according to the estimated peptide amount) following the ZipTip manufacturer's procedure. The peptide-depleted solution is then returned back to the original container of the gel piece, a microtube, for a second

extraction. The peptides are desalted and eluted in approximately 0.7–1.5 µL of α-cyano-4-hydroxycinnamic acid (CHCA) MALDI matrix dissolved in a 0.1% TFA solution of a 4:1 (v:v) mixture of ACN and water, which is sufficient to create three MALDI spots per sample using the dried-droplet technique. The first two spots are used directly for MALDI-MS and data-dependent MS/MS analysis, and the remaining spot is saved as a reserve for mass-targeted MALDI MS/MS. In the ZipTip elution dried-droplet technique, the sample is eluted and mixed with MALDI matrix by repeatedly drawing and expelling (ca. 20 times) the smallest possible amount of the eluting solution into a microcentrifuge tube while avoiding passing air bubbles through the ZipTip. MALDI matrix is dissolved in a solution with an increased acetonitrile content (0.1% TFA solution of a 4:1 (v:v) mixture of ACN and water) with respect to the final 1:1 ratio usually recommended for spotting, to prevent matrix precipitation upon mixing with the aqueous desalting solution wetting the ZipTip. The higher than 50% (v/v) content of organic solvent in the spotting solution enables spotting smaller volumes (0.5 µL or less) as organic solvents spread easily on metal surfaces, and it also reduces drying time. Proper crystallization exhibits milky amorphous spots with equally sized rounded crystals visible under a microscope, evenly distributed on the plate without clumping. Touching the plate surface with a ZipTip during spotting is not recommended as it often results in uneven crystallization and clumping. As a general guidance, sample transfer and solvent evaporation should be avoided during peptide extraction or peptide losses will occur. This was demonstrated in a study by Speicher et al. (11) using radiolabeled proteins by systematically evaluating peptide recoveries after in-gel trypsin digestion. It was found that approximately 80% of the labeled tryptic peptides could be extracted from gel bands containing 1–10 pmol of protein, and at least 70% could be extracted at 200- to 500-fmol levels. It was demonstrated that although 70–85% of tryptic peptides could be extracted from gels over a range of conditions and protein concentrations, further extraction attempts resulted in substantial additional losses. Even minimal handling resulted in adsorption loss of about 10–15% of extracted peptides. Adsorptive losses to plastic surfaces were particularly high, sometimes greater than 50%, and were variable if extracts were partially dried in a SpeedVac to concentrate the sample or to remove ACN. Although the ZipTip purification procedure is labor-intensive, it provides consistently good MS data quality and sensitivity in MALDI MS. CHCA MALDI matrix solutions can elute peptides from C18 ZipTips directly onto a MALDI plate, and the thorough mixing of peptides with the matrix creates excellent matrix crystallization conditions. It eliminates extraneous salt adduct MS peaks due to the thorough desalting performed before sample elution. This method provides good peptide recovery (50–70%), and it concentrates peptides in a small solvent volume providing significantly improved sensitivity of MS analysis.

9. There are numerous reports of additional spotting techniques, MALDI matrices, and additives that improve the limit of detection under specific MALDI-MS conditions. For example, in a modification of the common dried-droplet technique, 0.5–1 µL of sample solution is deposited on the MALDI plate and then mixed by aspiration with a comparable volume of matrix solution before drying the droplet in the air or nitrogen (15). Since this technique decouples sample deposition from the matrix addition step, additional sample preparation can be performed directly on the MALDI plate, including on-plate enzymatic digestion and MALDI-MS calibrant addition. Dried sample spots can be desalted directly on the MALDI plate using a crystal washing procedure that might improve MALDI-MS spectra by reducing extraneous adducts and ionization suppression. Vacuum drying has been reported to produce uniform crystals in a dried-droplet method (16). Application of matrix additives, e.g., ammonium monobasic phosphate or ammonium dibasic citrate, reduces sodium and potassium metal ion cluster interferences, thus improving peptide ion intensity and signal-to-noise ratio for peptide samples spotted at low femtomole level (17). A thin-layer spotting technique is applied to achieve enhanced sensitivity of peptide detection, for instance, for sample concentrations around 0.1 pmol/µL or lower (18, 19). The technique requires a polished MALDI plate without etched sample miniwells. Spotting and drying of the matrix and sample solutions are performed separately. Because the matrix and sample layers are thin, the number of laser shots per desorption spot needs to be reduced to below 20. In an optimized thin-layer spotting protocol, the CHCA matrix (20 g/L) and nitrocellulose (5 g/L) are introduced as a thin layer onto a MALDI target in acetone/2-propanol (1:1; v:v). An equivalent sample volume is introduced onto the matrix surface, the solvent evaporated, and the sample is desalted by surface washing using a small amount of water (19). Peptides stored on a spotted MALDI plate are, in general, relatively stable and can be reanalyzed days or weeks later, especially if the plate is kept in a dry and oxygen-free environment. Tryptophan and methionine are, of all amino acid residues, the most prone to oxidation which may occur during MALDI-MS sample preparation. Therefore, in order to increase protein coverage, appropriate amino acid residue modifications can be included in the search parameters during software-assisted database searches of the

MALDI-MS and MS/MS spectra of proteins and peptides. The phenomenon of artifactual oxidation was recently demonstrated by SL Cohen (14), whereby oxidation artifacts were found on methionine and tryptophan residues.

10. A fully spotted stainless steel MALDI target plate should contain several standard calibration spots which are needed for mass calibration. Ideally, there are three MALDI spots (technical replicates) made per sample in a sample spot set. Depending on the fluorescence intensity of the gel spot, either one or two spots per sample are analyzed automatically using the MALDI MS and automated MS/MS data acquisition options in which the 10 top-intensity MS peaks are subjected to the subsequent MS/MS analysis. If two spots are analyzed, the MS/MS precursor ion masses from the first spot are excluded from the second spot analysis by using an "N first precursors to skip" option in the Interpretation Method of the instrument-control software. Importantly, a MALDI-MS spectrum of a "digestion blank" is acquired before the MS/MS data acquisition begins, and the top background ions are placed on a permanent exclusion list to stop the MALDI MS/MS software from acquiring their MS/MS spectra. The MS/MS exclusion list should also contain major human keratin and trypsin autolysis peaks. The following set of instrument parameters can be applied to enhance MS and MS/MS data quality for weaker 2D DIGE spots processed using the dried-droplet MALDI spotting method: total laser shots per spectrum/desorption spot is ~8,000/200 for the MS and ~8,000/100 for the MSMS; laser intensity is fixed at the level set to produce approximately $1-6 \times 10^4$ ion counts; and MS/MS acquisition stop conditions are in effect after 10–15 consecutive subacquisitions from desorption spots fail to pass acceptance criteria (S/N above 4–10) or after an accumulated spectrum reaches S/N of 50–80 with a minimum of 8–10 peaks above the threshold level and a minimum of 15–20 subacquisitions accepted before the S/N test stops. These MS/MS parameter ranges are set to preserve sample while providing spectral quality by stopping the system from acquiring poor quality MS/MS spectra which would result in no peptide identification, and to stop data acquisition as soon as a good quality MS/MS spectrum has been collected. In order to preserve samples further, a more laborious manual stepwise procedure can be applied, in which the peptide mass fingerprint (MS spectrum) is used to obtain an initial precursor MS/MS ion list for targeted MS/MS verification of the top protein candidates. The best possible quality MS spectrum is usually acquired first, followed by the MS/MS spectra, if possible, in the order of their importance to protein identification. If faint gel spot peptide extracts fail to produce clear top

protein candidates after the peptide mass fingerprint acquisition, then isoelectric point and molecular weight range information obtained from the 2D gel can help restrict the search parameters. Should this approach also fail, one can compare the MS spectra obtained for all the protein and blank gel spots in the data acquisition set searching for small and unique MS peaks to be used as precursors for MS/MS experiments (if available, AB Sciex Peak Explorer software can be of assistance in this task). Using a "Link Spectrum Traces" option allows resizing the mass-to-charge axis range for all the displayed spectra facilitating the spectral comparison and analysis. Once a preliminary protein identity emerges from the initial MS and MS/MS experiments and from the analysis of the location of the spots on the 2D gel, the remaining MALDI spots are used to increase the protein coverage and to differentiate between different protein forms. On the MALDI 4800 TOF/TOF instrument, MS/MS data can be acquired either with or without the collision gas with similar results.

11. Both MS and MS/MS spectra are acquired during the initial automated MALDI run. High-resolution MS spectrum of a single protein digest can be used for protein identification. This is called peptide mass fingerprinting and is independent of the protein identification procedure that relies on matching experimental MS/MS peptide spectra to the in silico theoretical MS/MS spectra generated for each protein digest in a database. Proteomic "Aldente" software is frequently used in our laboratory for this task (20). First, MALDI-MS peak lists are background-subtracted in a Microsoft Excel spreadsheet to remove MS contaminant peaks (see Notes 9 and 10). Then the resulting MS peak lists are submitted to Aldente, initially for further contaminant identification (usually human keratins), followed by the protein identification to give an initial protein and peptide report. This report is then compared with the protein identification data obtained from the automated data acquisition MS/MS experiments processed by the Protein Pilot or Mascot software in search for additional MS peptide peaks lacking their MS/MS counterparts. These MS peaks proposed by Aldente are used to acquire more MS/MS data using the remaining MALDI spot(s) in order to maximize protein coverage and thus protein identification probability. The final cumulative data report in Excel shows a list of peptides and proteins with their probability scores. The probability scores depend on the proteomic software platform, for example, Protein Pilot characterizes quality of every match between a MS/MS spectrum of an identified peptide and its theoretical barcode-like spectrum with a peptide probability score 0–99. The protein probability score corresponds to a product of the

individual probabilities assigned to the peptide MS/MS spectra; therefore, it is important for the overall identification success rate of protein identification to allow, at the minimum, two high-quality MS/MS unique peptide matches characterized with a peptide probability score above 95%.

Acknowledgments

The research described here was possible thanks to financial support from the Systems-Proteomics Center—University of North Carolina at Chapel Hill. We would like to thank Dr. Roger Madison (Duke University Medical Center) for helping with the acquisition and analysis of human BDNF.

Disclaimer: The research described in this article was reviewed by the National Health and Environmental Effects Research Laboratory, United States Environmental Protection Agency, and approved for publication. Approval does not signify that the contents necessarily reflect the views of the Agency, nor does mention of trade names or commercial products constitute endorsement or recommendation for use.

References

1. Diez R, Herbstreith M, Osorio C, Alzate O (2009) 2-D Fluorescence Difference Gel Electrophoresis (DIGE) in neuroproteomics. In: Alzate O (ed) Neuroproteomics, CRC Press, Boca Raton
2. Parker CE, Warren MR, Mocanu V (2009) Mass Spectrometry for proteomics. In: Alzate O (ed) Neuroproteomics, CRC Press, Boca Raton
3. Rial-Otero R, Carreira RJ, Cordeiro FM, Moro AJ, Santos HM, Vale G, Moura I, Capelo JL (2007) Ultrasonic assisted protein enzymatic digestion for fast protein identification by matrix-assisted laser desorption/ionization time-of-flight mass spectrometry. Sonoreactor versus ultrasonic probe. J Chromatogr A 1166(1–2):101–107
4. Pramanik BN, Mirza UA, Ing YH, Liu YH, Bartner PL, Weber PC, Bose AK (2002) Microwave-enhanced enzyme reaction for protein mapping by mass spectrometry: A new approach to protein digestion in minutes. Protein Science 11:2676–2687
5. Lopez-Ferrer D, Petritis K, Hixson KK, Heibeck TH, Moore RJ, Belov ME, Camp II DG, Smith RD (2008) Application of pressurized solvents for ultrafast trypsin hydrolysis in proteomics: proteomics on the fly. J Proteome Res 7:3276–3281
6. Lazarev AV, Rejtar T, Dai S, Karger BL (2009) Centrifugal methods and devices for rapid in-gel digestion of proteins. Electrophoresis 30:966–973
7. Lanne B, Potthast F, Höglund Å, Brockenhuus von Löwenhielm H, Nyström AC, Nilsson F, Dahllöf B (2001) Thiourea enhances mapping of the proteome from murine white adipose tissue. Proteomics 1:819–828
8. Pierce in-gel digestion protocol: http://www.piercenet.com/files/1468dh4.pdf
9. Panfilov O, Lanne B (2002) Peptide mass fingerprinting from wet and dry two-dimensional gels and its application in proteomics. Analytical Biochemistry 307:393–395
10. Trypsin profile IGD kit for in-gel digests: http://www.sigmaaldrich.com/etc/medialib/docs/Sigma/Bulletin/pp0100bul.Par.0001.File.tmp/pp0100bul.pdf
11. Speicher KD, Kolbas O, Harper S, Speicher DW (2000) Systematic analysis of peptide recoveries from in-gel digestions for protein identifications in proteome studies. Journal of Biomolecular Techniques 11:74–86

12. Winnik WM, Alzate O, Bruno M, Burgoon L, Ge Y, Klinefelter GR, Kodavanti PR, Robinette JB, Suarez J, Wallace K (2010) Improved method for identification of low abundance proteins using 2D-gel electrophoresis, MALDI-TOF and TOF/TOF. Conference proceedings, American Society for Mass Spectrometry, Salt Lake City, UT, May 2010
13. Shevchenko A, Tomas H, Havli J, Olsen JV, Mann M (2007) In-gel digestion for mass spectrometric characterization of proteins and proteomes. Nature Protocols 1:2856–2860
14. Cohen SL (2006) Ozone in ambient air as a source of adventitious oxidation. A mass spectrometric study. Anal Chem 78:4352–4362
15. Moseley III MA, Sheeley DM, Blackburn RK, Johnson RL, Merrill BM (1998) "Mass Spectrometry in Drug Discovery", in Mass Spectrometry of Biological Materials, 2nd edition, ed. by Barbara Larsen and Charles N. McEwen, Marcel Dekker, Inc., NY, p.162
16. Weinberger SR, Boernsen KO, Finchy JW, Roberstson V, Musselman BD (1993) Proceedings of the 41th Annual ASMS Conference on Mass Spectrometry and Allied Topics; San Francisco, May 31-June 5, p. 775a
17. Zhu X, Papayannopoulos IA (2003) Improvement in the detection of low concentration protein digests on a MALDI TOF/TOF workstation by reducing alpha-cyano-4-hydroxycinnamic acid adduct ions. J Biomol Tech. 14:298–307
18. Vorm O, Roepstorff P, Mann M (1994) Improved Resolution and Very High Sensitivity in MALDI TOF of Matrix Surfaces Made by Fast Evaporation. Anal Chem 66:3281–3287
19. Shevchenko A, Wilm M, Vorm O, Mann M (1996) Mass Spectrometric Sequencing of Proteins from Silver-Stained Polyacrylamide Gels. Anal Chem 68:850–858
20. http://www.genebio.com/products/phenyx/aldente/index.html

Chapter 6

Synthesis and Validation of Cyanine-Based Dyes for DIGE

Michael E. Jung, Wan-Joong Kim, Nuraly K. Avliyakulov, Merve Oztug, and Michael J. Haykinson

Abstract

The application of difference gel electrophoresis (DIGE), in particular its most common "minimal labeling" variety, utilizes N-hydroxysuccinimide esters of Cy2, Cy3, and Cy5 dyes, which are commercially available. We describe methods for the efficient synthesis of all three dyes from relatively inexpensive and commercially available precursors in only a few steps and with relatively high yields. In model DIGE experiments, the newly synthesized dyes proved to be indistinguishable from commercially available ones and have been shown to be stable for years while stored under argon as dry solids or after being dissolved in N,N-dimethylformamide.

Key words: DIGE, Difference gel electrophoresis, Benzoxazolium Cy2, Propyl Cy3, Ethyl Cy5, Cy dye synthesis

1. Introduction

Difference gel electrophoresis (DIGE) is the latest in a series of major advances in the field of two-dimensional (2D) gel electrophoresis, which was originally introduced by O'Farrel (1). The key feature of the DIGE method relies on labeling of protein samples before electrophoresis with fluorescent dyes (originally using two N-hydroxysuccinimide (NHS) esters of Cy3 and Cy5) followed by protein separation of the combined samples on the same 2D gel (2). Multiplexed labeled proteins comigrate in the gel ensuing full coregistration of spectrally different gel images, thus minimizing gel-to-gel technical variation commonly associated with regular 2D gels. The introduction of a third Cy2-based dye allowed the implementation of an internal standard for DIGE, which further improved the method. The internal standard concept is based on labeling with Cy2 a pooled sample consisting of equal aliquots of all samples. The same amount of the Cy2-labeled standard is then loaded on all the gels in the experiment along with individual samples labeled with Cy3 and Cy5. The ratios of Cy3-to-Cy2 and Cy5-to-Cy2 can then be calculated and normalized against the Cy2-labeled internal standard (3). This approach resulted in further elimination of technical noise in DIGE experiments and enabled the analysis of the relative abundances for each protein spot resolved on gels with high precision and unprecedented statistical power (4, 5).

The DIGE method was originally introduced in 1997 (2); in a few years, it was commercialized by GE Healthcare, which offered a full line of reagents and equipment necessary to successfully implement the technology, including the dyes for labeling. Besides being commercially available, the dyes could be synthesized using known procedures (2, 6, 7). In this chapter, we provide detailed procedures for the efficient synthesis of all three dyes used in DIGE, benzoxazolium dye Cy2, and two indocyanine dyes, propyl Cy3 and ethyl Cy5, and their NHS esters, from commercially available precursors. We also validated the newly synthesized dyes in DIGE experiments by comparing them to commercially available dyes and evaluated the effects of long-term and short-term storage of the synthesized dyes on DIGE output.

2. Materials

2.1. Dye Synthesis

The following reagents were obtained from Sigma-Aldrich: 2-methylbenzoxazole, bromoethane, 1,2-dichlorobenzene, 4-(bromomethyl) phenylacetic acid, N,N'-diphenylformamidine, N,N'-disuccini-midyl carbonate (DSC), 2,3,3-trimethyl-3H-indole, 1-bromopropane, 6-bromohexanoic acid, malondialdehyde bis(dimethyl acetal), 1,3,3-trimethyl-2-methyleneindoline.

Other reagents and common solvents were obtained from Fisher Scientific: anhydrous N,N-dimethylformamide (DMF), diethyl ether, acetonitrile, acetic anhydride, dichloromethane, hexane, methanol, absolute ethanol, anhydrous triethylamine, anhydrous pyridine, glacial acetic acid, anhydrous sodium acetate, and HCl. Aniline was obtained from Alfa Aesar.

2.2. DIGE Experiments

The following reagents and materials were supplied by GE Healthcare: urea, thiourea, CHAPS, 2D Quant kit, CyDye fluor minimal dyes Cy2, Cy3, and Cy5, and IPG strips. Other reagents were obtained from Sigma-Aldrich (DMF, SDS, DTT, iodoacetamide, L-lysine, magnesium acetate), and Fisher Scientific (glycerol, isopropanol).

All 2D electrophoresis equipment, the IPGPhor II (first-dimension isoelectric focusing) and the DALT 12 apparatus (second-dimension SDS-PAGE) as well as the DeCyder v. 6.5 software for DIGE analysis were obtained from GE Healthcare.

3. Methods

3.1. Dye Synthesis

General notes

All reactions were carried out under an atmosphere of nitrogen, and all commercial reagents were used as provided by the manufacturers. ^{1}H and ^{13}C NMR data were obtained on a Bruker 400 MHz spectrometer. ^{1}H NMR and ^{13}C NMR data are reported in parts per million (d) downfield from tetramethylsilane. The following abbreviations are used: s (singlet), d (doublet), t (triplet), q (quartet), m (multiplet), and br (broad). Infrared spectra were recorded on a Thermo Nicolet Avatar 370 FT infrared spectrophotometer as a liquid film or as a thin crystalline film. All IR data are reported in wavenumbers (per cm). Thin-layer chromatography (TLC) was performed using Merck silica gel 60F254 0.2 mm alumina-backed plates. Visualization was accomplished using ultraviolet light or one of the following stains: anisaldehyde, phosphomolybdic acid, and potassium permanganate. Flash chromatography was carried out using ICN Biomedicals silica gel 60 (230–400 mesh). Mass spectrometry analysis was carried out using electrospray ionization (ESI) on an IonSpec FT mass spectrometer.

3.1.1. Cy2 Synthesis

The synthesis of the Cy2, the benzoxazolium dye, and its NHS ester (see Fig. 1) began with alkylation of the commercially available 2-methylbenzoxazole (1) with two different alkyl halides: alkylation with ethyl bromide gave the ethyl salt (2) (8) in 22% yield (this compound is also commercially available), while alkylation with the commercially available 4-bromomethylphenylacetic acid (3) gave an 80% yield of the salt (4) (9). Condensation of the ethyl salt (2) with diphenylformamidine in the presence of excess acetic anhydride as solvent afforded the acetanilidylvinyl indolium salt (5) in 93% yield (10–13). This compound was then reacted with the other benzoxazolium salt (4) in ethanol in the presence of triethylamine to give the desired dye (6) in 75% yield as a deep yellow powder. The dye was easily converted into the NHS ester by treatment with DSC in the presence of pyridine to give the activated dye (7) in 86% yield. The spectroscopic data were in agreement with the structures assigned (see Note 1).

A. *N-Ethyl-2-methylbenzoxazolium bromide* (2). A mixture of 2-methylbenzoxazole (1) (Aldrich, 2.24 g, 0.01 mol) and bromoethane (Aldrich, 2.5 mL, 0.03 mol) in 1,2-dichlorobenzene (Aldrich) was heated at 110°C for 24 h. The solution was cooled to room temperature, and the residue obtained was filtered and washed with diethyl ether. The solid obtained was dried under vacuum to give the salt (2) as a white powder (0.60 g, 22%) (14, 15). This compound is also commercially available from various suppliers (Aldrich, Alfa Aesar, TCI, etc.). IR (neat): 3,084, 3,047, 2,974, 2,929, 2,859, 2,729, 1,593,

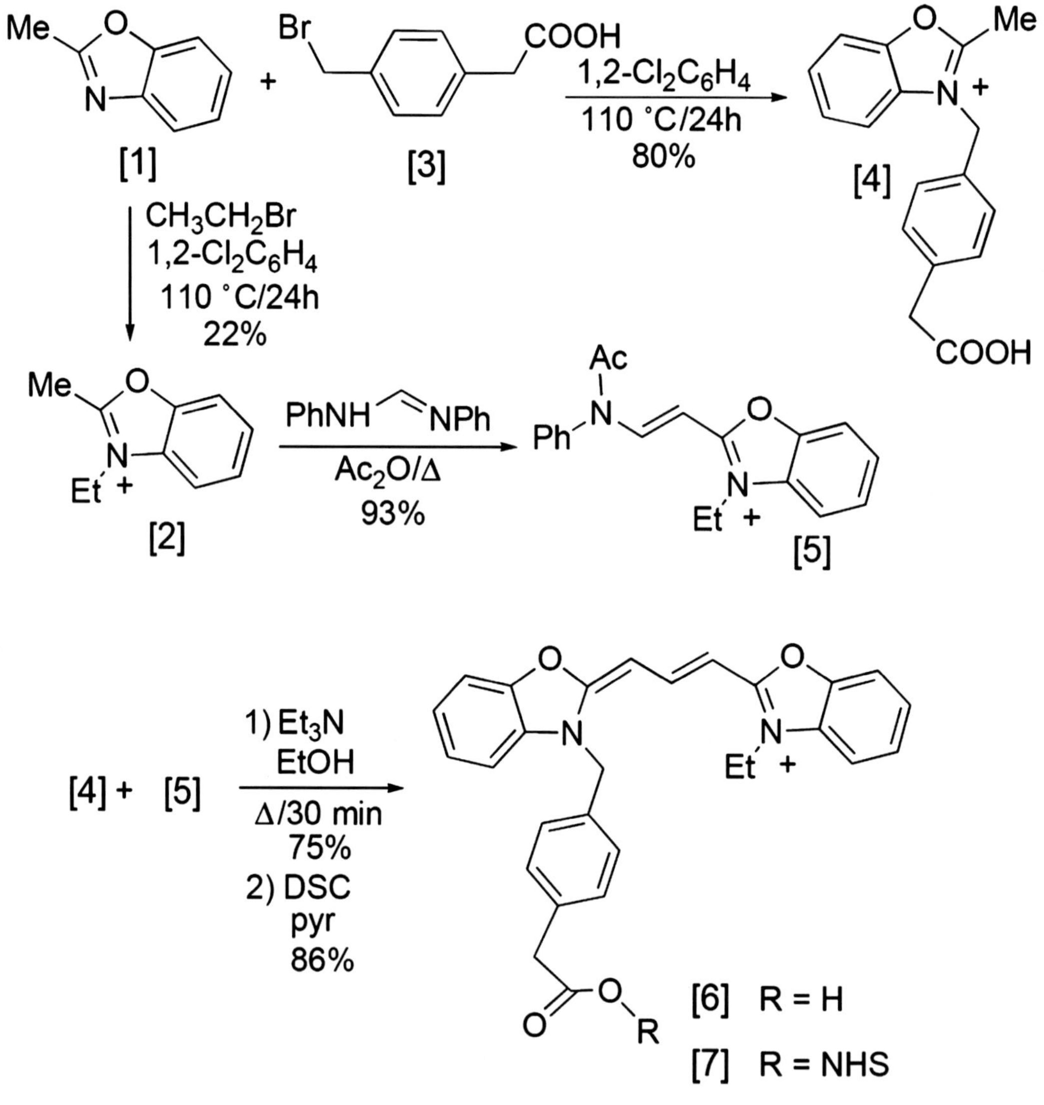

Fig. 1. Synthesis of the benzoxazolium Cy2 dye and its *N*-hydroxysuccinimide (NHS) ester.

1,462, 1,388, 1,188, 1,147, 1,025, 759/cm. ^{1}H NMR (400 MHz, DMSO-d_6): δ 8.09–8.23 (2H, m), 7.71–7.80 (2H, m), 4.60 (2H, q, *J*=7.3 Hz), 3.02 (3H, s), 1.43 (2H, t, *J*=7.3 Hz). ^{13}C NMR (100 MHz, DMSO-d_6): δ 169.03, 147.91, 129.73, 129.04, 128.13, 115.00, 113.42, 42.38, 13.84, 13.40.

B. *N-((4-Carboxymethyl)phenylmethyl)-2-methylbenzoxazolium bromide* (**4**). A mixture of 2-methylbenzoxazole (**1**) (Aldrich, 2.24 g, 0.01 mol) and 4-(bromomethyl)phenylacetic acid (**3**) (Aldrich, 3.22 g, 0.01 mol) in 1,2-dichlorobenzene (Aldrich) was heated at 110°C for 12 h. The solution was cooled to room temperature, and the residue obtained was filtered and washed with acetonitrile. The solid obtained was dried under vacuum to give the salt **(4)** as a light yellow powder (3.20 g, 80%) (9). IR (neat): 3,014, 1,733, 1,579, 1,456, 1,360, 1,226, 1,164, 753/cm. ^{1}H NMR (400 MHz, DMSO-d_6): δ 7.13 (4H, s), 7.01 (1H, br t, J=7.8 Hz), 95 (1H, br d, J=8.1 Hz), 85 (1H, br d, J=7.8 Hz), 68 (1H, br t, J=7.6 Hz), 5.20 (1H, d, J=15.0 Hz), 4.19 (1H, d, J=15.0 Hz), 3.48 (2H, s), 3.17 (3H, s). ^{13}C NMR (100 MHz DMSO-d_6): δ 173.18, 170.49, 153.38, 1375, 133.98, 130.15, 130.04, 129.54, 129.46, 128.38, 119.71, 117.15, 50.79, 40.84, 22.21.

C. *2-(2-Phenylacetamido-E-1-ethenyl)-N-ethylbenzoxazolium salt* **(5)**. A mixture of *N,N'*-diphenylformamidine (Aldrich, 0.38 g, 1.98 mmol) and the salt **(2)** (0.26 g, 1.65 mmol) in acetic anhydride (10 mL) was refluxed for 30 min. The solution was cooled to room temperature, the solvent was removed under reduced pressure, and the residue was purified by flash chromatography on silica gel (dichloromethane/hexane/methanol = 5:1:1) to give the salt **(5)** as a light yellow powder (0.41 g, 93%) (9). IR (neat): 3,084, 3,064, 2,978, 1,719, 1,646, 1,613, 1,589, 1,491, 1,466, 1,413, 1,372, 1,319, 1,252, 1,151, 1,004, 755, 702/cm. ^{1}H NMR (400 MHz, CDCl$_3$): δ 9.19 (1H, d, J=13.8 Hz), 7.26–7.83 (9H, m), 5.33 (1H, d, J=13.8 Hz), 4.44 (2H, q, J=7.4 Hz), 2.02 (3H, s), 1.31 (3H, t, J=7.4 Hz).^{13}C NMR (100 MHz, CDCl$_3$): δ 173.25, 169.92, 162.86, 147.68, 1488, 138.12, 131.30, 130.86, 129.62, 128.57, 127.85, 113.93, 112.21, 87.14, 42.36, 23.45, 13.79.

D. *2-(3-(3-(4-(2-carboxyethyl)phenylmethyl)-2(3H)-benzoxazolylidene)-1-propenyl)-3-ethylbenzoxazolium, Cy2* **(6)**. A mixture of the enamide **(5)** (0.10 g, 0.32 mmol), the salt **(4)** (0.92 g, 0.32 mmol), and dry triethylamine (0.1 mL) in absolute ethanol (10 mL) was refluxed for 30 min. The solvent was removed under reduced pressure and the crude residue was purified by flash chromatography on silica gel (dichloromethane/hexane/methanol = 5:1:1) to give the dye **(6)** as a deep yellow powder (0.11 g, 75%). IR (neat): 3,408, 2,927, 1,710, 1,609, 1,565, 1,508, 1,461, 1,394, 1,347, 1,280, 1,201, 1,154, 1,116, 1,083, 978, 906, 748/cm. ^{1}H NMR (400 MHz, CDCl$_3$): δ 8.38 (1H, dd, J=13.2, 13.3 Hz), 7.16–7.48 (12H, m), 5.90 (1H, d, J=13.3 Hz), 5.88 (1H, d, J=13.2 Hz), 5.22 (2H, s), 4.12 (2H, br q, J=7.2 Hz), 3.51 (2H, s), 1.30 (3H, t, J=7.2 Hz). ^{13}C NMR (100 MHz CDCl$_3$): δ 174.31, 1637, 162.25, 161.74, 148.00, 1488, 1473, 131.23, 130.93, 130.58, 130.38, 127.17, 1233, 1223, 125.59, 125.38, 111.01, 110.86, 110.76, 110.53, 85.95, 85.26, 47.41, 45.52, 39.60, 12.98.

E. *2-(3-(3-((4-(2-((2,5-Dioxo-1-pyrrolidinyl)oxy)-2-oxoethyl)phenyl)methyl)-2(3H)-benzoxazolylidene)-1-propenyl)-3-ethylbenzoxazolium, Cy2-NHS ester* **(7)**. Anhydrous pyridine (0.1 mL) and DSC (Aldrich, 21 mg, 0.08 mmol) were added to a stirred solution of the dye (6) (25 mg, 0.05 mmol) in dry DMF (Fisher, 2 mL) under nitrogen. The reaction mixture was stirred at 60°C for 1.5 h. After evaporation of the solvent, the deep yellow residue was purified by column chromatography on silica gel (dichloromethane/hexane/methanol = 5:1:1) to give the pure Cy2-NHS ester (7) (26 mg, 86%) as an orange powder. IR (neat): 2,924, 2,851, 1,736, 1,565, 1,507, 1,461, 1,395, 1,348, 1,280, 1,201, 1,115, 1,082, 747/cm. ^{1}H NMR (400 MHz, CDCl$_3$): δ 8.40 (1H, t, J = 13.2 Hz), 7.22–7.49 (12H, m), 52 (1H, d, J = 13.2 Hz), 40 (1H, d, J = 13.2 Hz), 5.49 (2H, s), 4.31 (2H, br q, J = 7.1 Hz), 3.85 (2H, s), 2.78 (4H, s), 1.46 (3H, t, J = 7.1 Hz). ^{13}C NMR (100 MHz, CDCl$_3$): δ 168.99, 1651, 162.18, 161.73, 1495, 1482, 132.87, 132.02, 131.37, 130.82, 130.13, 128.36, 1209, 125.28, 125.08, 110.98, 110.90, 110.76, 110.66, 87.15, 849, 37.19, 25.60, 13.34. MS (EI): m/z (%) = 550 (100), 453 (6), 304 (6).

3.1.2. Cy3 Synthesis

The synthesis of the propyl Cy3 dye and its NHS ester (see Fig. 2) began with the alkylation of the commercially available 2,3,3-trimethyl-3H-indole (8) with propyl bromide to give the 1-propyl-2-methyleneindoline (9) (16, 17) in an unoptimized yield of 44%. Condensation with commercially available N,N'-diphenylformamidine in the presence of excess acetic anhydride as solvent afforded the acetanilidylvinyl indolium salt (10) in 87% yield. The second component of the dye was prepared by alkylation of trimethylindole (8) with 6-bromohexanoic acid in dichlorobenzene at 110°C for 12 h to give 67% yield of the methylindolium salt (11) (2, 18–20). This compound was then reacted with the acetanilidylvinyl indolium salt (10) in ethanol in the presence of triethylamine to give the desired dye (12) in 85% yield as a red powder (14, 15). The dye was easily converted into the NHS ester by treatment with DSC in the presence of pyridine to give the activated dye (13) in 96% yield. Again, all the pertinent spectroscopic data, especially high-field NMR and mass spectrometry, were in agreement with the structures assigned (see Note 1).

A. *3,3-Dimethyl-2-methylene-1-propylindoline* (9). A mixture of 2,3,3-trimethyl-3H-indole (8) (Aldrich, 0.2 g, 1.25 mmol) and 1-bromopropane (Aldrich, 2.28 mL, 0.025 mol) in 1,2-dichlorobenzene (Aldrich) was heated at 110°C for 24 h. The solution was cooled to room temperature, and the residue obtained was filtered and washed with a mixture of acetonitrile/diethyl ether (1/1). The solid obtained was dried under vacuum to give the 3,3-dimethyl-2-methylene-1-propylindoline (9) as a light red powder (0.11 g, 44%) (see Note 1). IR (neat): 2,966, 2,925, 1,617, 1,601, 1,474, 1,454, 1,356, 1,290,

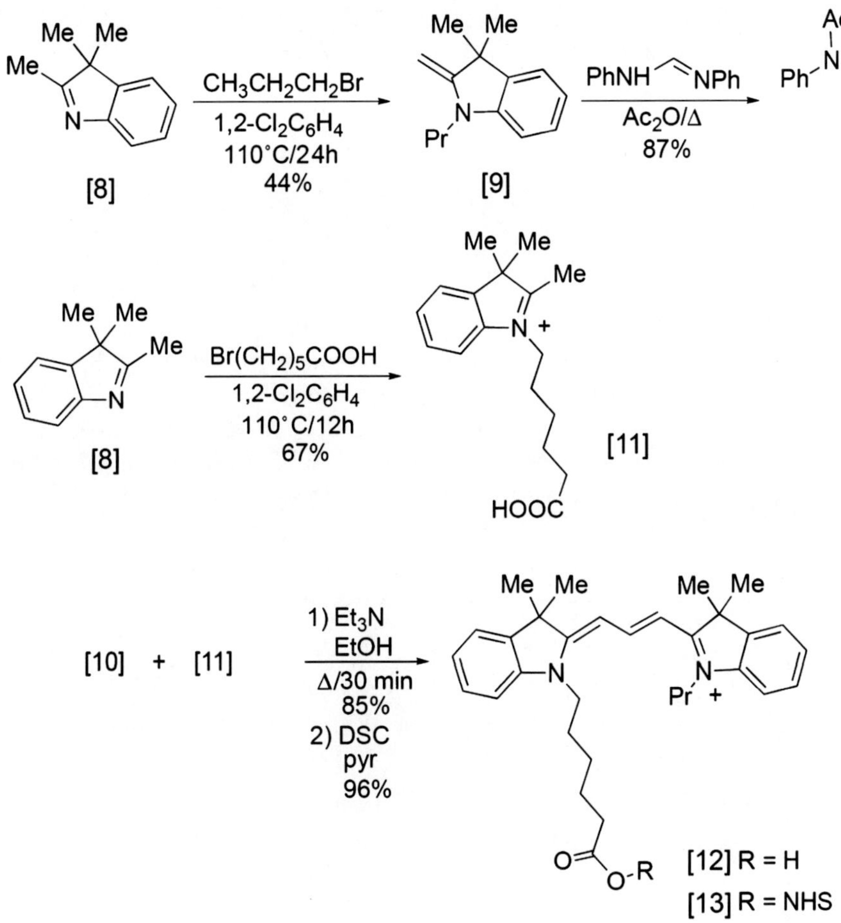

Fig. 2. Synthesis of the propyl Cy3 dye and its NHS ester.

1,119, 931, 767/cm. ^{1}H NMR (400 MHz, D$_2$O): δ 7.66 (1H, m), 7.60 (1H, m), 7.45–7.50 (2H, m), 4.65 (2H, s), 4.30 (2H, t, J=7.4 Hz), 1.86 (2H, m), 1.42 (6H, s), 0.87 (3H, t, J=7.4 Hz). ^{13}C NMR (100 MHz, D$_2$O): δ 141.77, 140.92, 129.75, 128.99, 123.26, 115.15, 54.33, 49.20, 21.72, 20.85, 10.18.

B. *2-(2-Phenylacetamido-E-1-ethenyl)-3,3-dimethyl-1-propylindolium salt* **(10)**. A mixture of *N,N'*-diphenylformamidine (Aldrich, 0.35 g, 1.78 mmol) and 3,3-dimethyl-2-methylene-1-propylindoline **(9)** (0.30 g, 1.48 mmol) in acetic anhydride (10 mL) was refluxed for 30 min. The solution was cooled to room temperature, the solvent was removed under reduced pressure, and the residue was purified by flash chromatography on silica gel (dichloromethane/hexane/methanol = 5:1:1) to give the salt **(10)** as a light yellow powder (0.45 g, 87%). IR (neat): 2,965, 2,926, 1,680, 1,638, 1,603, 1,580, 1,553, 1,492,

1,369, 1,311, 1,200, 1,130, 996, 757/cm. ^{1}H NMR (400 MHz, DMSO-d_6): δ 9.10 (1H, d, J=14.2 Hz), 7.05–7.70 (9H, m), 5.34 (1H, d, J=14.2 Hz), 4.06 (2H, t, J=7.1 Hz), 2.05 (3H, s), 1.70 (6H, s), 1.60 (2H, m), 0.67 (3H, t, J=7.4 Hz). ^{13}C NMR (100 MHz CDCl$_3$): δ 162.80, 154.24, 142.85, 139.56, 129.48, 128.41, 125.42, 123.32, 121.96, 119.42, 109.13, 94.35, 47.84, 45.19, 29.26, 27.86, 20.15, 11.45.

C. *1-(5-Carboxypentyl)-2,3,3-trimethyl-3H-indolium bromide* (**11**). A mixture of 2,3,3-trimethyl-3*H*-indole (**8**) (Aldrich, 0.2 g, 1.25 mmol) and 6-bromohexanoic acid (Aldrich, 0.36 g, 1.80 mmol) in 1,2-dichlorobenzene was heated at 110°C for 12 h. The solution was cooled to room temperature, and the residue obtained was filtered and washed with a mixture of acetonitrile/diethyl ether (1/1). The solid obtained was dried under vacuum to give the product (**11**) as a light red powder (2.3 g, 67%) (2, 18–20). IR (neat): 3,405, 2,936, 1,724, 1,624, 1,460, 1,392, 1,168, 767/cm. ^{1}H NMR (400 MHz, DMSO-d_6): δ 7.93–7.97 (1H, m), 7.79–7.87 (1H, m), 7.56–7.64 (2H, m), 4.43 (2H, t, J=7.7 Hz), 2.82 (3H, s), 2.19 (2H, t, J=7.2 Hz), 1.81 (2H, m), 1.52 (2H, m), 1.50 (6H, s), 1.35 (2H, m). ^{13}C NMR (100 MHz, DMSO-d_6): δ 1998, 174.77, 142.33, 141.51, 129.85, 129.40, 123.99, 115.97, 54.62, 47.90, 33.83, 27.41, 25.87, 24.48, 22.47, 14.51.

D. *2-(3-(1-(5-Carboxypentyl)-1,3-dihydro-3,3-dimethyl-2H-indol-2-ylidene)-1-propenyl)-3,3-dimethyl-1-propyl-3H-indolium, propyl Cy3* (**12**). A mixture of the salt (**11**) (0.08 g, 0.28 mmol), enamide (**10**) (0.10 g, 0.28 mmol), and dry triethylamine (0.1 mL) in absolute ethanol (10 mL) was refluxed for 30 min. The solvent was removed under reduced pressure, and the crude residue was purified by flash chromatography on silica gel (dichloromethane/hexane/methanol = 5:1:1) to give the dye propyl Cy3 (**12**) as a deep red powder (0.12 g, 85%) (16, 17). IR (neat): 3,407, 2,969, 2,934, 2,876, 2,734, 2,673, 1,557, 1,453, 1,427, 1,242, 1,192, 1,130, 1,030, 930, 754/cm. ^{1}H NMR (400 MHz, DMSO-d_6): δ 8.31 (1H, dd, J=13.4, 13.5 Hz), 7.61 (2H, d, J=7.4 Hz), 7.36–7.47 (4H, m), 7.26 (2H, t, J=7.1 Hz), 54 (1H, d, J=13.5 Hz), 53 (1H, d, J=13.4 Hz), 4.08 (2H, t, J=7.2 Hz), 3.43 (2H, t, J=7.2 Hz), 2.14 (2H, t, J=7.2 Hz), 1.75 (2H, m), 1.66 (12H, s), 1.52 (2H, m), 1.40 (2H, m), 1.19 (2H, m), 0.84 (3H, t, J=7.4 Hz). ^{13}C NMR (100 MHz DMSO-d_6): δ 174.20, 161.01, 1445, 143.11, 141.37, 137.32, 129.11, 127.92, 1213, 122.98, 122.21, 118.45, 105.73, 102.72, 74.11, 49.36, 44.06, 43.30, 30.26, 27.94, 27.90, 226, 20.89, 19.44, 11.79, 11.4.

E. *2-(3-(1,3-Dihydro-3,3-dimethyl-1-propyl-2H-indol-2-ylidene)-1-propenyl)-1-(6-((2,5-dioxo-1-pyrrolidinyl)oxy)-6-oxohexyl)-3,3-dimethyl-3H-indolium, propyl Cy3-NHS ester* (**13**). Anhydrous

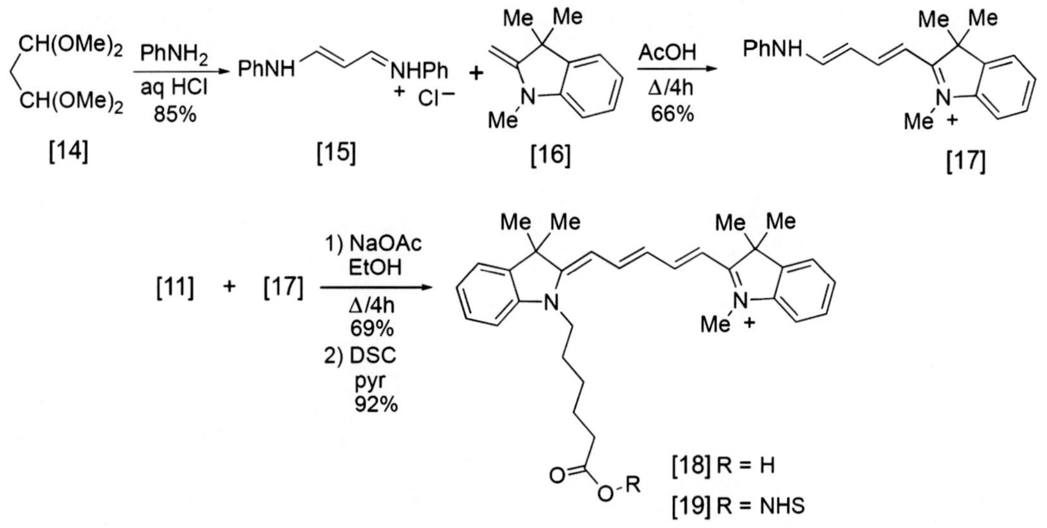

Fig. 3. Synthesis of the methyl Cy5 dye and its NHS ester.

pyridine (0.1 mL) and DSC (21 mg, 0.08 mmol) were added to a stirred solution of the dye (**12**) (27 mg, 0.05 mmol) in dry DMF (2 mL) under nitrogen. The reaction mixture was stirred at 60°C for 1.5 h. After evaporation of the solvent, the deep red residue was purified by column chromatography on silica gel (dichloromethane/hexane/methanol = 5:1:1) to give the pure propyl Cy3-NHS ester (**13**) (31 mg, 96%) as a red powder (2, 6). IR (neat): 2,924, 2,853, 1,737, 1,555, 1,456, 1,428, 1,371, 1,248, 1,196, 1,158, 1,116, 1,019, 930, 796, 579, 680/cm. ^{1}H NMR (400 MHz, DMSO-d_6): δ 8.32 (1H, dd, J=13.4, 13.4 Hz), 7.61 (2H, d, J=7.6 Hz), 7.38–7.47 (4H, m), 7.26 (2H, t, J=7.4 Hz), 48 (2H, d, J=13.5 Hz), 4.09 (4H, m), 2.74 (4H, s), 2.66 (2H, t, J=7.2 Hz), 1.74 (2H, m), 1.69 (2H, m), 1.66 (12H, s), 1.48 (2H, m), 1.19 (2H, m), 0.95 (3H, t, J=7.4 Hz). ^{13}C NMR (100 MHz, DMSO-d_6): δ 174.20, 173.40, 169.17, 168.61, 151.50, 142.30, 140.52, 129.11, 128.86, 125.04, 122.03, 111.14, 110.96, 105.17, 48.90, 440, 44.20, 30.70, 29.71, 28.20, 27.21, 25.82, 25.62, 24.53, 21.17, 11.30. MS (EI): m/z (%) = 582 (100), 485 (22).

3.1.3. Cy5 Synthesis

For the synthesis of the methyl Cy5 dye and its NHS ester (see Fig. 3), the three-carbon spacer (**15**) had to be prepared. Condensation of commercially available malondialdehyde bis(dimethyl acetal) (**14**) with aniline under acidic conditions afforded the anilino anilinium salt (**15**) in 85% yield (2, 21). The reaction of (**15**) with the commercially available 1,3,3-trimethyl-2-methyleneindoline (**16**) in refluxing acetic acid afforded the anilinobutadienyl salt (**17**) in 66% yield (2, 22). Finally, the reaction of the activated indolium salt (**17**) with the methylindolium salt (**11**) synthesized earlier (see Subheading 3.1.2, step C and Fig. 2) in ethanol in the presence of sodium acetate afforded the desired dye (**18**) in 69% yield as a blue powder. The dye was easily converted into the NHS ester by treatment with commercially available DSC in the presence of pyridine to give the activated dye (**19**) in 92% yield. All spectroscopic data, especially high-field NMR and mass spectrometry, were in agreement with the structures assigned (see Note 1).

A. *N-((1E)-3-(Phenylimino)prop-1-enyl)benzenamine hydrochloride* (**15**). A solution of distilled water (140 mL), HCl (10 mL), and aniline (Alfa Aesar, 7.4 mL, 0.08 mol) was added dropwise to a solution of distilled water (171 mL), HCl (8.5 mL), and malondialdehyde bis(dimethyl acetal) (**14**) (Aldrich, 10.5 mL, 0.06 mol) with stirring at 50°C. The precipitate was isolated by filtration to give malondialdehyde dianil hydrochloride (**15**) (12 g, 85%) as an orange powder (2, 21). IR (neat): 3,425, 1,642, 1,620, 1,580, 1,492, 1,343, 1,273, 1,194, 749, 683/cm. ^{1}H NMR (400 MHz, DMSO-$d6$): δ 12.72 (2H, d, J=13.2 Hz), 8.92 (2H, t, J=12.4 Hz), 7.40 (8H, m), 7.20 (2H, m), 50 (1H, t, J=11.5 Hz). ^{13}C NMR (100 MHz, DMSO-$d6$): δ 158.89, 139.18, 130.33, 1231, 117.86, 99.09.

B. *2-(4-Phenylamino-1E,3E-butadien-1-yl)-1,3,3-trimethy-lindolium chloride* (**17**). A mixture of malondialdehyde dianil hydrochloride (**15**) (1 g, 4.49 mmol) and 1,3,3-trimethyl-2-methyleneindoline (**16**) (Aldrich, 0.93 mL, 4.49 mmol) in glacial acetic acid (10 mL) was refluxed for 4 h. The solution was cooled to room temperature, the acetic acid was removed under reduced pressure, and the residue was purified by flash chromatography on silica gel (dichloromethane/hexane/methanol = 5:1:1) to give the product (**16**) as a red powder (0.9 g, 66%) (22). ^{1}H NMR (400 MHz, DMSO-$d6$): δ 8.18 (1H, d, J=9.8 Hz), 7.52 (1H, t, J=13.3 Hz), 74–7.36 (9H, m), 48 (1H, dd, J=14.1, 9.9 Hz), 5.55 (1H, d, J=12.5 Hz), 3.24 (3H, s), 1.55 (6H, s). ^{13}C NMR (100 MHz, DMSO-$d6$): δ 129.41, 129.17, 128.04, 125.24, 121.71, 121.36, 120.41, 118.18, 107.27, 468, 28.38.

C. *2-(5-(1-(5-Carboxypentyl)-1,3-dihydro-3,3-dimethyl-2H-indol-2-ylidene]-1,3-pentadienyl]-1,3,3-trimethyl-3H-indolium, methyl Cy5* (**18**). A solution of the anil (**16**) (0.20 g, 0.65 mmol), the acid (**11**) synthesized earlier in Subheading 3.1.2, step C (0.18 g, 0.65 mmol), and anhydrous sodium acetate (0.11 g, 0.79 mmol) in absolute ethanol (50 mL) under nitrogen was refluxed for 4 h. The solid was purified by flash chromatography on silica gel (dichloromethane/methanol = 5:1) to give the methyl Cy5 dye (**18**) (0.22 g, 69%) as a blue powder (2). IR (neat): 3,406, 2,925, 1,716, 1,575, 1,470, 1,425, 1,371, 1,335, 1,217, 1,146, 1,016, 1,040, 923, 796, 750, 708/cm. ^{1}H NMR (400 MHz, DMSO-$d6$): δ 8.30

(2H, t, *J*= 13.1 Hz), 7.58 (2H, d, *J*= 7.4 Hz), 7.34 (4H, m), 7.21 (2H, m), 53 (1H, t, *J*= 12.3 Hz), 25 (2H, dd, *J*= 13.8, 13.9 Hz), 4.06 (2H, t, *J*= 7.2 Hz), 3.56 (3H, s), 2.15 (2H, t, *J*= 7.2 Hz), 1.69 (2H, m), 1.64 (12H, s), 1.52 (2H, m), 1.32 (2H, m). ^{13}C NMR (100 MHz DMSO-*d6*): δ 175.03, 173.69, 172.95, 154.47, 143.21, 142.47, 141.54, 141.47, 128.80, 125.83, 125.16, 125.05, 122.89, 122.76, 111.49, 103.77, 103.51, 55.39, 49.30, 43.67, 34.16, 31.56, 27.61, 27.44, 27.15, 212, 24.73.

D. *2-(5-(1,3-Dihydro-1,3,3-trimethyl-2H-indol-2-ylidene)-1,3-pentadienyl]-1-(6-((2,5-dioxo-1-pyrrolidi-nyl)oxy]-6-oxohexyl]-3,3-dimethyl-3H-indolium*, methyl Cy5-NHS ester **(19)**. Anhydrous pyridine (0.1 mL) and DSC (23.6 mg, 0.09 mmol) were added to a stirred solution of the acid **(18)** (0.36 g, 0.06 mmol) in dry DMF (2 mL) under nitrogen. The reaction mixture was stirred at 60°C for 1.5 h. After evaporation of the solvent, the deep blue residue was purified by column chromatography on silica gel (dichloromethane/hexane/methanol = 5:1:1) to give the pure NHS ester of the dye **(19)** (0.33 g, 92%) as a blue powder (2, 6). IR (neat): 2,926, 1,733, 1,496, 1,481, 1,456, 1,372, 1,336, 1,217, 1,182, 1,149, 1,097, 1,040, 1,016, 924, 796, 756, 708, 668/cm. ^{1}H NMR (400 MHz, DMSO-*d6*): δ 8.30 (2H, t, *J*= 13.1 Hz), 7.58 (2H, d, *J*= 7.4 Hz), 7.36 (4H, m), 7.20 (2H, m), 53 (1H, t, *J*= 12.4 Hz), 25 (2H, t, *J*= 14.2 Hz), 4.05 (2H, t, *J*= 7.2 Hz), 3.56 (3H, s), 2.77 (4H, s), 2.64 (2H, t, *J*= 7.2 Hz), 1.68 (2H, m), 1.65 (2H, m), 1.64 (12H, s), 1.44 (2H, m). ^{13}C NMR (100 MHz, DMSO-*d6*): δ 173.74, 173.26, 172.93, 170.71, 169.35, 154.49, 151.10, 143.21, 142.44, 141.54, 141.47, 129.11, 128.81, 125.83, 125.19, 125.05, 122.87, 122.76, 111.50, 103.44, 49.31, 49.03, 43.63, 31.54, 30.44, 27.61, 27.43, 284, 25.90, 25.66, 24.37. MS (EI): *m/z* (%) = 580 (100).

3.2. Validation of Newly Synthesized Dyes in DIGE Experiments

To validate the newly synthesized dyes, which will be called LAB dyes further, we applied them in a regular DIGE experiment and compared them to commercially available dyes from GE Healthcare, which further will be described as COM dyes.

For the experimental model, we chose a comparison between two strains of *Escherichia coli* K-12 MG1655: a wild-type (WT) and a Fis⁻ mutant strain (both strains were kindly provided by Daniel Yoo and Reid Johnson, Department of Biological Chemistry, UCLA School of Medicine). Fis is a small nucleoid-associated protein, the expression levels of which undergo dramatic changes during cell growth: it has been shown that if stationary phase cells are transferred into a fresh rich medium such as LB (1% Tryptone, 0.5% yeast extract, 1% NaCl, pH 7.3), in about 1 h Fis levels increase from about 100 copies per cell to tens of thousands of copies (23). Fis has been implicated in transcriptional regulation of multiple genes (24); hence, differential expression of many different proteins may also be expected when WT *E. coli* and Fis⁻ mutant cells are analyzed using the DIGE method. This study is designed to compare individual protein abundances calculated by the DIGE method after using either LAB or COM dyes for protein labeling.

3.2.1. Sample Preparation

Five independently grown *E. coli* WT and five Fis⁻ mutant cultures were grown in LB for 15 h. A stationary overnight culture was diluted 1/50 in prewarmed LB medium and incubated for 45 min at 37°C as Fis protein level is known to reach its maximum at this time (23). Immediately after removing the cell cultures from the incubator, they were quickly chilled on ice and harvested by centrifugation at $10,000 \times g$ for 10 min at 4°C. The pellets were washed twice with 20 mM Tris–HCl, pH 8.0, 5 mM magnesium acetate. Cell pellets (from approximately 10^{10} cells) were resuspended in the lysis buffer containing 7 M urea, 2 M thiourea, 4% CHAPS, 20 mM Tris–HCl, pH 8.8, and sonicated on ice for six cycles of 30 s each at high-power setting with 1 min of interruption between the cycles (see Note 2). The samples were centrifuged at $14,000 \times g$ for 15 min at 4°C, and the protein concentration was determined using 2D Quant kit.

3.2.2. Labeling with Cy Dyes

A total of 50 μg of each protein sample was labeled with 400 pmol of either Cy3 or Cy5 dyes, standard conditions for a so-called minimal labeling. A pool of 25-μg aliquots collected from each sample was labeled with 400 pmol Cy2 dye/every 50 μg of protein and used as an internal standard for each gel. The labeling reaction was stopped after 30-min incubation on ice by adding 1 μL of 10 mM L-lysine/400 pmol dye and incubated on ice for additional 10 min. The rehydration solution containing 7 M urea, 2 M thiourea, 4% CHAPS, 5% glycerol, 10% isopropanol, 1% DTT, and 0.5% IPG 4–7 buffer was added to the labeled protein samples to a final volume of 450 μL. The samples were incubated for 20–30 min at room temperature (RT) and then centrifuged at $12,000 \times g$ for 5 min at RT.

3.2.3. IEF, SDS-PAGE, and Scanning

Protein samples (450 μL) were loaded overnight onto a 24-cm pH 4–7 IPG strips (GE Healthcare) using an IPGPhor II apparatus (GE Healthcare). Isoelectric focusing (IEF) was performed at 20°C at 50 μA for a total of 80,000 Vh. After the IEF, IPG strips were incubated at RT in equilibration buffer containing 50 mM Tris–HCl, pH 8.8, 6 M urea, 30% glycerol, 2% SDS, and 10 mg/mL DTT for 15 min with gentle shaking, and then in the buffer containing 50 mM Tris–HCl, pH 8.8, 6 M urea, 30% glycerol, 2% SDS, and 40 mg/mL iodoacetamide for another 15 min. IPG strips were then rinsed in the 1× SDS running buffer and placed on top of the 12.5% SDS-PAGE gels and run at 25 V for 1 h, 50 V for another hour, and then at 2 W/gel overnight at 24°C in an Ettan DALT 12 apparatus (GE Healthcare). Next morning, gels were scanned, and fluorescently labeled proteins were visualized by the Typhoon Trio Variable Mode Imager (GE Healthcare) using the following parameters: Cy2 dye-labeled proteins using 488 nm excitation wavelength and 520BP40 emission filter (520 nm center, 40 nm bandpass), Cy3 dye-labeled proteins using 532 nm excitation wavelength and 580BP30 emission filter, and Cy5-labeled proteins using 633 nm excitation and 670BP30 emission filter. All gels were scanned at 100-μm resolution, the PMT (photomultiplier tube) voltage was adjusted to keep all the recognized grayscale image values below 80,000, which are within the linear range of the scanner. Images were cropped using ImageQuant v.5.2 software (GE Healthcare) to remove areas outside of the gel image.

3.2.4. Image Analysis

Image analysis was performed using the DeCyder software v. 6.5 (GE Healthcare). The same software parameters were applied for the analysis of all images in the automatic batch mode. The estimated number of spots was set in the DeCyder DIA module at 2,500 in all cases. Gel matching was performed within the DeCyder BVA module automatically without land-marking or manual matching to minimize the operator impact on the analysis. The following filtering criteria were applied for further protein-of-interest (POI) selection: spot presence in all spot maps, average ratios of ≥ 1.4 and ≤ -1.4, t test ≤ 0.01, and spot volume of $\geq 1.0e+05$ and $\leq 1.0e+07$.

3.2.5. Validation Results

In order to determine how similarly the newly synthesized (LAB) and commercially available dyes (COM) behave in DIGE, two identical DIGE experiments were conducted using LAB dyes and COM dyes. For each experiment, ten protein samples comprising five WT and five Fis$^-$ samples were used for labeling ($N=5$). The dyes were "swapped" to minimize the potential effects of individual dyes: three samples in the WT group were labeled with Cy3, and two with Cy5, while three Fis$^-$ group samples were labeled with Cy5, and two with Cy3. In both cases (LAB and COM), the Cy2 dye was used to label the common internal standard consisting of equal aliquots from all ten samples. Cy3 and Cy5 samples were paired randomly in order to avoid bias related to loading, and the Cy2-labeled internal standard was loaded onto each gel along with a Cy3/Cy5 sample pair. Although experiments with LAB and COM dyes were run independently and conducted on two separate days, spot patterns were found to be remarkably similar (a portion of two gels representing each experiment is shown in Fig. 4). A nonsupervised spot detection was performed on both sets of gels representing five WT and five Fis$^-$ samples labeled with two different dyes. As shown in Table 1, the number of detected spots for both LAB- and COM-labeled samples is very similar across all gels representing both LAB and COM experiments with the differences of 4.5 and 6% in CV values.

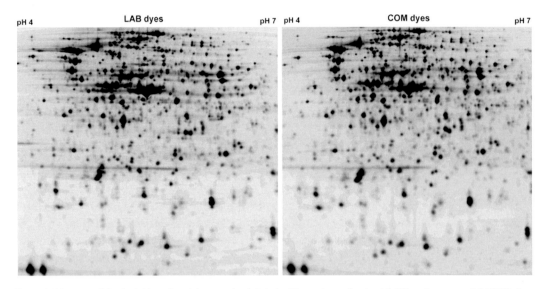

Fig. 4. Gel images of *Escherichia coli* protein samples labeled with newly synthesized (LAB) and commercial (COM) dyes. Spot patterns produced by using the two different dyes for labeling are remarkably similar.

Table 1
Comparison of spot maps for samples labeled with LAB and COM dyes

Sample	Number of detected spots (per gel, 1 through 5)	Mean	Standard deviation	CV (%)
LAB	2,172 (1), 2,143 (2), 2,243 (3), 2,090 (4) 1,990 (5)	2,128	95	4.5
COM	2,000 (1), 2,071 (2), 2,166 (3), 2,376 (4), 2,204 (5)	2,163	143	6

To check for similarities, individual spots average ratio measurements and an automated POI selection filter was then applied, and 37 protein spots were selected that satisfied the selection criteria, as described in Subheading 3.2.4 above. Out of 37 spots passing the filter, 27 were found to be upregulated in Fis⁻ cells, and 10 protein spots were downregulated in Fis⁻ cells, as compared to WT (see Note 3).

Comparison of average ratios in protein abundance for all 37 spots plotted in Fig. 5 shows that selected spots in both LAB- and COM-labeled samples are very similar. As shown in Table 2, the difference in average ratios for the majority of protein spots in both LAB and COM samples is small, as 29 spots (80% of total) differ less than 5% from each other, while the rest of the spots differ by no more than 15%. Paired Student's t-test analysis performed on

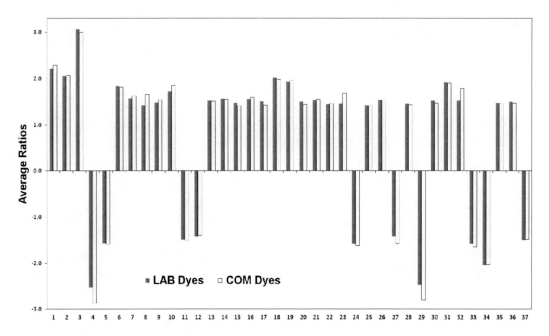

Fig. 5. Comparison of average ratios of individual protein measurements in samples labeled with newly synthesized (LAB) and commercial (COM) dyes. Spot numbers are shown at the *bottom* of the graph, while standardized log abundances for proteins selected are shown on the *y*-axis.

Table 2
Comparison of average ratios for individual spots in samples labeled with LAB and COM dyes

Deviation from mean average ratio (%)	≤1	1–5	5–10	10–15	≥15
Number of spots (total = 37)	8	21	2	6	0

this set has also confirmed that there is no statistical difference between the corresponding average ratios for LAB- and COM-labeled protein spots.

3.3. Dye Storage and Stability

3.3.1. Aliquoting the Dyes for a Long-Term Storage

Newly synthesized NHS dyes were weighed into regular 1.5-mL Eppendorf tubes (0.5 mg dye/tube) and stored as dry solids. The tubes were filled with argon gas, closed tightly, and stored at −20°C. The use of argon presumably reduces the amount of oxygen and water vapors in contact with the dye and therefore extends the dyes' storage life (see Subheading 3.3.3 below and Note 4).

3.3.2. Dyes Dilution and Storage

For everyday work, dyes were dissolved in anhydrous DMF at a final concentration of 2 nmol/μL. This solution was distributed into smaller aliquots (usually 25–30 μL) and stored in 0.5-mL Eppendorf tubes at –80°C. Before closing, the tubes were filled with argon gas. These aliquots were taken out of the –80°C freezer as necessary for creating working dye solutions for protein labeling. Before returning the remaining aliquots to the freezer, the tubes were filled with argon gas again.

3.3.3. Long-Term Storage and Dye Stability

To evaluate the effects of long-term storage of dry dyes on their performance in DIGE, as well as the effect of dilution in DMF on dye stability, we compared two groups of dye samples. The first group of newly synthesized dyes (Cy2, Cy3, and Cy5) was placed in storage immediately after synthesis and stored at –20°C as dry solids for 5 years under argon atmosphere, as described in Subheading 3.3.1. These tubes were taken out of the freezer and dissolved in fresh anhydrous DMF right before being used in a DIGE experiment. This dilution is further called NEW dilution, and it was the same dilution used to compare newly synthesized dyes with commercially available ones, as described in Subheading 3.2 above. The second group of new dyes was diluted in DMF at 2 nmol/μL at the time of dye synthesis and then stored in DMF for 5 years, as described in Subheading 3.3.2. This sample is further called OLD dilution and was used to compare to the freshly made NEW dilution in DIGE experiments.

OLD and NEW dyes dilutions were used for protein labeling in the same type of experiment, as described in Subheading 3.2 above, except that it was done on a slightly smaller scale: instead of five pairs of samples, only three WT and three Fis⁻ protein samples were used ($N=3$). As previously, after labeling, IEF and SDS-PAGE separation, protein spots were analyzed using DeCyder software in an automated fashion. Spot detection and automatic matching was performed without manual intervention, and a protein filter was applied as described in Subheading 3.2.4. Out of 25 protein spots that satisfied the selection criteria, 16 are upregulated, and 9 are downregulated in Fis⁻ cells, as compared to WT cells (see Fig. 6).

Average ratios in protein abundance for all 25 spots plotted in Fig. 6 shows that selected spots in both OLD and NEW dilution-labeled samples are very similar. As presented in Table 3, the majority of protein spots (21 spots, 84% of total) in both OLD and NEW dilution samples differ by less than 10% from each other.

In summary, three dyes used in DIGE, benzoxazolium dye Cy2, propyl Cy3, and ethyl Cy5, and their NHS esters were synthesized from commercially available precursors. Newly synthesized dyes were tested in DIGE experiments and were shown to be indistinguishable from commercially available dyes at the level of individual protein abundance measurements. Synthesized dyes

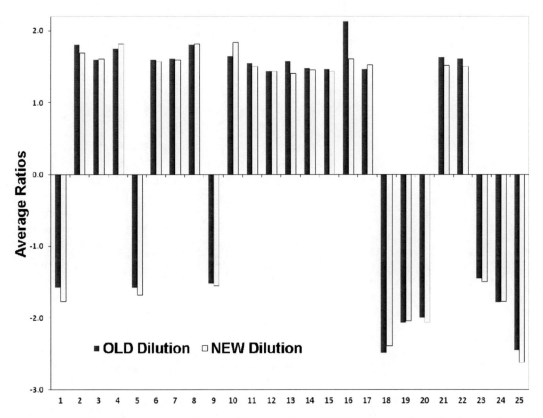

Fig. 6. Comparison of average ratios of individual protein measurements in samples labeled with NEW and OLD dyes dilutions (see text for details). Spot numbers are shown at the *bottom* of the graph, while standardized log abundances for proteins selected are shown on the *y*-axis.

Table 3
Comparison of average ratios for individual spots in samples labeled with OLD and NEW dye dilutions

Deviation from mean average ratio (%)	≤1	1–5	5–10	10–15	≥15
Number of spots (total = 25)	5	11	5	3	1

were shown to be stable for years when stored as dry solids under argon, the dyes are equally stable when stored in anhydrous DMF in the argon atmosphere.

4. Notes

1. We have no information on the counterions of these final quaternary ammonium salt dyes, but they are almost certainly not acetate since no peak for an acetate group appears in the proton NMR of the final compounds.

2. Charged urea/thiourea degradation products may cause protein carbamylation, and therefore it is advisable to deionize urea and thiourea before use. To prepare 25 mL of the deionized buffer, add 10.5 g of urea and 3.8 g thiourea to 11 mL of milli-Q water. Fill a small column with 7–9 g of ion-exchange resin AG501X8 (Bio-Rad) and wash it with milli-Q water. After water removal, pass the urea/thiourea solution through the column 3–4 times until the solution conductivity is around 0.2–0.4 µS. Add 1 g of CHAPS, 0.33 mL of 1.5 M Tris–HCl, pH 8.8, and water to make a 25-mL solution and then filter it through a 0.45-µm filter. For longer-term storage, distribute the lysis buffer into 1.5-mL Eppendorf tubes and store at −80°C.

3. The identifications of these and other proteins affected by Fis at the early logarithmic phase of *E. coli* cell growth will be published elsewhere.

4. Dye precursors, such as **(6)**, **(12)**, and **(18)** (see Figs. 1–3), could be successfully stored as dry solids at −20°C under argon, so when more NHS dyes are needed, one last reaction could be performed to convert the acid form of the dye into its NHS ester.

Acknowledgments

Wan-Joong Kim and Nuraly K. Avliyakulov contributed equally to this study. We thank the Dean's Office of the David Geffen School of Medicine at UCLA and Senior Associate Dean Leonard Rome for the generous and continuing support of this work.

References

1. O'Farrel PH (1975) High resolution two-dimensional electrophoresis of proteins. J Biol Chem, 250: 4007–4021.
2. Unlu M, Morgan ME, Minden JS (1997) Difference gel electrophoresis. A single gel method for detecting changes in protein extracts. Electrophoresis 18: 2071–2077.
3. Alban A, David SO, Bjorkesten L, Andersson C, Sloge E, Lewis S, Currie I (2003) A novel experimental design for comparative two-dimensional gel analysis: two-dimensional difference gel electrophoresis incorporating a pooled internal standard. Proteomics 3: 36–44.
4. Lilley KS, Friedman DB (2004) All about DIGE: quantification technology for differential-display 2D-gel proteomics. Expert Rev Proteomics 1: 401–409.

5. Karp NA, Lilley KS (2005) Maximising sensitivity for detecting changes in protein expression: experimental design using minimal CyDyes. Proteomics 5: 3105–3115.
6. Tonge R, Shaw J, Middleton B, Rowlinson R, Rayner S, Young J, Pognan F, Hawkins E, Currie I, Davison M (2001) Validation and development of fluorescence two-dimensional differential gel electrophoresis proteomics technology. Proteomics 1: 377–39.
7. Jung ME, Kim W-J (2006) Practical syntheses of dyes for difference gel electrophoresis. Bioorg Med Chem 14: 92–97.
8. Jedrzejewska B, Kabatc J, Pietrzak M, Paczkowski J (2003) Hemicyanine dyes: synthesis, structure and photophysical properties. Dyes Pigments 58: 47–58.
9. Hamilton AL, Birch MN, Hatcher MJ, Bosworth N, Scott B (1999) Energy transfer assay method and reagent. PCT Int Appl WO/1999/064519.
10. Schouten JA, Ladame S, Mason SJ, Cooper MA, Balasubramanian SG (2003) Quadruplex-specific peptide–hemicyanine ligands by partial combinatorial selection. J Am Chem Soc 125: 5594–5595.
11. Gorb LT, Romanov NN, Fedotov KV, Tolmachev AI (1981) Meso-ionic compounds with a nitrogen bridging atom. Polymethine dyes of the thiazolo[3,2-a]quinolinium 1-oxide series. Khim Geterotsikl Soedin, 481–484.
12. Abramenko PI, Zhiryakov VG (1975) Polymethine dyes, derivatives of 6-furo[2,3-b] pyridine. Khim Geterotsikl Soedin, 475–479.
13. Bailey J, Elvidge JA (1973) Synthesis and properties of dyes containing the pyrano[2,3-d] pyrimidine nucleus. J Chem Soc, Perkin Trans, 1: 823–828.
14. Cummins WJ, West RM, Smith JA (1999) Cyanine dyes. PCT Int Appl WO/1999/005221.
15. Jackson P, Cummins WJ, West R, Smith JA, Briggs MSJ (1998) Analysis of carbohydrates. PCT Int Appl WO/1998/015829.
16. Wurthner F (1999) DMF in acetic anhydride: A useful reagent for multiple-component syntheses of merocyanine dyes. Synthesis, 2103–2113.
17. Durr H, Ma Y, Cortellaro G (1995) Preparation of photochromic molecules with polymerizable organic functionalities. Synthesis: 294–298.
18. Lee LG, Woo SL, Head DF, Dubrow RS, Baer TM (1995) Near-IR dyes in three-color volumetric capillary cytometry: Cell analysis with 633- and 785-nm laser excitation. Cytometry 21: 120–128.
19. Mader O, Reiner K, Egelhaaf H-J, Fischer R, Brock R (2004) Structure-property analysis of pentamethine indocyanine dyes identification of a new dye for life-science applications. Bioconjugate Chem 15: 70–78.
20. Reichardt C, Engel HD (1988) An improved method for the synthesis of 1,3,3-trialkyl-2-alkylideneindolines. Chem Ber 121: 1009–1011.
21. Shiobasa Y, Ishida S (1960) Malonaldehyde dianil. Yamanouchi Pharmaceuticals Patent, Japan JP35 017020.
22. Kiprianov AI, Buryak VYu (1972) Cyanine dyes with two conjugated chromophores. XVIII. Effect of steric hindrance on the absorption spectra of bis(hemicyanines) from isomeric phenylenediamines with methyl groups on the benzene rings. Zh Org Khim 8: 1707–1712.
23. Ball CA, Osuna R, Ferguson KC, Johnson RC (1992) Dramatic changes in Fis levels upon nutrient upshift in *Escherichia coli*. J Bacteriol 174: 8043–805.
24. Bradley MD, Beach MB, de Koning AP, Pratt TS, Osuna R (2007) Effects of Fis on *Escherichia coli* gene expression during different growth stages. Microbiology 153: 2922–2940.

Part II

Methods

Chapter 7

2D DIGE Saturation Labeling for Minute Sample Amounts

Georg J. Arnold and Thomas Fröhlich

Abstract

The 2D DIGE technique, based on fluorophores covalently linked to amino acid side chain residues and the concept of an internal standard, has significantly improved reproducibility, sensitivity, and the dynamic range of protein quantification. In saturation DIGE, sulfhydryl groups of cysteines are labeled with cyanine dyes to completion, providing a so far unraveled sensitivity for protein detection and quantification in 2D gel-based proteomic experiments. Only a few micrograms of protein per 2D gel facilitate the analysis of about 2,000 analytes from complex mammalian cell or tissue samples. As a consequence, 2D saturation DIGE is the method of choice when only minute sample amounts are available for quantitative proteome analysis at the level of proteins rather than peptides.

Since very low amounts of samples have to be handled in a reproducible manner, saturation DIGE-based proteomic experiments are technically demanding. Moreover, successful saturation DIGE approaches require a strict adherence to adequate reaction conditions at each step. This chapter is dedicated to colleagues already experienced in 2D PAGE protein separation and intends to support the establishment of this ultrasensitive technique in proteomic workgroups. We provide basic guidelines for the experimental design and discuss crucial aspects concerning labeling chemistry, sample preparation, and pitfalls caused by labeling artifacts. A detailed step-by-step protocol comprises all aspects from initial sample preparation to image analysis and statistical evaluation. Furthermore, we describe the generation of preparative saturation DIGE gels necessary for mass spectrometry–based spot identification.

Key words: Quantification, Protein, Proteomics, 2D DIGE, Saturation labeling, CyDyes

1. Introduction

Current proteomic workflows use chromatographic or electrophoretic methods for separation of proteins or peptides from complex mixtures before identification in a mass spectrometer. Two-dimensional (2D) gel electrophoresis is a well-established technique, facilitating the separation of around 2,000 protein analytes in a standard gel. A special benefit of this method is the convenient detection and quantification of protein isoforms or modifications which differ in the molecular mass or the isoelectric point. Quantification of proteins is generally performed at the protein level, whereas mass spectrometry–based quantification usually occurs at the peptide level. Proteome analysis based on 2D polyacrylamide gel electrophoresis (PAGE) has been substantially improved by the introduction of 2D DIGE (1) with respect to reliability, accuracy, dynamic range, and reproducibility of protein spot quantification. In 2D DIGE approaches, fluorophores are covalently attached to an amino acid side chain group before electrophoretic separation. In the frequently applied 2D DIGE minimal labeling concept, three different CyDyes (Cy5, Cy3, and Cy2) are available, which are balanced with respect to charge and attached to the ε-amino group of lysines and free N-terminal residues. Typically, only a few percent of the molecules from each protein species are labeled, and on each 2D gel, around 50 μg of protein sample per CyDye are co-separated. In 2003, an extremely sensitive modification of the DIGE concept was developed (2). Since all accessible sulfhydryl residues of cysteines are labeled to completion in this modification, it is referred to as "saturation labeling." Two CyDyes (Cy3 and Cy5) are available for this system, which are coupled via a maleimide linker to sulfhydryl groups after chemical protein reduction (see Fig. 1). The total protein amount required for 2D gel electrophoresis of complex cell lysates could thereby be reduced by two orders of magnitude down to the lower microgram range. This high sensitivity has made accessible many new areas in biomedicine and biology to quantitative 2D gel-based proteomic studies, e.g., analysis of samples from microdissections (3, 4), glomerular cell preparations (5), membrane protein preparations (6), or mammalian oocytes (7).

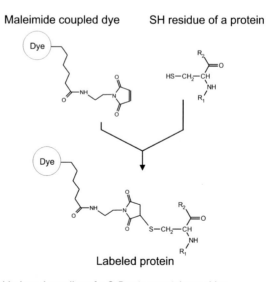

Fig. 1. Maleimide-based coupling of a CyDye to a cysteine residue.

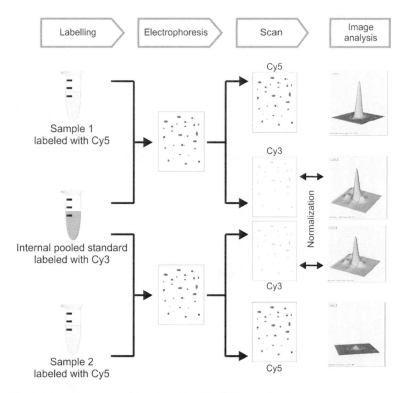

Fig. 2. Schematic representation of a saturation DIGE experiment.

In a typical saturation DIGE experiment (see Fig. 2 for graphical overview), one dye is used for the labeling of individual samples, while the other is used as a label for the so-called internal pooled standard (IPS), which consists of aliquots from each of the samples used within the study. As a consequence, each individual spot present in any of the samples is also represented by a spot in the IPS. Cysteine coupling of Cy3 and Cy5 does not affect protein IP, and the dyes are almost perfectly size-matched. Furthermore, fluorescence emission characteristics of Cy3 and Cy5 are different, and therefore, a co-separation of a Cy5-labeled sample and a Cy3-labeled IPS on each gel is possible. After electrophoresis, two images from each gel, representing the sample and IPS, respectively, are generated by a fluorescence scanner. Further image analysis is performed with a dedicated software, e.g., DeCyder (GE Healthcare). Being present in the same amount on each gel, the IPS is a perfect reference to normalize for inter-gel variances with respect to spot coordinates and spot intensities. For each spot on each gel, the intensity ratio between sample and IPS is calculated. After matching of individual gels based on the IPS readout image, statistical analysis can be performed to calculate significant spot intensity differences among individual samples or sample groups and calculate an abundance ratio for the corresponding proteins.

The saturation labeling concept is especially useful when sample amounts are very limited, i.e., only a few micrograms of total protein are available per individual sample. If sample amount is not limiting, the DIGE minimal labeling approach offers clear advantages with respect to gel number (only half the number of gels is necessary) and robustness of the labeling reaction.

As a consequence from the concept of labeling sulfhydryl groups to completion, saturation DIGE requires a strict stoichiometry between reactive sulfhydryl groups present in the lysate and CyDyes.

The labeling reaction must be carefully optimized for each kind of sample, since the sulfhydryl content may vary substantially between proteins from different biological sources, notably also within different cells, tissues, or body fluids from one distinct organism. In addition, the sample preparation procedure used may influence labeling efficiency since different procedures may lead to different amounts of substances interfering with the labeling reaction.

For successful proteomic analyses based on 2D DIGE saturation labeling, the following preconditions must be fulfilled: (1) availability of a reliable and robust protocol to lyse the biological material in a very reproducible manner, a challenging task especially with minute sample amounts; (2) availability of a reproducible and robust protocol for protein quantification from scarce samples; (3) the ability to conduct 2D PAGE in an excellent manner with respect to reproducibility and separation power in both dimensions; and (4) in cases where protein identification is obligatory, the possibility to generate enough sample(s) for at least one preparative gel containing a total protein amount in the order of 100–500 µg, depending on sample complexity and intensity of relevant spots, for mass spectrometry–based protein identification.

This chapter describes:

1. Special considerations for sample preparation in saturation DIGE.
2. Optimization of the saturation dye labeling reaction.
3. Design of a 2D DIGE study.
4. Evaluation of a 2D DIGE experiment.
5. Identification of protein spots from a preparative saturation DIGE gel.

It is assumed that the reader is familiar with 2D PAGE using immobilized pH gradients in the first and SDS PAGE in the second dimension. Therefore, no descriptions about 2D PAGE in general or 2D PAGE technical equipment are given.

2. Materials

2.1. Cell Lysis

1. Cell lysis buffer: 30 mM Tris, 7 M urea, 2 M thiourea, 4% (w/v) CHAPS, pH 8.0. Adjust using 1 M HCl and store in aliquots at –20°C. The buffer is stable for at least 3 months.

2.2. Reconstitution of Dye and Protein Labeling

1. Cy3 and Cy5 saturation dyes for cysteine labeling (GE Healthcare Bio-Sciences, Uppsala, Sweden). Store at –70°C.
2. *N,N*-Dimethylformamide (DMF) anhydrous >99.8%. Do not use longer than 3 months after opening.
3. Reconstituted solutions of Cy3 and Cy5 saturation dyes in DMF at a 2 mM concentration.
4. 2 mM Tris-(2-carboxyethyl)phosphinehydrochloride (TCEP) solution: dissolve 2.8 mg TCEP in 5 mL of water. Aqueous TCEP solution is unstable and must be prepared fresh.
5. Sample buffer I: 7 M urea, 2 M thiourea, 4% (w/v) CHAPS. Store in aliquots at –20°C. This buffer is stable for at least 6 months.
6. Sample buffer II: 7 M urea, 2 M thiourea, 4% (w/v) CHAPS, 130 mM dithiothreitol (DTT), 2% (v/v) Pharmalyte™ pH 3–10 (GE Healthcare Bio-Sciences, Uppsala, Sweden; Cat. No. 17-0456-01). Prepare fresh by addition of Pharmalyte and DTT to sample buffer I.
7. Protein Determination Reagent (e.g., USB, code 30098) for determination of protein concentration.
8. Appropriate laser scanner (e.g., from the Typhoon series; GE Healthcare Bio-Sciences, Uppsala, Sweden).
9. Appropriate software for image analysis and spot quantification (e.g., DeCyder 2D Differential Analysis software; GE Healthcare Bio-Sciences, Uppsala, Sweden).

3. Methods

3.1. Sample Preparation for Saturation DIGE

Biological samples contain highly complex mixtures of completely different biochemical macromolecules such as DNA/RNA, proteins, and lipids. The distribution among these substance classes is dependent on the biological source and may therefore require special steps to avoid interference with 2D gel electrophoresis in general or 2D DIGE saturation labeling, e.g., removal of excess amounts of lipids or nucleic acids. In addition, some samples may contain especially high amounts of hydrophobic proteins, the solubilization of which may require special detergents, whose compatibility has to be assured by separate experiments. In saturation DIGE, the user typically handles very limited sample amounts, leading to additional challenges with respect to reproducibility. Consequently, all steps which are not absolutely essential should be avoided to facilitate the highest possible degree of reproducibility. It is more effective to consciously disregard some protein analytes (e.g., hardly soluble proteins) in favor of a reproducible procedure than to end up with lysates differing in their protein composition within biological replicates. While the former will simply lead to a lack of information about distinct individual proteins, the latter might lead to a complete failure of the experiment as a consequence of abundance differences introduced by the procedure. Here, we describe a sample preparation procedure based on the buffer recommended by GE Healthcare. Note that this buffer must not contain a disulfide-reducing agent (see Note 1).

1. Carefully wash your tissue or cells at least 4 times, e.g., with PBS (phosphate buffered saline), at 4°C to remove blood and any contaminants contained in the culture media. Carryover of serum or media proteins can have disastrous consequences and limit the scientific outcome of the experiment (see Note 2).

2. Centrifuge after each washing step for 10 min at $600 \times g$, 4°C, and remove the supernatant carefully and completely. Even traces of remaining wash buffer may significantly reduce the concentration of chaotropic agents in the lysis buffer and affect extraction and solubilization of proteins. After the last centrifugation step, remove even the last microliter of remaining wash buffer by aspiration with a very thin glass capillary or a pipette tip used for ultrathin gel loading.

3. Add lysis buffer to end up with a total protein concentration of >0.55 μg/μL and <5.0 μg/μL for analytical gels (see Notes 3 and 4).

4. Check if the protein amount to be analyzed per 2D gel is contained in a volume which can be readily applied to your isoelectric focusing device (see Note 4).

5. Apply a lysis procedure (e.g., sonication, freeze thaw) suitable for the sample analyzed.

6. Check for solubilization after centrifugation at $10,000 \times g$ for 5 min at 4°C. Ideally, no pellet at all is visible after completion of the lysis procedure (see Note 5).

7. Transfer the supernatant to a fresh microfuge tube. Check if the pH is still in the range of 7.8–8.2 (see Note 6).

8. If the pH is below 7.8, titrate to pH 8.0 by stepwise addition of lysis buffer adjusted to pH 9.5 instead of 8.0.

9. If the pH is above 8.2, use lysis buffer adjusted to pH 6.5 accordingly. pH readjustment using lysis buffer is preferable to adjustment by NaOH or HCl, since the concentration of the chaotropic salts is maintained at the optimal level.

3.2. Determination of Total Protein Concentration

1. Carefully determine the total protein concentration of the lysate using a detergent compatible assay (saturation DIGE manufacturer GE Healthcare recommends Protein Determination Reagent USB, code 30098). Do not change the assay throughout the saturation DIGE experiment since assays of different type and manufacturer may lead to different results.

2. If the concentration matches the recommended range (>0.55 µg/µL, <5 µg/µL), but a dilution is required for technical reasons, this may be performed by addition of lysis buffer.

3. In contrast, if the concentration is not within the recommended range, prepare a new lysate using an adequate sample amount and/or lysis buffer volume.

3.3. Optimization Experiment: Development of a Sample-Specific Labeling Protocol

The maleimide derivatives of saturation CyDyes react with free sulfhydryl groups, and consequently, proteins to be included in the analysis must contain at least one cysteine residue. Additional cysteines lead to a proportional increase in fluorescence signal intensity of the affected protein. If present in excess amount, CyDyes also react with primary amine groups, leading to one or even more additional labels of the protein at lysine residues. Since the positive charge of each affected lysine residue is eliminated, the pI of the molecule becomes more acidic. As a consequence, two or even more derivatives of the protein are generated, leading to additional 2D gel spots along the horizontal (pI) axis of the affected protein. This effect is referred to as "overlabeling." In contrast, an insufficient concentration of CyDye will fail to label all cysteine residues in the lysate to completion. Since sulfhydryl groups are not charged at neutral or mild acidic or basic pH, no horizontal shift will occur. Instead, especially in cysteine-rich small proteins, the lower molecular mass of unlabeled cysteine residues may be resolved by 2D gel electrophoresis, leading to additional spots in the vertical (molecular mass) axis. The stoichiometric conditions facilitating complete labeling of all cysteine residues without affecting lysine residues have to be determined individually in a titration experiment, since (1) content of sulfhydryl-containing biological compounds (e.g., glutathione) may vary considerably between organisms, organs, cells, and body fluids (see Note 7) and (2) individual sample preparation conditions may lead to the introduction of exogenous interfering chemical compounds (traces of reducing agents, free primary amines, etc.). Whenever the biological source, the cell culture conditions, or the sample preparation protocol is changed, the optimization step must be repeated. Within one experimental series, conditions should not be changed at all with respect to these parameters.

Reduction of Cys–Cys bonds in proteins is essential to make them accessible to CyDye labeling. This is performed using the agent TCEP, not containing a sulfhydryl group, and the amount of TCEP necessary is proportional to the amount of CyDye necessary for optimal labeling. The molar ratio of TCEP:dye is always kept at 1:2 to ensure efficient labeling.

The protocol developed in the following procedure is used for all analytical labeling experiments throughout the saturation DIGE experiment.

For the labeling of 5 μg protein (see Note 8) in the recommended lysis buffer at a concentration of 0.55–5 μg/μL, typically 2 nmol TCEP and 4 nmol dye have to be added. Within this range of protein concentration, no adjustments must be made with respect to the reaction kinetic.

Careful optimization following the scheme in Table 1 is strongly recommended. Since this is a laborious task comprising six 2D gels, and since changes in the sample preparation will require a new optimization experiment, the lysis protocol should have been successfully established before the optimization experiment. To check for appropriate lysis conditions, label a sample following the protocol outlined for gel number 3 in Table 1. If necessary, optimize your lysis protocol and eventually your 2D electrophoresis conditions until the spot pattern obtained perfectly meets your needs with respect to number of resolved protein spots visible, spot resolution, and absence of extensive streaking.

The IPS would generally represent the individual samples best; however, not all biological samples might already be available when

Table 1
Pipetting scheme, amounts, and final concentrations for the optimization experiment of the sample labeling

Gel number	Volume (μL) of TCEP solution (2 mmol/L)	Amount of TCEP (nmol)	Final concentration of TCEP (mmol/L)	Volume (μL) of CyDye solution (2 mmol/L)	Amount of CyDye (nmol)	Final concentration of CyDye (mmol/L)
1	0.50	1.00	0.095	1.00	2.00	0.190
2	0.75	1.50	0.133	1.50	3.00	0.267
3	1.00	2.00	0.167	2.00	4.00	0.333
4	1.25	2.50	0.196	2.50	5.00	0.392
5	1.50	3.00	0.222	3.00	6.00	0.444
6	2.00	4.00	0.267	4.00	8.00	0.533

performing the optimization experiment. Under these circumstances, use the most typical sample available instead. If all samples are already available, prepare the IPS by mixing aliquots comprising equal protein amounts of all samples integrated in the saturation DIGE study. In both cases, follow the scheme of Table 1, using six 5 µg aliquots of the sample for Cy3 labeling and another six 5 µg aliquots for Cy5 labeling:

1. Transfer a 5 µg protein aliquot into a sterile microfuge tube.
2. Adjust the volume to 9 µL using the cell lysis buffer.
3. Add the required volume of freshly prepared 2 mM TCEP solution according to Table 1 to the protein solution of step 1. Discard any unused material.
4. Mix vigorously by pipetting.
5. Spin down the sample in a microcentrifuge.
6. Repeat steps 4 and 5 to make sure the solution is really homogeneous and is located completely at the bottom of the microfuge tube.
7. Incubate at 37°C for 60 min in the dark.
8. Add the required volume of 2 mM CyDye according to Table 1.
9. Mix vigorously by pipetting.
10. Spin down the sample in a microcentrifuge.
11. Repeat steps 9 and 10 to make sure the solution is really homogeneous and is located completely at the bottom of the microfuge tube.
12. Incubate at 37°C for 30 min in the dark.
13. Stop the reaction by adding a volume equal to the reaction volume of sample buffer II.
14. Mix vigorously by pipetting.
15. Spin down the sample in a microcentrifuge.
16. Samples are ready for 2D PAGE analysis and can be stored frozen in the dark for up to 1 month, preferably at –70°C.

Before 2D PAGE analysis, combine corresponding Cy3 and Cy5 samples for each stoichiometric condition. Analyze each mixture (see Note 9) on the 2D gel electrophoresis equipment designated for the final study. For isoelectric focusing conditions of analytical saturation DIGE gels, the protocol given in Table 2 has been elaborated by the manufacturer for 24-cm pH 3–10 strips, which should be followed whenever possible.

For the SDS PAGE in the second dimension, make sure to use low fluorescent glass plates.

From each of the six gels, generate two fluorescence readout images, using a scanner suitable for Cy3 and Cy5 dyes. Laser scanners

Table 2
Parameters recommended by GE Healthcare for isoelectric focusing of analytical saturation DIGE samples

Step	Power (V)	Ramp	Duration (h)
1	300	Step and hold	3.0
2	600	Gradient	1.5
3	1,000	Gradient	2.0
4	8,000	Gradient	2.5
5	8,000	Step and hold	3.0
6	500	Step and hold	Up to 48.0

Current should be below 25 µA/strip, and temperature should be adjusted to 25°C

(e.g., from the Typhoon series; GE Healthcare) usually provide better resolution and signal-to-noise ratios than camera-based imaging instruments. Generate overlay images of corresponding readouts using appropriate software tools (e.g., ImageQuant).

3.4. Evaluation of the Optimization Experiment

To decide which labeling condition gives the best labeling results, carefully inspect the overlay images with respect to the following parameters:

1. All spots should be overlaid between the Cy3 and Cy5 image; nonoverlaid spots are an indication for insufficient amounts of CyDye.

2. No significant horizontal spot "arrays" should be present: horizontal spot arrays are an indication for overlabeling. They can be distinguished from natural protein isoforms differing in their pI by their disappearance at lower CyDye concentrations.

3. No significant vertical spot "arrays" should be present: vertical spot arrays are an indication for under-labeling. Labeling-induced vertical spot arrays can be distinguished from natural proteins of identical pH but slightly different molecular mass by their disappearance at higher CyDye concentrations. Small proteins are more suitable for detection of under-labeling, since the CyDye dependant differences in molecular mass can be resolved on the gel in the 10–30 kDa range. Figure 3 shows an example of under-labeling and the disappearance of under-labeled spots upon an increase in CyDye amount.

4. Usually, this optimization experiment will lead to at least one gel without the obstructions mentioned above. If this is not the case, carefully check all solutions and buffers and the sample preparation procedure. Cross-check your protein concentration

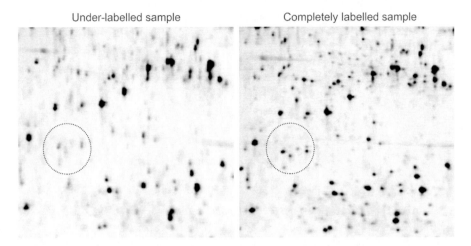

Fig. 3. Cy3 readout of an under-labeled and optimally labeled analytical DIGE saturation gel. The area characteristic for an under-labeled artifact is *circled*.

with another assay. If indeed no errors can be detected at all, prepare a new sample and repeat the procedure. Only if this does not solve the problem, further increase or reduce the amount of CyDyes beyond the limits of Table 1.

5. Check the signal-to-noise ratio of the weakest spots to be integrated into the quantification. This can be performed using ImageQuant software: select an area containing a single spot, read out the integrated signal intensity from all pixels selected (signal), and compare it with the same area selected in a spot-free neighboring area (noise). As a rule of thumb, the signal-to-noise ratio should exceed 10 to facilitate reliable quantifications. Optimize the parameters of your scanner or imaging instrumentation when necessary.

6. In a gel that is free from under- and overlabeling artifacts, check for spots which are not co-separated and therefore appear in the Cy3 or Cy5 color rather than in the overlay color. These spots must be noted and merged manually in the DeCyder 2D Differential Analysis software in the main experiment, prior to statistical analysis.

7. Since aliquots of one single sample were labeled, the ratio of Cy5 intensity and Cy3 intensity of each spot in a gel should, in theory, be exactly 1.0. Visualize the images using the DeCyder DIA software and check for proteins, which exhibit differential intensities between the two dyes (visible by color change in an overlay image). This is usually due to Cy5 quenching effects with some highly labeled proteins. Using the recommended experimental design with the incorporated IPS, this phenomenon will be compensated in the DeCyder evaluation of the final experiment. However, systematic deviation from a

1:1 ratio affecting a large number of spots may indicate an irreproducible labeling procedure and requires corresponding improvements.

The optimized labeling protocol established here is used for all labeling reactions for the saturation DIGE experiment. It should be applied without any further changes in either the labeling procedure or the sample preparation steps.

3.5. 2D DIGE Saturation Labeling of Analytical Samples

A 2D DIGE gel of mammalian samples typically facilitates the detection and quantification of around 1,500–2,000 spots with a sufficient signal-to-noise ratio of the fluorescence readout. As for all multianalyte systems, a statistical evaluation of a sufficient number of replicates is mandatory to correct for random deviations of spot intensities between samples and controls. Reference (8) gives an overview about the number of biological replicates necessary from the statistical point of view. As a rule of thumb, six biological replicates (see Note 10) should be prepared for the final experiment. It is recommended to prepare at least one additional sample to facilitate completion of the experiment even if, for whatever reasons, one sample proves to be inadequate upon analysis. This additional sample must be contained in the IPS. The DIGE saturation approach facilitates all conceivable comparisons between all samples which are contained in the IPS, making the design of the experiment very straightforward (see Note 11):

1. Prepare and label all samples to be integrated into the 2D DIGE saturation experiment according to the optimized labeling protocol (see Subheading 3.3) with Cy5.

2. Prepare an amount of IPS which is sufficient for the planned number of samples (or gels) plus an additional 50%, in case some of the gels fail (see Note 12). Prepare the IPS by combining aliquots containing the same amounts of protein from each sample.

3. Label the IPS with Cy3 according to the optimized labeling protocol.

4. Combine each sample with an aliquot of IPS and subject the mixture to 2D electrophoresis.

5. Immediately after electrophoresis, scan the gel to generate Cy5 and Cy3 readouts.

3.6. Image Analysis and Spot Quantification

In order to detect spots differing in intensity between several gels of different samples, dedicated software tools are available on the market. The software packages usually perform (1) the detection and quantification of all spots in all gels, (2) the inter-gel matching of these spots, and finally (3) the calculation of spot intensity ratios between gels from different sample groups. For saturation DIGE experiments, it is recommended to use software packages supporting the concept of co-migration of proteins labeled with different dyes as well as the concept of IPS. In this section, image analysis with the frequently used DeCyder 2D V7.0 software package (GE Healthcare, Uppsala, Sweden) is described. The focus of this section is not to replace the software manual but to briefly outline the principle procedure.

3.6.1. Image Loading

1. Take into account that DeCyder uses an Oracle database (Oracle, Redwood Shores, CA, USA) installed on a dedicated server to store images and all kinds of related data. Therefore, be sure that you have a valid DeCyder account and appropriate privileges to create your own projects and workspaces.
2. Login into DeCyder and generate a new project using the "Organizer" tool to which images and data should be linked to.
3. If you want the images to be automatically grouped and named, the DeCyder gel name recommendations should be followed. The filename has to consist of a gel number (e.g., Gel 01), a description of the function of the image (key terms: "Standard," "Control," "Treated"), and the dye used. Instead of the key terms, other words can be used but must be in brackets. For example, the following two image files produced by the Typhoon scanner will be grouped into a Gel named Gel 01:

   File 1: Gel 01 Standard Cy3.gel.

   File 2: Gel 01 (Sample name) Cy5.gel.
4. Start the Image Loader and select the project into which the gel images are to be imported.
5. Add gel files to the import list.
6. If the files do not match the recommended names, perform manual grouping as described in the manual.
7. Crop the images using the DeCyder image editor tool (or ImageQuant, before importing the files into DeCyder) to remove areas which contain information of no interest.
8. Ensure that only relevant spots remain inside the image and that the patterns of the different images are similar.
9. Double-check settings (dye chemistry, assignment of fluorophores) and change these settings if required.
10. Import the selected image files.

3.6.2. Differential In-Gel Analysis

The differential in-gel analysis (DIA) module processes images from *individual* gels and performs (1) spot co-detection, (2) spot quantification and (3) normalization, and (4) in-gel intensity ratio calculation of the spots from the Cy5 and Cy3 samples.

To perform a DIA analysis:

1. Open the DIA module within the DeCyder software.
2. Create a new DIA workspace and select all images of one gel.

3. Perform spot detection using the "DeCyder spot detection algorithm 6.0" with a number of 10,000 for "estimated number of spots." This number is an overestimation and incorporates the false positive detection of dust particles and other disturbing nonprotein-related artifacts.

4. To remove false positives, use the exclusion filter "Volume <30,000." Other filters are usually not needed for good quality gels.

5. Visually inspect if spot detection leads to a satisfactory result. A method to control spot detection quality is to uncheck "Excluded spots" in the "Spot Display Properties" window and to ensure that the broad majority of spots are detected and the majority of very weak spots (typically false positives) are not detected.

6. Save the DIA workspace to the corresponding DeCyder project.

3.6.3. Biological Variation Analysis

The biological variation analysis (BVA) module performs gel-to-gel spot matching and allows quantitative comparisons between several groups:

1. Create a BVA workspace.

2. Import the DIA workspaces which are to be included in the BVA analysis.

3. Add experimental groups in the "Experimental Design View" of the "Spot Map View."

4. Assign the unassigned gel images to the corresponding experimental group.

5. Select the most representative gel as master gel for matching.

6. Match the gels with "warping" enabled.

7. Switch to "Match mode" and check the matching quality manually. If warping has been performed it is helpful to display warp grids. In addition, color overlay images can be displayed, where the master image within every gel is presented in yellow and the standard image is presented in blue.

8. For gels showing areas with unsatisfactory matching, set landmarks in the corresponding areas. Alternatively, correct matches in the corresponding area can be set manually (see Note 13).

9. Rematch the gels.

3.6.4. Batch Processing

When multiple gels have to be analyzed, it is much faster to use the Batch Processor. After setup, the Batch Processor performs all DIA and subsequent BVA analyses automatically:

1. First, import the cropped gel images into the corresponding project as described in Subheading 3.6.1.

2. Open the Batch Processor module.

3. Add the images by selecting Menu item → "File" → "Add DIA Batch item."
4. Select the images which are to be analyzed.
5. Enter 10,000 as the number of estimated spots.
6. Check the "Include in BVA batch list" box.
7. Set the "Spot exclusion filter" to <30,000 for the volume.
8. Click "OK" to proceed.
9. Set up the experimental groups in the "BVA item settings" window.
10. Select the most representative gel as master gel for matching.
11. Save the batch to the corresponding project.
12. Process the batch.

3.6.5. Statistical Analysis

The BVA module allows some basic statistical analyses including:

1. The determination of average intensity ratios between two groups and the calculation of Student's t test p values.
2. The statistical analysis between all groups using One-Way ANOVA (ANalysis Of VAriance).
3. Calculation of Two-Way ANOVA p values (statistical analysis between the two conditions in an experiment with two independent factors, e.g., time-dose studies).
4. The application of a false discovery rate (FDR) correction.

Perform the statistical analysis with the BVA module as follows:

1. Make sure the gel images are in the correct experimental groups.
2. Switch to Protein mode.
3. Define the statistics which should be calculated (Menu item "Process" → "Protein statistics"). For comparisons of two groups, choose "Average Ratio" and "Student's t test" and select the groups which are to be compared in the "Population 1" and "Population 2" text box.
4. If more than two groups are compared, select additionally the One-Way ANOVA. Should a Two-Way-ANOVA be required, please refer to the DeCyder software manual.
5. If FDR correction is required, check the corresponding box (see Note 14).

3.6.6. Protein Filtering

To select protein spots of interest in most cases, filtering must be performed. If the Student's t test was applied, filtering of (1) average spot ratios between two groups and (2) p values should be applied. When more than two groups are analyzed, filtering of ANOVA p values is appropriate. Commonly, the filter is set to

p values <0.05 if FDR correction was performed. Moreover, it is recommended to filter for spots which are present in at least three-quarter of the gels:

1. Filter settings can be set in the BVA module within "Protein view" (Menu item "Process" → "Protein Filter").
2. Make sure the "Assign Protein of Interest" box is checked.
3. To display only the spots assigned as "Protein of interest," set the view properties to "Protein of interest" (Menu item "View" → "Properties" → Tab "Protein Table" → "Protein Table Filter" → "Protein of Interest").

3.6.7. Extended Data Analysis

The Extended Data Analysis (EDA) module of the DeCyder software is a tool for multivariate analysis of protein expression data generated with the BVA module. In addition to the univariate analyses (Student's t test, One-Way ANOVA and Two-Way ANOVA) performed with the BVA module, the following analyses can be performed:

- Principal component analysis (PCA).
- Hierarchical clustering.
- K-means clustering.
- Self-organizing maps.
- Gene shaving.
- Discriminant analysis.

To describe all the statistical methods provided by the EDA module would clearly go beyond the scope of this chapter. Nevertheless, we want to point out that, especially, the PCA analysis has shown to be a powerful tool to find outlier gels in large gel sets. Furthermore, hierarchical clustering has demonstrated in many experiments to be a powerful tool to estimate the grade of difference between experimental groups and the grade of similarity of samples within an experimental group.

3.7. Preparative 2D DIGE Saturation Labeling for Protein Identification

For mass spectrometry–based identification of relevant proteins, the amount of protein present on an analytical saturation DIGE gel (5 μg total protein) is usually not sufficient. Therefore, a preparative gel, typically containing 100–500 μg of total protein, has to be prepared. To reproduce exactly the spot pattern of the analytical gel, it is essential to label the protein solution with saturation dye Cy3 (see Note 15). The stoichiometry of protein and TCEP/CyDye determined in the optimization experiment has to be maintained in the preparative labeling procedure. Since the saturation labeling procedure for a preparative gel is designed to give a final volume of 450 μL, cup loading of the sample for isoelectric focusing is not possible. Instead, in-gel rehydration sample loading must be used, and 450 μL is the volume that should be loaded when using 24-cm IEF strips. The procedure for labeling the preparative sample differs from the procedure for analytical gels in several aspects and is therefore described separately:

1. Prepare a protein lysate (see Notes 15 and 16) using the same lysis procedure as used for the analytical protein amounts. The desired amount of protein (usually 100–500 μg) must be contained in a final volume of 250 μL of lysis buffer.
2. Transfer 250 μL of this solution to a microcentrifuge tube.
3. Calculate the amount of TCEP necessary by scaling up the amount determined in the optimization experiment (see Subheading 3.3, step 3), using the formula:

   amount of TCEP(nmol) = (amount of protein to be labeled / amount of protein used in the optimization experiment) * optimal amount of TCEP in optimization experiment

   (e.g., 500 μg protein/5 μg protein * 2 nmol TCEP = 200 nmol TCEP).
4. Freshly dissolve the calculated amount of TCEP in 10 μL of water.
5. Transfer 10 μL of the TCEP solution to the protein lysate.
6. Mix vigorously and thoroughly by pipetting, taking into account the high viscosity of concentrated protein solutions (see Note 17).
7. Spin down the sample in a microcentrifuge.
8. Repeat steps 6 and 7 to make sure the solution is really homogeneous and is located completely at the bottom of the microcentrifuge tube.
9. Incubate at 37°C for 60 min in the dark.
10. Calculate the required amount of Cy3 by scaling up the amount determined in the optimization experiment (see Subheading 3.3, step 3), using the formula:

    amount of Cy3(nmol) = (amount of protein to be labeled / amount of protein used in the optimization experiment) * optimal amount of Cy3 in optimization experiment

    (e.g., 500 μg protein/5 μg protein * 4 nmol Cy3 = 400 nmol Cy3).

11. Calculate the volume of the Cy3 solution (20 mmol/L, i.e., 20 nmol/µL) containing the amount of Cy3 calculated in step 10, using the formula:

    volume of 20 mmol/L Cy3 solution (µL) = amount of Cy3 (nmol) / 20 nmol/µL

    (e.g., 400 nmol Cy3/20 nmol/µL = 20 µL)

12. Add the calculated volume of 20 mM Cy3 to the reaction mixture.
13. Mix vigorously by pipetting (see Note 17).
14. Spin down the sample in a microcentrifuge.
15. Repeat steps 13 and 14 to make sure the solution is really homogeneous and is located completely at the bottom of the microcentrifuge tube.
16. Incubate at 37°C for 30 min in the dark.
17. Calculate the required volume of sample buffer I to adjust the total volume of the labeling reaction mixture to 445.5 µL, using the formula:

    volume of sample buffer I (µL) = 445.5 µL − 250 µL − 10 µL − volume of Cy3 calculated in step 11

18. Stop the reaction by adding the volume of sample buffer I calculated in step 17 to the reaction mixture.
19. Mix vigorously by pipetting.
20. Add 4.5 µL of Pharmalytes (pI range corresponding to pI range of IEF strips used). Total volume is now 450 µL, corresponding to the maximal volume applicable.
21. Add 9.0 mg of solid DTT for a final DTT concentration of 65 mM DTT.
22. Mix vigorously by pipetting.
23. Spin down the sample in a microcentrifuge.
24. The sample with a total volume of 450 µL is now ready for isoelectric focusing. The sample can be stored frozen in the dark for up to 1 month, preferably at −70°C.

3.8. Preparative 2D Gel Separation

It is essential to use the same type of electrophoretic devices as used for the analytical gels. Otherwise, an unambiguous assignment of corresponding spots between the analytical and preparative gels may be impossible. For isoelectric focusing of preparative IEF strips, use the parameters given in Table 3.

Appropriate precautions are strongly recommended in all steps to avoid contamination with human keratin (see Note 18).

Table 3
Parameters recommended by GE Healthcare for isoelectric focusing of preparative saturation DIGE samples

Step	Power (V)	Ramp	Duration (h)
1	150	Step and hold	4.0
2	300	Step and hold	3.0
3	1,000	Gradient	6.0
4	8,000	Gradient	1.0
5	8,000	Step and hold	6.0
6	500	Step and hold	Up to 48.0

Current should be below 75 µA/strip, and temperature should be adjusted to 25°C

3.8.1. Visualization and Assignment of Spots

Generate a fluorescence Cy3 image from the preparative gel by following the instructions of the scanner's manufacturer. Use the same instrument as used for the analytical gels. Make sure no distortions of the gel can occur between the start of scanning and the end of spot picking, e.g., by using an appropriate gel frame.

Carefully match the spots in the preparative gel with the spots from an analytical gel, preferably using the readout representing the IPS. For spots to be identified by mass spectrometry, a very careful manual inspection for correct matching is indispensable. Print out a true to scale image of the Cy3 readout and mark all spots to be picked for identification.

3.8.2. Spot Picking

Spots can either be picked by a spot-picking robot or manually. For spot picking by a robot, the following procedure is recommended:

1. Export the *x*- and *y*-coordinates of relevant spots from DeCyder and import these into the software of the spot-picking robot. Refer to the robot's manual for this procedure.
2. After picking, perform a rescan of the Cy3 image to control if spots have been hit correctly by the spot excision device.
3. In addition, a subsequent Coomassie staining may be helpful (see Notes 19 and 20).

For manual spot picking, the following procedure is recommended:

1. Stain the gel with colloidal Coomassie (freshly prepared and keratin-free) to generate visible spots (see Notes 19 and 20).
2. Carefully align the gel to the true to scale printout, using characteristic spots visible in both the Cy3 and the Coomassie stained gel image.
3. Pick spots using your preferred spot-picking device.

3.9. Spot Identification by Mass Spectrometry

With the picked spots, proceed according to standard procedures for in-gel digestion, peptide elution, and peptide purification. A final ZipTip purification is highly recommended. Analyze spots by ESI or MALDI tandem mass spectrometry. For data analysis, include a modification for each cysteine residue for the Cy3 fluorophore (Monoisotopic mass addition: 672.30 Da; Average mass addition: 672.83 Da).

4. Notes

1. If the use of a different lysis buffer is inevitable, strictly avoid the inclusion of thiols (e.g., dithiothreitol (DTT), dithioerythritol (DTE), β-mercaptoethanol (β-MSH)) or primary amines (e.g., from Pharmalytes or carrier ampholytes added to the buffer), since these compounds act as competitors for the saturation dye.

2. Vascularized tissue frequently contains endogenous blood to a varying degree from sample to sample. This represents a considerable bias, possibly leading to false positive proteins erroneously characterized as proteins of different abundance. As a consequence, check each gel for the presence of a characteristic array of albumin spots in the 2D gel at a molecular mass of around 60 kDa. Presence of these spots requires a careful consideration whether protein spots present at different abundance in the samples may be the result of a different degree of vascularization in individual samples. Investigate appropriate databases to rule out typical serum proteins.

3. Total protein content may vary considerably between different biological materials. As a very rough first-order approximation, assume a total extractable protein amount of 100 pg per somatic mammalian cell from cell culture and 25 µg of protein/mg tissue sample. Thus, to prepare a lysate containing 5 µg protein, 50,000 cells or 0.20 mg tissue sample are required.

4. If protein spot identification from a preparative gel has to be performed, make sure lysis conditions can be scaled up to the required protein amount in a volume compatible with your isoelectric focusing device. Switching to different 2D PAGE equipment for preparative gels must be avoided, since deviations in the spot pattern might impair the correct spot allocation between analytical and preparative gels. In addition, beware of protein concentrations exceeding 5 µg/µL, since lysates may get significantly depleted of poorly soluble proteins.

5. In cases were substantial amounts of sample have remained undissolved, add fresh lysis buffer and repeat the lysis procedure once or twice. Repeated extraction of the pellet with fresh

buffer accelerates solubilization kinetics, though overall solubility of a protein remains unaffected. Choose your lysis buffer volumes appropriately so the minimum protein concentration is >0.55 mg/mL.

6. pH of volumes as low as 0.5 μL can be measured using a narrow range pH indicator strip. However, since the color references tables of the manufacturer refer to indicator strips completely saturated with the solvent to be measured, a significant color deviation may occur when only submicroliter volumes are applied. To avoid erroneous pH measurement, apply on the same indicator strip in the appropriate field an equal amount of fresh lysis buffer, adjusted to pH 8.0 with a calibrated pH meter. The resulting color of the original buffer is your unbiased reference. Should readjustment of your sample pH be necessary, proceed in the same manner, preparing a fresh reference spot each time.

7. Assuming a protein concentration of 5 μg/μL in the protein lysate and 1% of all amino acids represent cysteine, the concentration of cysteine-derived SH groups in the lysate equals 420 μmol/L. Calculating with a glutathione concentration of 5 mmol/L in mammalian tissue and 25 μg protein extracted/mg tissue, the concentration of glutathione-derived SH groups in the lysate is 1,000 μmol/L, e.g., in the same range as cysteine-derived sulfhydryl groups. Therefore, varying amounts of GSH or other sulfhydryl-containing compounds in individual samples may considerably alter the overall sulfhydryl content in the lysate.

8. Labeling procedures for analytical gels are described for a total protein amount of 5 μg, as recommended by the manufacturer (GE Healthcare). According to our experience, the amount can be further decreased if necessary. In any case, develop the sample labeling protocol with the same amount of protein as will be used in the final experiments.

9. According to our experience, not all the material is necessary to produce excellent gel images. Consider to use only half the amount (e.g., 2.5 μg Cy3 and 2.5 μg Cy5 labeled material per gel) in order to be prepared for a potential repetition of the 2D PAGE analysis.

10. If all experimental procedures are carried out carefully, the variations introduced by the biological system will usually exceed the variations introduced by the experimental setup. Thus, it is much more effective to perform biological replicates than technical replicates, since the biological replicates necessarily comprise all the technical variations, whereas technical replicates do not comprise any biological variations.

11. As an example, analyzing effector-induced proteome changes at four time points (t1, t2, t3, t4) in two cell lines (A, B), six

comparisons within each cell line A and B (t1A vs. t2A, t1A vs. t3A, t1A vs. t4A, t2A vs. t3A, t2A vs. t4A, t3A vs. t4A, and corresponding comparisons in cell line B) plus 16 comparisons between cell lines can be performed (t1A vs. t1B, t2A vs. t1B, t3A vs. t1B, t4A vs. t1B, t1A vs. t2B, t2A vs. t2B, t3A vs. t2B, t4A vs. t2B, t1A vs. t3B, t2A vs. t3B, t3A vs. t3B, t4A vs. t3B, t1A vs. t4B, t2A vs. t4B, t3A vs. t4B, t4A vs. t4B).

12. For reasons of reproducibility, it should be avoided to have more than one IPS preparation in the experimental setup. Even minor deviations in labeling efficiency may complicate or even impair a proper evaluation.

13. It should be noted that for reproducible gels, the time needed for matching is usually low.

14. If thousands of spots are compared using Student's t test or ANOVA for statistical significance, many of these spots may be statistically different in intensity, from which several may have achieved this significance by chance alone. The FDR correction adjusts the Student's t test or ANOVA p values for each spot to keep the overall error rate as low as possible.

15. In most cases, the preparative sample should have the same composition as the IPS. This ensures that each spot which was identified as differentially abundant in the quantitative analysis is present in a detectable amount in the preparative gel. Otherwise, individual spots may be missed, when they are present only in a subgroup of samples which was not contained in the preparative gel. However, if all spots relevant for identification are increased in a subgroup of samples, the preparative gel is preferentially prepared using this subgroup.

16. If not enough sample material is available for a preparative gel, it has been suggested to use sample material most similar to the samples analyzed. For instance, when analyzing laser microdissection samples of tumors, surrounding healthy tissue could be used for preparing the preparative gel (or vice versa, if tumor material is abundant and healthy tissue is limiting). In any case, this must be regarded a less than ideal solution, and we strictly recommend a verification experiment to ensure that the correct protein has been identified. Verification could be performed by analyzing the corresponding tissues by Western blotting when appropriate antibodies are available. Alternatively, an antibody-independent verification could be performed by selected reaction monitoring (SRM) mass spectrometry (MS), which is sensitive, specific, and precise and requires low amounts of protein in the same range used for analytical saturation DIGE gels. A description of the SRM MS method is available in this book series (9).

17. This step is very critical since the reaction may be incomplete if the solution is not homogenous. Incomplete reaction will

result in incomplete labeling, leading to a different image as compared to the analytical gel and probably impairing a correct spot assignment between analytical and preparative gels.

18. Keratin contaminations cause problems in mass spectrometry for two reasons: (1) in human samples, exogenously introduced human keratin cannot be discriminated from endogenous keratin potentially contained in the sample; (2) tryptic peptides derived from exogenous keratin are potent competitors in peptide ionization. The resulting ion suppression may impair the identification of weak spots. Beware that keratin contaminations in solvents or solutions cannot be rendered innocuous by autoclaving. Use carefully prepared fresh solutions for gel casting, gel staining, etc., in keratin-free labware.

19. As a consequence of labeling cysteine residues to completion, spot intensity is determined not only by the amount of protein contained in the spot but in addition by the number of cysteine residues per molecule. Spot intensities obtained by staining procedures, e.g., Coomassie or Sypro Ruby®, do not at all reflect the cysteine content but are predominantly dependent on the protein amount. Intense spots in the Cy3 image may therefore appear as weak spots in the Coomassie gel and vice versa, and some spots visible in the Coomassie gel may not be visible in the Cy3 gel when no cysteine residue is present in the protein. Therefore, the CyDye image may differ considerably from the Coomassie image of the same gel. A spot assignment in the preparative gel should therefore always be based on the comparison of analytical and preparative CyDye images.

20. Take into account that after Coomassie staining, a further Cy3 readout is impaired by Coomassie-derived fluorescence signals and absorption effects.

Acknowledgments

The authors would like to extend their appreciation to the Deutsche Forschungsgemeinschaft (DFG) for ongoing research grants within the research unit FOR 1041, previous grants within FOR 478 and GRK 1029, the Federal Ministry for Education and Research (BMBF) for ongoing research grants within PHANOMICS, FUGATO and FUGATO-plus (REMEDY), the European Union for a research grant within the Plurisys consortium, and the European Science Foundation (ESF) for funding of the "Stressflea" consortium. The authors wish to acknowledge their appreciation to Patrick Bolbrinker, Myriam Demant, Daniela Deutsch, and Julia Korte for critical reading of the manuscript and to Miwako Kösters for technical assistance.

References

1. Unlu, M., Morgan, M. E., and Minden, J. S. (1997) Difference gel electrophoresis: a single gel method for detecting changes in protein extracts, *Electrophoresis,* **18,** 2071–7.
2. Kondo, T., Seike, M., Mori, Y., Fujii, K., Yamada, T., and Hirohashi, S. (2003) Application of sensitive fluorescent dyes in linkage of laser microdissection and two-dimensional gel electrophoresis as a cancer proteomic study tool, *Proteomics,* **3,** 1758–66.
3. Sitek, B., Sipos, B., Alkatout, I., Poschmann, G., Stephan, C., Schulenborg, T., Marcus, K., Luttges, J., Dittert, D. D., Baretton, G., Schmiegel, W., Hahn, S. A., Kloppel, G., Meyer, H. E., and Stuhler, K. (2009) Analysis of the Pancreatic Tumor Progression by a Quantitative Proteomic Approach and Immunhistochemical Validation, *J Proteome Res.,* **4,** 1647–56.
4. Kondo, T., and Hirohashi, S. (2006) Application of highly sensitive fluorescent dyes (CyDye DIGE Fluor saturation dyes) to laser microdissection and two-dimensional difference gel electrophoresis (2D-DIGE) for cancer proteomics, *Nat Protoc* **1,** 2940–56.
5. Sitek, B., Potthoff, S., Schulenborg, T., Stegbauer, J., Vinke, T., Rump, L. C., Meyer, H. E., Vonend, O., and Stuhler, K. (2006) Novel approaches to analyse glomerular proteins from smallest scale murine and human samples using DIGE saturation labelling, *Proteomics,* **6,** 4337–45.
6. Helling, S., Schmitt, E., Joppich, C., Schulenborg, T., Mullner, S., Felske-Muller, S., Wiebringhaus, T., Becker, G., Linsenmann, G., Sitek, B., Lutter, P., Meyer, H. E., and Marcus, K. (2006) 2-D differential membrane proteome analysis of scarce protein samples, *Proteomics,* **6,** 4506–13.
7. Berendt, F. J., Frohlich, T., Bolbrinker, P., Boelhauve, M., Gungor, T., Habermann, F. A., Wolf, E., and Arnold, G. J. (2009) Highly sensitive saturation labelling reveals changes in abundance of cell cycle-associated proteins and redox enzyme variants during oocyte maturation in vitro, *Proteomics* **9**, 550–64.
8. Karp, N. A., and Lilley, K. S. (2007) Design and analysis issues in quantitative proteomics studies, *Proteomics, 7 Suppl 1,* 42–50.
9. Fröhlich T. and Arnold G.J. (2011) Quantifying Attomole Amounts of Proteins from Complex Samples by Nano-LC and Selected Reaction Monitoring (SRM), *Methods Mol Biol.,* **790,** 141–164.

Chapter 8

Proteomic Analysis of Redox-Dependent Changes Using Cysteine-Labeling 2D DIGE

Hong-Lin Chan, John Sinclair, and John F. Timms

Abstract

Redox-modification of proteins plays an important role in the regulation of protein function and cellular physiology and in pathological conditions such as oncogenic activation, inhibition of tumor suppression, and ischemia reperfusion injury. This occurs, at least in part, through the reduction or oxidation of cysteine groups in these proteins resulting in the modulation of their activities. Herein, we focus on the development of a pair of cysteine-labeling iodoacetylated cyanine dyes (ICy3/5) for two-dimensional difference gel electrophoresis (2D DIGE) to monitor redox-dependent changes on cysteine residues. The method is applied to a cellular model of human mammary luminal epithelial cells treated with H_2O_2 to induce oxidative stress. Differences in labeling are caused either by differential protein expression or from the loss or gain of reactive thiol groups of cysteines in response to oxidative stress. Proteins displaying differential labeling would then be picked for MS-based identification. In summary, this cysteine-labeling 2D-DIGE approach provides an MS-compatible and reproducible technique for identifying alterations in the expression and redox-modification of free thiol-containing proteins.

Key words: Thiol-reactive cyanine dyes, Two-dimensional difference gel electrophoresis, Redox proteomics, Mass spectrometry

1. Introduction

Two-dimensional gel electrophoresis (2DE) is one of the most widely used proteomic separation methods and has been employed for the analysis of differential protein expression in many different biological sample types (1, 2). However, as most users realize, 2DE and the methods commonly used for in-gel protein visualization are inherently variable and many replicate gels must be run before significant differences in protein expression can be ascribed with accuracy. Moreover, these protein visualization methods often have narrow linear dynamic ranges of detection, making them unsuitable for the analysis of biological samples where protein copy numbers vary enormously. A significant improvement in the ability to use gel-based methods for protein quantitation and detection was achieved with the introduction of two-dimensional difference gel electrophoresis (2D DIGE), which enables co-detection of several samples on the same 2DE gel, so avoiding gel-to-gel variation (3–7).

Several chemical moieties have been found to be potential regulators of cellular redox status. One of these, the free thiol group (R-SH) of cysteine residues is a potent nucleophilic agent and can undergo a number of redox-induced modifications under physiological and pathological conditions. Modifications of R-SH include the formation of protein disulfides and mixed disulfides (e.g., with glutathione and free cysteine) and oxidation to the sulfenic (R-SOH), sulfinic (R-SO$_2$H), and sulfonic (R-SO$_3$H) acids depending on the oxidative capacity of the oxidant (8). The R-SH group can also be modified by reactive nitrogen species to give the S-nitrosylated form (R-SNO), while oxidized forms can be glutathionylated for active secretion from cells or interconverted between forms by various enzyme activities (see Fig. 1).

Numerous studies have combined 2DE with cysteine thiol labeling to study redox-dependent protein changes. Maleimides, iodoacetic acid, iodoacetamide, and other chemicals have been modified with labels (biotin, fluorophores, radionuclides) to study

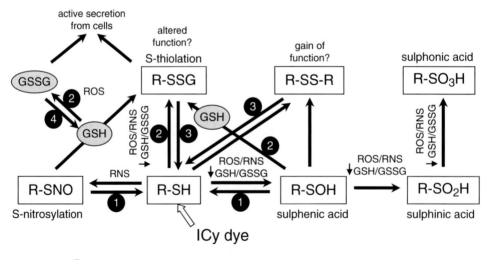

Fig. 1. Mechanisms of thiol-dependent cellular redox regulation. *GSH* reduced glutathione; *GSSG* oxidized glutathione; *ROS* reactive oxygen species; *RNS* reactive nitrogen species.

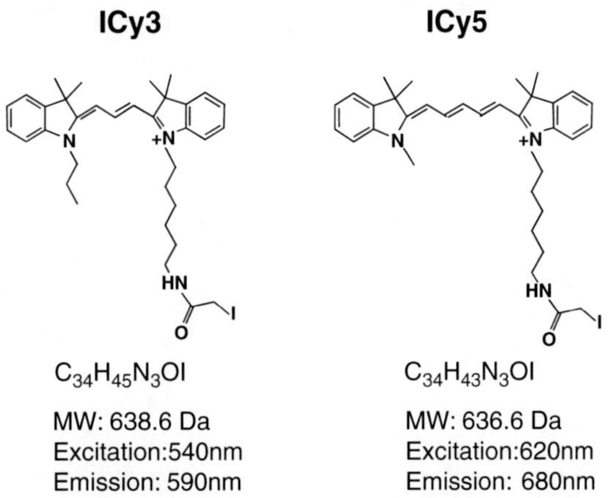

ICy3   **ICy5**

$C_{34}H_{45}N_3OI$ $C_{34}H_{43}N_3OI$

MW: 638.6 Da   MW: 636.6 Da
Excitation: 540nm   Excitation: 620nm
Emission: 590nm   Emission: 680nm

Fig. 2. Structure of the matched ICy3 and ICy5 iodoacetylated cyanine dyes. Chemical formulas, molecular weights, and excitation/emission wavelengths are shown.

changes in the redox status of proteins in 2D gels (9, 10). For example, N-(biotinoyl)-N'-(iodoacetyl) ethylenediamine was used for detecting selenium metabolite targets with labeled proteins detected using HRP-streptavidin and chemiluminescence (11). In another approach to monitor protein thiol oxidation in cultured cells, reduced thiols were first blocked with N-ethylmaleimide, then any oxidized thiols were reduced with dithiothreitol and subsequently labeled with 5-iodoacetamidofluorescein prior to 2DE and fluorescence detection (12). Such labeling approaches, especially using biotinylated derivatives, also allow affinity enrichment of the labeled pool of proteins for improved sensitivity of gel- and MS-based analyses.

A major drawback of these single labeling methods is reduced quantitative accuracy and precision due to the inherent variation of 2DE. To overcome this, we have developed a pair of iodoacetylated cyanine dyes (ICy3/5; see Fig. 2) for cysteine-labeling 2D DIGE to monitor redox-dependent changes on protein thiols and have tested the method in a model cell system of human mammary luminal epithelial cells exposed to H_2O_2 (13) and in plasma preparations disinfected by UVC irradiation (14). It is expected that any differences in labeling are caused by changes in the content of reactive thiol groups. Proteins displaying differential labeling on 2D gels are then picked for identification by MALDI-TOF MS peptide mass fingerprinting or LC-MS/MS with further validation of changes carried out by 1D and 2D immunoblotting. As the outlined protocol directly measures the labeling of free thiols between protein samples, changes in the expression levels of thiol-containing proteins are also expected to give rise to an altered ICy dye signal. For this reason, we recommend running a lysine-labeling experiment with NHS-cyanine dyes in parallel to aid in the discrimination of expression changes vs. redox-dependent changes. A detailed protocol for this is provided elsewhere in this volume.

Modifications of this technique have used BODIPY FL-N-(2-aminoethyl) maleimide and BODIPY TMR C_5-maleimide for dual labeling of reduced and oxidized cysteines to monitor changes in the thiol redox state of proteins in cells cultured at different oxygen concentrations (15). Similarly, commercial Cy3 and Cy5 maleimides have been also used for differential redox labeling (16), although the reproducibility and specificity of maleimides for thiol labeling has been called into question (17). It is important to note that the ICy dyes used in the protocol here carry a net charge of +1 and would be expected to cause a shift in the pI of labeled proteins on a 2D gel adding complexity to the analysis. This problem has recently been circumvented with the synthesis of a pair of sulfonated iodoacetamido-cyanine dyes which have a net charge of zero (17).

2. Materials

2.1. Cell Culture and Hydrogen Peroxide Treatment

1. Cells: human mammary luminal epithelial cells (HMLECs) (18, 19).
2. RPMI-1640 growth medium: RPMI-1640 medium (with 25 mM HEPES and L-glutamine), supplemented with 10% fetal bovine serum (FBS), 2 mM L-glutamine, 100 μg/mL penicillin-streptomycin (all from Gibco/Invitrogen), 5 μg/mL hydrocortisone and 5 μg/mL insulin.
3. Solution of 0.25% trypsin and 1 mM EDTA (Gibco/Invitrogen).
4. Hydrogen peroxide treatment: 30% hydrogen peroxide solution diluted to 0.5 mM final concentration in growth media.

2.2. Preparation of ICy Dye-Labeled Samples for 2D DIGE

1. Iodoacetyl cyanine dyes 3 and 5 (ICy3 and ICy5) (see Fig. 2 and Note 1): From lyophilized powder (stored at −20°C), reconstitute to 10 mM stock by dissolving in the appropriate volume of anhydrous N,N-dimethylformamide (DMF). Keep stock solutions in dark at −20°C.
2. 2D lysis buffer: 8 M urea, 4% (w/v) CHAPS, 1 mM EDTA 10 mM Tris–HCl pH 8.3. To make 100 mL, dissolve 48 g of urea in 50 mL of distilled H_2O. Add 4 g CHAPS, 0.5 g NP-40, 0.1 mL of 1 M EDTA, and 0.67 mL of 1.5 M Tris pH 8.8 solution. This should give a final pH of 8.3. Make up to final volume, aliquot, and store at −20°C. Do not heat (see Notes 2 and 3).
3. DTT solution: 1.3 M DTT in H_2O. To make 10 mL, dissolve 2 g DTT in distilled H_2O and make to 10 mL. Aliquot and store at −20°C. Do not heat.
4. Ampholines/Pharmalyte mix: Mix equal volumes of Ampholines (pH 3.5–10) and Pharmalyte (pH 3–10). Store at 4°C. These broad pH range IPG buffers can be replaced with narrow-range buffers depending on the first dimension pH range.

5. Bromophenol blue: 0.05% (w/v) bromophenol blue in H$_2$O. To make 10 mL, weigh 5 mg bromophenol blue and make to 10 mL with distilled H$_2$O. Filter and store at room temperature.

2.3. Preparation of 2D-Gels, Imaging, and Image Analysis

1. Bind-Silane solution: For twelve 24×20 cm plates, mix 12 μL of PlusOne Bind-Silane (GE Healthcare), 300 μL glacial acetic acid, 12 mL ethanol and 2.7 mL ddH$_2$O.
2. PlusOne Repel-Silane solution (GE Healthcare).
3. Immobiline DryStrip pH 3–10 NL gel strips (GE Healthcare).
4. Mineral oil.
5. Equilibration buffer: 6 M urea, 30% (v/v) glycerol, 50 mM Tris–HCl pH 6.8, 2% (w/v) SDS. To make 200 mL, dissolve 72 g urea in 100 mL distilled H$_2$O. Add 60 mL of 100% glycerol, 10 mL of 1 M Tris pH 6.8 solution, and 4 g SDS. Dissolve all powders and adjust volume to 200 mL with ddH$_2$O. Aliquot and store at −20°C.
6. Ammonium persulfate (APS): Prepare a 10% solution in ddH$_2$O and store at 4°C for no more than a month.
7. Gel preparation: 30% acrylamide/bisacrylamide solution, 1.5 M Tris–HCl, pH 8.8, 10% SDS, 10% APS, N,N,N,N'-tetramethylethylenediamine (TEMED).
8. Agarose overlay: 0.5% (w/v) low-melting point agarose in SDS-PAGE running buffer. To make 200 mL, melt 1 g of agarose in 200 mL of 1× SDS-PAGE running buffer in a microwave on low heat. Add bromophenol blue solution to give a pale blue color.
9. SDS-PAGE running buffer (10×): Tris-glycine SDS buffer 10× (Severn Biotech LTD). Store at room temperature.
10. Ettan DALT*twelve* Large Vertical Electrophoresis System (GE Healthcare).
11. Typhoon 9400 Variable Mode Imager and ImageQuant software (GE Healthcare).
12. DeCyder software (v7.0) (GE Healthcare).

2.4. Post-Staining and Spot Excision

1. Colloidal CBB fixing solution: 35% (v/v) ethanol, 2% (v/v) phosphoric acid in ddH$_2$O.
2. Colloidal CBB staining solution: 34% (v/v) methanol, 17% (w/v) ammonium sulfate, and 3% (v/v) phosphoric acid in ddH$_2$O (see Note 4).
3. Coomassie Blue G-250.
4. Bio-Rad GS-800 scanning densitometer and QuantityOne software (Bio-Rad Laboratories Inc. USA.) or Typhoon 9400 Variable Mode Imager and ImageQuant software (GE Healthcare).
5. Ettan Spot Picker (GE Healthcare).

2.5. In-Gel Digestion

1. 50 and 100% HPLC-grade acetonitrile (ACN). Store at room temperature.
2. 5 mM ammonium bicarbonate (ABC) pH 8.0 in ddH$_2$O. Store at 4°C or prepare a 100 mM stock, aliquot, and store at −20°C.
3. 10 mM dithiothreitol (DTT) in 5 mM ABC. Prepare fresh.
4. 50 mM iodoacetamide (IAM) in 5 mM ABC. Prepare fresh.
5. 5% trifluoroacetic acid (TFA) in 50% ACN. Prepare fresh.
6. 10 ng/μL sequencing-grade modified trypsin (Promega) in 5 mM ABC pH 8.0. Prepare 500 ng-stocks in buffer provided (50 mM acetic acid). Store at −20°C.

2.6. Matrix-Assisted Laser Desorption/Ionization Time-of-Flight Mass Spectrometry

1. Matrix solution: saturated aqueous 2,5-dihydroxybenzoic acid (Bruker Daltonics).
2. Peptide external calibration standard (Bruker Daltonics).
3. Ultraflex matrix-assisted laser desorption/ionization time-of-flight (MALDI-TOF)/TOF mass spectrometer (Bruker Daltonics).
4. MTP AnchorChip target (Bruker Daltonics).
5. FlexControl software (Bruker Daltonics).
6. FlexAnalysis software (Bruker Daltonics).

2.7. Validation by Immunoblotting

1. Transfer-Blot tank (Bio-Rad).
2. Transfer buffer: 195 mM glycine, 25 mM Tris–HCl, pH 7.4, 20% (v/v) methanol. Prepare 10× transfer buffer in ddH$_2$O without methanol and store at 4°C. Dilute to a 1× solution prior to transfer, adding 20% (v/v) methanol.
3. Tris-buffered saline with Tween-20 (TBS-T): 50 mM Tris–HCl pH 8.0, 150 mM NaCl, and 0.1% Tween-20 (Sigma).
4. Polyvinylidene fluoride membrane (PVDF) (Immobilon P, Millipore, Bedford, MA), and 3-MM chromatography paper (Whatman, Maidstone, UK).
5. Blocking buffer: 5% w/v low-fat milk in TBS-T.
6. Enhanced chemiluminescence (ECL) reagents (Perkin-Elmer Life Sciences) and Fuji RX X-ray film (Genetic Research Instrumentation).
7. Primary antibody: appropriate primary antibodies chosen for validation of MALDI-TOF results. Appropriate working concentrations of the primary antibodies should be determined beforehand.
8. Secondary antibody: anti-mouse or anti-rabbit IgG-HRP linked antibodies (GE Healthcare). Prepare freshly at a dilution of 1:5,000 in TBS-T.

3. Methods

3.1. Tissue Culture

1. Culture HMLECs in 15-cm tissue culture dishes in RPMI-1640 growth media at 37°C in a 10%-CO_2-humidified incubator.
2. Split cells approximately 1:5 when confluent. Do not over-split.

3.2. Protein Quantification

1. HMLEC cells at ~80% confluence are washed twice with ice-cold 0.5× PBS and then 1,000 μL of 2D-lysis buffer added per plate. Place dishes immediately on ice.
2. Scrape cells and collect them in labeled tubes.
3. Homogenize by passage through a 25-gauge needle 10 times. Vortex and remove insoluble material by centrifugation (13,000×g/10 min/4°C) and transfer supernatant to fresh tubes.
4. Determine protein concentration using the Coomassie Protein Assay Reagent. Make a 5 mg/mL stock of BSA in 2D-lysis buffer and prepare serial dilutions of 0, 0.25, 0.5, 1.0, 2.5, and 5.0 mg/mL to make a standard curve. Use a 96-well flat-bottomed assay plate and make triplicate measurements for the BSA standards and four replicates for the experimental samples. For this, add 2 μL of sample per well and 198 μL of assay reagent and mix without introducing bubbles. Use a plate reader at a wavelength of 595 nm and calculate protein concentrations using the standard curve (see Note 5).

3.3. Cell Treatment and Lysis

1. Add H_2O_2 solution to growth media of HMLEC cells at ~80% confluence to a final concentration of 0.5 mM with gentle swirling or leave untreated. Leave treated cells for 2, 20, or 240 min.
2. Cells are washed twice with ice-cold 0.5× PBS and lysed with 1,000 μL 2D lysis buffer per plate containing ICy3/5 at 80 pmoL/μg protein (see Note 6).
3. Scrape cells and collect them in labeled tubes.
4. Homogenize by passage through a 25-gauge needle 10 times. Vortex and remove insoluble material by centrifugation (13,000×g/10 min/4°C) and transfer supernatant to fresh tubes.
5. The ICy3/5-labeled proteins are subsequently incubated on ice in the dark for 1 h and quenched with DTT at a 65 mM final concentration.
6. Mix 150 μg ICy3- and 150 μg ICy5-labeled samples to give 300 μg total protein. Volumes are adjusted to 450 μL with 2D-lysis buffer containing 65 mM DTT and 9 μL carrier Ampholines/Pharmalyte (1:1; v/v) (pH 3–10) and 1 μL of

Table 1
Example of differential labeling, mixing, and gel loading for comparison of thiol reactivity and protein expression under four treatment conditions in triplicate using cysteine-labeling 2D DIGE

	ICy3 (μg pool)	ICy5
Gel 1	150	150 μg condition 1 (0 min H_2O_2)
Gel 2	150	150 μg condition 1 (0 min H_2O_2)
Gel 3	150	150 μg condition 1 (0 min H_2O_2)
Gel 4	150	150 μg condition 2 (2 min H_2O_2)
Gel 5	150	150 μg condition 2 (2 min H_2O_2)
Gel 6	150	150 μg condition 2 (2 min H_2O_2)
Gel 7	150	150 μg condition 3 (20 min H_2O_2)
Gel 8	150	150 μg condition 3 (20 min H_2O_2)
Gel 9	150	150 μg condition 3 (20 min H_2O_2)
Gel 10	150	150 μg condition 4 (240 min H_2O_2)
Gel 11	150	150 μg condition 4 (240 min H_2O_2)
Gel 12	150	150 μg condition 4 (240 min H_2O_2)

bromophenol blue solution added. The mixed samples are allocated appropriately for separation on triplicate 2D-gels as shown in Table 1. This scheme controls for dye bias, although labeling combinations are interchangeable so long as each gel is loaded with samples labeled with distinct dyes. This experiment generates 24 images for matching, cross-comparison, and statistical analysis in the biological variation analysis (BVA) module of the DeCyder software.

7. Rehydrate Immobiline DryStrip pH 3–10 NL gel strips with samples for at least 12 h in the dark at room temperature in a rehydration tray (passive rehydration method). Strips should be covered with mineral oil.

3.4. Preparation of 2D-Gels, Imaging, and Image Analysis

1. For SDS-PAGE casting, an Ettan DALT*twelve* system (24 cm) is employed.

2. Prior to gel casting, treat low-fluorescence glass plates with 1 mL of fresh Bind-Silane solution per plate, wiping over one surface with a lint-free tissue. Leave plates to dry for a minimum of 1.5 h (see Note 7 and Fig. 3).

3. Treat the clean and dry inner surface of the other plate with 1 mL Repel-Silane to ensure easy separation after electrophoresis.

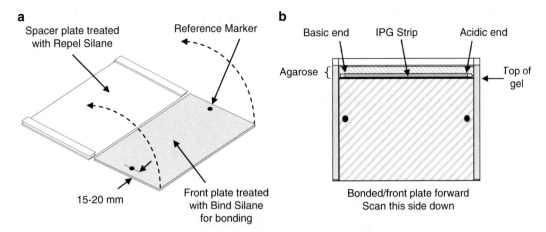

Fig. 3. (a) Treatment of plates for bonding and reference marker positioning. (b) Casting and loading of second dimension gels based on the Ettan Dalt 24-cm strip format.

Apply Repel-Silane solution to a lint-free tissue and wipe over the surface. Remove excess Repel-Silane by wiping with a clean tissue, rinse with ethanol, then with distilled H_2O. Leave to dry for 10 min (see Notes 7 and 8 and Fig. 3).

4. Stick fluorescent reference markers to the bonded surface of the plates. These should be placed half-way down the plates and 15–20 mm in from each edge. These markers are used as references for determining picking coordinates for automated spot picking using the Ettan Spot Picker (see Fig. 3).

5. Assemble plates in casting chamber according to the manufacturer's guidelines.

6. Prepare 12.5% acrylamide/bisacrylamide resolving gel solution and pour into assembled casting chamber leaving approximately 1.5 cm at the top for the IPG strip (~100 mL resolving gel solution per gel). Overlay carefully with water or butanol. Leave gel to polymerize for at least 4 h.

7. Perform isoelectric focusing (IEF) for a total of 80 kVh in the dark according to the manufacturer's instructions.

8. Equilibrate strips for 15 min in equilibration buffer containing 65 mM DTT for reduction and then 15 min in the same buffer containing 240 mM iodoacetamide for alkylation.

9. Rinse strips in 1× SDS-PAGE running buffer and place onto the top of second dimension gels in melted 0.5% agarose overlay, with the basic end of the strip toward the right hand side when the bonded plate is facing backward.

10. Run second dimension gels until the dye front has completely run off to avoid fluorescence signals from bromophenol blue and free dye. For the Ettan DALT*twelve* system, this can be

achieved by running 12.5% gels for approximately 16 h at 2.2 W per gel.

11. Images are best acquired directly after the 2DE by scanning gels between glass plates using a Typhoon 9400 Imager or a similar device. Ensure that both outer plate surfaces are clean and dry before scanning and that the bonded plate is the lower plate on the scanner bed.

12. Perform an initial low-resolution scan (1,000 μm) for one gel on the Cy3 and Cy5 channels with the photomultiplier tube (PMT) voltages set low (e.g., 500 V). The excitation/emission wavelengths for fluorescence detection using the Typhoon 9400 are 532/580 nm for ICy3 and 633/680 nm for ICy5. An image is then built up by the scanner for each channel and is converted to grayscale pixel values.

13. Using ImageQuant software for the Typhoon 9400, establish maximum pixel values in various user-defined, spot-rich regions of each image, and adjust the PMT voltages for a second low-resolution scan to give similar maximum pixel values (within 10%) on each channel and without saturating the signal from the most intense peaks. Repeat scans may be required until values are within 10% for the two channels (see Note 9).

14. Once set for the first gel, use the same PMT voltages for the whole set of gels scanning at 100-μm resolution. A 24- × 20-cm gel image takes approximately 10 min to acquire per channel and two gels are scanned simultaneously. Images are generated as gel files.

15. Crop overlayed images in ImageQuant and import into DeCyder Batch Analysis software for subsequent BVA analysis, according to the DeCyder software user manual. An example of using ICy to monitor H_2O_2-induced redox-proteome alterations in HMLECs is shown in Fig. 4. A number of proteins showed a rapid increase in labeling (e.g., spots 25 and 26), suggesting generation of new free thiols, for example, via scission of disulfides. On the other hand, a number of proteins showed a rapid decrease in labeling (e.g., spots 24 and 27), implying oxidation of free thiols that would not be labeled with the ICy dyes. In some cases, the change in ICy labeling recovered over time, indicating reversible modification (e.g., spot 27) (see Note 10).

3.5. Post-Staining and Spot Excision

1. After ICy fluorescence image scanning, gels are immersed in fixing solution and incubated overnight with gentle shaking. Fixed and bonded gels can now be stored for several months at 4°C by sealing in plastic bags with 1% (v/v) acetic acid.

2. For post-staining, the colloidal Coomassie Brilliant Blue (CCB) G-250 staining method is modified from that of (20). Fix gels

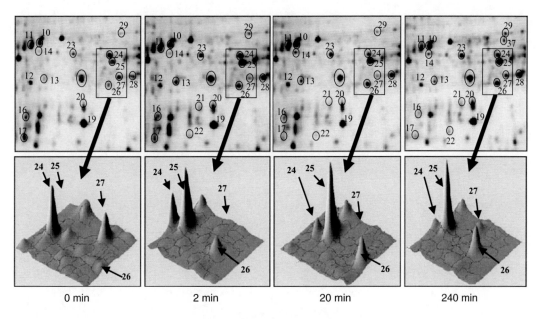

Fig. 4. Multiplex 2D DIGE analysis to monitor oxidant-dependent thiol reactivity in HMLECs. Sections of overlaid 2D DIGE images and 3D fluorescence profiles of HMLEC cell lysates from untreated cells (0 min) and cells treated with 0.5 mM H_2O_2 for 2, 20, and 240 min are shown. Individual lysates were labeled with ICy5 and run on each gel against a standard pool of all samples labeled with ICy3.

in colloidal CBB fixing solution for at least 3 h on a shaking platform. Wash 3 times for 30 min with ddH_2O and incubate in CCB staining solution for 1 h. Add one crushed CCB tablet (250 mg)/500 mL of staining solution (i.e., 0.5 g/L) and leave to stain for 3–5 days. No destaining step is required to visualize proteins. Stained gels can be scanned on the Typhoon 9400 imager using the red laser and no emission filters. Alternatively, the stained images can be scanned on a Bio-Rad GS-800 densitometer (see Note 11).

3. Align post-stained and fluorescence gel images to identify spots of interest for picking. Alignment and spot identification can be carried out by comparing images in DeCyder BVA or using Adobe Photoshop to overlay images.

4. For automated spot picking, process post-stained images in DeCyder BVA software and create a pick list for the spots of interest by comparing with the results of the BVA analysis. To facilitate sample tracking and later data matching with MS results, the post-stained image can be imported and matched within the current experimental BVA workspace. The advantage is that any spot picked according to the post-stained image will have the same master spot number as in the BVA quantitative analysis. Define the positions of the two reference markers in

DeCyder (left then right) and export the pick list coordinate file (.txt) to the spot picker controller. Subsequently, open the imported pick list and align the Ettan Spot Picker with the reference markers according to the manufacturer's instructions. Pick and collect spots in 96-well plates, drain the water, and store at −20°C prior to MS analysis. Alternatively, excision of spots from the post-stained gel can be done manually with a gel-plug cutting pipette. The gel is best submerged under 1–2 mm of distilled water, and picking performed in a dedicated clean area.

3.6. In-Gel Digestion

1. Ideally samples for MS analysis should be prepared in a clean room or other clean area to avoid keratin and other contamination.

2. Shake gel pieces in ddH$_2$O for 15 min. Replace water with 50% ACN and shake for a further 15 min. Repeat this step 3 times until gel pieces are completely destained (see Note 12).

3. Remove the 50% ACN and dry in a speed vacuum for 15–20 min (see Note 13).

4. Reduce the samples by adding sufficient 10 mM DTT (in 5 mM ABC pH 8.0) to cover the gel pieces and incubating for 45 min at 50°C, with gentle shaking.

5. Remove the DTT solution and alkylate by adding enough 50 mM IAM (in 5 mM ABC pH 8.0) to cover the gel pieces and incubating for 1 h at room temperature in the dark.

6. Remove the IAM solution and wash the gel pieces twice with 50% ACN for 15 min each.

7. Dry the gel pieces in a speed vacuum for approximately 15–20 min (see Note 13).

8. Digest the samples with trypsin. From a 500-ng/μL stock of trypsin in buffer, dilute 100 times (to 5 ng/μL) to provide sufficient volume for all samples (10 μL of trypsin at 50 ng per sample). Allow the trypsin solution to soak into the gel piece and then add sufficient 5 mM ABC pH 8.0 to cover the gel piece. Place samples in an incubator or rocking heater block at 37°C and leave to digest overnight.

9. Briefly spin the samples and collect the supernatant and transfer to new siliconized tubes. Add sufficient 50% ACN/5% TFA to cover the gel pieces and agitate briefly to aid peptide extraction. Remove the supernatant and pool together with the first. Repeat this step twice.

10. Speed vacuum to dryness. Samples can be stored at −20°C or be directly analyzed by MS.

11. Resuspend peptides in 5 μL of 0.1% formic acid by gently shaking prior to MALDI-TOF MS (or ESI MS/MS) analysis.

3.7. Protein Identification by MALDI-TOF MS and Data Analysis

1. 0.5 µL of tryptic digest is mixed with 1 µL of saturated aqueous 2,5-dihydroxybenzoic acid matrix solution and spotted onto a target plate and dried.

2. Mass spectra are acquired on an Ultraflex TOF/TOF mass spectrometer in the reflector mode. The spectrometer is calibrated using the peptide external calibration standard, and internal calibration is carried out using trypsin autolysis peaks at m/z 842.51 and m/z 2,211.10.

3. Peaks in the mass range of m/z 500–5,000 are used to generate a peptide mass fingerprint that is searched against the updated NCBInr database using Mascot Peptide Mass Fingerprint software (Matrix Science, London, UK; http://www.matrixscience.com/search_form_select.html). The following parameters are used for the search: *Homo sapiens* (or relevant taxonomy); tryptic digest with a maximum of one missed cleavage; carbamidomethylation of cysteine; protein N-terminal acetylation, methionine oxidation, and glutamine to pyroglutamate as variable modification; and a mass tolerance of ±50 ppm (see Note 14).

4. A positive identification is accepted based on a significant Mascot MOWSE score ($p<0.05$), at least 6 peptide masses matching a particular protein, matched peptides covering >20% of the matched protein sequence and general agreement between the observed and theoretical molecular weight on 2D gels.

3.8. Validation by Immunoblotting

1. Separate samples by 1D SDS-PAGE and transfer electrophoretically to PVDF membrane using transfer buffer in a transfer tank (Bio-Rad Transfer-Blot cell). Wet the PVDF membrane in 100% methanol for 1 min, then in transfer buffer for 5 min and place on top of the gel without air bubbles. Sandwich membrane and gel between two pieces of Whatman paper soaked in transfer buffer. Remove air bubbles with a plastic pipette and fix firmly in the tank cassette. Insert the cassette into the transfer tank containing transfer buffer and make sure the membrane is located between the gel and the anode. Connect the tank to an appropriate power supply and transfer at 350 mA for 5 h.

2. Once transfer is complete, remove the "sandwich" from the cassette and disassemble. The colored molecular weight markers should be clearly visible on the membrane if the transfer has worked.

3. Incubate the membrane in 500 mL of 5% low-fat milk in TBS-T for 1 h at room temperature on a rocking platform. The membrane can also be left in blocking buffer over night at 4°C.

4. Discard the blocking buffer and rinse the membrane with TBS-T prior to the addition of a dilution of primary antibody.

5. Incubate with the appropriate primary antibody for 2 h at room temperature or overnight at 4°C on a rocking platform.
6. Remove the primary antibody and wash the membrane 3 times for 15 min in TBS-T on a rocking platform.
7. Incubate with the appropriate HRP-conjugated secondary antibody for 1 h at room temperature on a rocking platform.
8. Remove the secondary antibody and wash the membrane 3 times for 15 min in TBS-T on a rocking platform.
9. Mix the two ECL reagents 1:1 and incubate immediately with the membrane for 1–2 min ensuring full coverage of the membrane.
10. Remove the membrane from the ECL reagents, drain excess fluid, enclose the membrane in Saran wrap, and tape it into an X-ray cassette.
11. In a darkroom, place an X-ray film on the top of the membrane and close the cassette. Leave to expose for an appropriate period to give a reasonable signal that is not saturated.
12. Scan the X-ray film on a Bio-Rad GS-800 densitometer and quantify the intensity with QuantityOne software.

4. Notes

1. The synthesis and purification of the ICy dyes are detailed in (13).
2. Urea decomposes to ammonium cyanate at temperatures above 30°C. The ammonium cyanate subsequently modifies the primary amino groups of proteins via carbamylation resulting in altered pI.
3. The lysis buffer should not contain thiourea, which competes with the ICy dyes. Thiourea is often used in 2-DE sample buffers to improve protein solubility.
4. Prepare the staining solution at least 1 h prior to use, considering that it takes the ammonium sulfate a while to dissolve completely.
5. It is recommended that at least four replicate assays are performed for each sample for accurate protein determination. Dilute concentrated samples with lysis buffer if necessary.
6. For ICy labeling, cells are lysed in the presence of dye to limit post-lysis thiol modification. Since ICy dyes interfere with the protein assay, protein concentrations are determined on replica lysates not containing the ICy dyes.

7. Low-fluorescence glass plates are used to reduce background in this experiment.

8. Only one plate in each set should be treated with Bind-Silane; treat the smaller, nonspacer plates if using Ettan DALT 24 cm gel plates. Bonding allows easier handling of gels during scanning, protein staining, storage, and importantly, for robotic spot excision (see Fig. 3).

9. By increasing PMT voltages, it is possible to increase the overall sensitivity of detection while saturating only a few of the signals from the more abundant proteins. This allows analysis of lower-abundance proteins that give reasonable signal-to-noise ratios.

10. In the protocol outlined, we are essentially measuring the change in R-SH reactivity upon oxidative insult. However, the method can be adapted to identify proteins whose thiols have become oxidized by the insult. In this case, the samples are first alkylated (e.g., with N-ethylmaleimide or iodoacetamide) to block existing R-SH groups, and then reduced (e.g., with DTT or TCEP) to regenerate R-SH groups from the oxidized forms prior to ICy dye labeling. Both alkylating and reducing agents must be removed after each step, either by protein precipitation or by size-exclusion columns.

11. Gels in containers should be placed on a rocking platform for staining. In addition, containers should be properly sealed to avoid evaporation.

12. The use of siliconized or low-bind tubes avoids protein or peptide loss due to adsorption on the tube walls and also avoids contaminants derived from the tube plastic.

13. Make sure the gel pieces are completely dried.

14. It is also possible to search for the ICy dye modifications and mono-isotopic masses of 512.36 Da $[ICy3-I+H]^+$ and 510.36 Da $[ICy5-I+H]^+$ should be added to searches as variable modifications. It should be noted, however, that we have had difficulty observing the modifications in digested gel extracts. We suspect that the poor recovery may be due to the lowered solubility and inefficient extraction of labeled peptides from gel pieces, since model peptides labeled in solution performed well in MALDI-TOF experiments.

References

1. Herbert, B. R., Harry, J. L., Packer, N. H., Gooley, A. A., Pedersen, S. K., and Williams, K. L. (2001) What place for polyacrylamide in proteomics?, *Trends Biotechnol 19*, S3–9.
2. Rabilloud, T. (1994) Two-dimensional electrophoresis of basic proteins with equilibrium isoelectric focusing in carrier ampholyte-pH gradients, *Electrophoresis 15*, 278–282.
3. Gharbi, S., Gaffney, P., Yang, A., Zvelebil, M. J., Cramer, R., Waterfield, M. D., and Timms, J. F. (2002) Evaluation of two-dimensional differential gel electrophoresis for proteomic expression analysis of a model breast cancer cell system, *Mol Cell Proteomics 1*, 91–98.
4. Tonge, R., Shaw, J., Middleton, B., Rowlinson, R., Rayner, S., Young, J., Pognan, F., Hawkins, E., Currie, I., and Davison, M. (2001) Validation and development of fluorescence two-dimensional differential gel electrophoresis proteomics technology, *Proteomics 1*, 377–396.
5. Unlu, M., Morgan, M. E., and Minden, J. S. (1997) Difference gel electrophoresis: a single gel method for detecting changes in protein extracts, *Electrophoresis 18*, 2071–2077.
6. Timms, J. F., and Cramer, R. (2008) Difference gel electrophoresis, *Proteomics 8*, 4886–4897.
7. Lai, T. C., Chou, H. C., Chen, Y. W., Lee, T. R., Chan, H. T., Shen, H. H., Lee, W. T., Lin, S. T., Lu, Y. C., Wu, C. L., and Chan, H. L. (2010) Secretomic and Proteomic Analysis of Potential Breast Cancer Markers by Two-Dimensional Differential Gel Electrophoresis, *J Proteome Res 9*, 1302–1322.
8. Schafer, F. Q., and Buettner, G. R. (2001) Redox environment of the cell as viewed through the redox state of the glutathione disulfide/glutathione couple, *Free radical biology & medicine 30*, 1191–1212.
9. Eaton, P. (2006) Protein thiol oxidation in health and disease: techniques for measuring disulfides and related modifications in complex protein mixtures, *Free radical biology & medicine 40*, 1889–1899.
10. Sheehan, D. (2006) Detection of redox-based modification in two-dimensional electrophoresis proteomic separations, *Biochem Biophys Res Commun 349*, 455–462.
11. Lee, J. S., Ma, Y. B., Choi, K. S., Park, S. Y., Baek, S. H., Park, Y. M., Zu, K., Zhang, H., Ip, C., Kim, Y. H., and Park, E. M. (2006) Neural network-based analysis of thiol proteomics data in identifying potential selenium targets, *Preparative biochemistry & biotechnology 36*, 37–64.
12. Cuddihy, S. L., Baty, J. W., Brown, K. K., Winterbourn, C. C., and Hampton, M. B. (2009) Proteomic detection of oxidized and reduced thiol proteins in cultured cells, *Methods Mol Biol 519*, 363–375.
13. Chan, H. L., Gharbi, S., Gaffney, P. R., Cramer, R., Waterfield, M. D., and Timms, J. F. (2005) Proteomic analysis of redox- and ErbB2-dependent changes in mammary luminal epithelial cells using cysteine- and lysine-labelling two-dimensional difference gel electrophoresis, *Proteomics 5*, 2908–2926.
14. Chan, H. L., Gaffney, P. R., Waterfield, M. D., Anderle, H., Peter Matthiessen, H., Schwarz, H. P., Turecek, P. L., and Timms, J. F. (2006) Proteomic analysis of UVC irradiation-induced damage of plasma proteins: Serum amyloid P component as a major target of photolysis, *FEBS Lett 580*, 3229–3236.
15. Lui, J. K., Lipscombe, R., and Arthur, P. G. (2010) Detecting changes in the thiol redox state of proteins following a decrease in oxygen concentration using a dual labeling technique, *J Proteome Res 9*, 383–392.
16. Hurd, T. R., Prime, T. A., Harbour, M. E., Lilley, K. S., and Murphy, M. P. (2007) Detection of reactive oxygen species-sensitive thiol proteins by redox difference gel electrophoresis: implications for mitochondrial redox signaling, *J Biol Chem 282*, 22040–22051.
17. Bruschi, M., Grilli, S., Candiano, G., Fabbroni, S., Della Ciana, L., Petretto, A., Santucci, L., Urbani, A., Gusmano, R., Scolari, F., and Ghiggeri, G. M. (2009) New iodo-acetamido cyanines for labeling cysteine thiol residues. A strategy for evaluating plasma proteins and their oxido-redox status, *Proteomics 9*, 460–469.
18. Harris, R. A., Eichholtz, T. J., Hiles, I. D., Page, M. J., and O'Hare, M. J. (1999) New model of ErbB-2 over-expression in human mammary luminal epithelial cells, *Int J Cancer 80*, 477–484.
19. White, S. L., Gharbi, S., Bertani, M. F., Chan, H. L., Waterfield, M. D., and Timms, J. F. (2004) Cellular responses to ErbB-2 overexpression in human mammary luminal epithelial cells: comparison of mRNA and protein expression, *Br J Cancer 90*, 173–181.
20. Neuhoff, V., Arold, N., Taube, D., and Ehrhardt, W. (1988) Improved staining of proteins in polyacrylamide gels including isoelectric focusing gels with clear background at nanogram sensitivity using Coomassie Brilliant Blue G-250 and R-250, *Electrophoresis 9*, 255–262.

Chapter 9

Analysis of Protein Posttranslational Modifications Using DIGE-Based Proteomics

Robert M. DeKroon, Jennifer B. Robinette, Cristina Osorio, Joseph S.Y. Jeong, Eric Hamlett, Mihaela Mocanu, and Oscar Alzate

Abstract

Difference gel electrophoresis (DIGE) is most often used to assess relative changes in the expression levels of individual proteins in multiple complex samples, and this information is valuable in making inferences about relative protein activity. However, a protein's activity is not solely dependent upon its expression level. A change in activity may also be influenced by myriad posttranslational modifications (PTMs), including palmitoylation, ubiquitination, oxidation, and phosphorylation. In this chapter, we describe the use of DIGE to determine specific PTMs by introducing specific labels or changes in pI and/or molecular weight.

Key words: Difference gel electrophoresis, Neuroproteomics, Oxidation, Palmitoylation, Phosphorylation, Posttranslational modification, Ubiquitination

1. Introduction

In eukaryotic systems, a protein's expression level is a widely accepted determination of its activity. However, protein activity is further regulated by posttranslational modifications (PTMs). Therefore, it is important to determine changes in PTM levels in conjunction with protein expression. Difference gel electrophoresis (DIGE) involves resolving proteins from multiple complex samples within the same gel and is thus a valuable technique for assessing relative protein expression. The high resolution of DIGE makes it possible to distinguish PTMs if they introduce changes to pI and/or relative mobility (Mr). Here, we present DIGE-based approaches for identifying PTMs in multiple complex samples by exploiting PTM-induced changes in pI and Mr. For example, by using phosphatases to remove the phosphate group from phosphorylated tyrosine, serine, and threonine amino acid residues and comparing the dephosphorylated lysate to an aliquot of the original sample, it is possible to identify which proteins contain these specific phosphorylated amino acids (1). Alternatively, with the development of labeling chemistries, it is increasingly possible to label specific PTMs directly with a reactive fluorophore. Here, we demonstrate oxidized protein labeling utilizing a probe that specifically reacts with carbonylated amino acids, a stable indicator of protein oxidation. Furthermore, PTMs may be determined by first modifying the PTM and subsequently labeling with existing chemistries. The ubiquitination and palmitoylation labeling protocols presented here, and the modified "biotin-switch" method to label S-nitrosylation (2), are examples of this strategy.

1.1. Experimental Design

Our preferred experimental design involves rotating the NHS-Cy3 and NHS-Cy5 labels within each experimental group (Fig. 1a); such that samples from each group are labeled with both fluorophores (3). In contrast, the fluorophore used to label the internal control (IC) remains constant (NHS-Cy2). The IC is a pooled sample containing an identical amount of protein of each sample in the study (3, 4).

However, in some cases, it is not possible to rotate the fluorophores. For example, the method for studying oxidized proteins described below uses an AlexaFluor 488 (AF-488) conjugated label that induces a shift in pI compared to that seen with NHS-Cy3 and NHS-Cy5. In addition, the AlexaFluor 647 label that is also available induces a different degree of shift. Therefore, a single-wavelength label for oxidation (AF-488) is used with all samples in

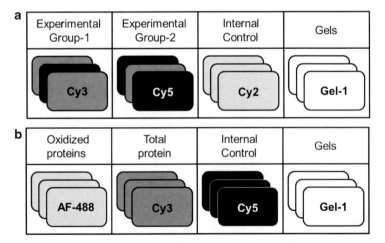

Fig. 1. Experimental design. (a) With our preferred design, the internal control (IC) is consistently labeled with NHS-Cy2. In contrast, NHS-Cy3 and NHS-Cy5 alternate between the experimental groups. (b) When labeling oxidized and total proteins, rotating the labels is not possible. The oxidized samples must remain labeled with AlexaFluor 488 (AF-488), the IC with NHS-Cy5, and the total protein samples with NHS-Cy3.

the study; NHS-Cy3 is used to label a separate aliquot of the same sample, and NHS-Cy5, the IC (Fig. 1b), with all three run on a single gel. To minimize the effects of any day-to-day variability, at least one sample from each experimental group is run at the same time, for example, four gels in the same tank. Furthermore, once complete, the oxidized and "total" protein data is analyzed separately but using the same ICs.

2. Materials

All materials are prepared with Milli-Q water (18 Ω) and stored at room temperature unless otherwise indicated. Analytical grade reagents are used whenever possible.

2.1. Sample Preparation

1. 1.5 M Tris–HCl buffer (pH 8.5): Add approximately 100 mL of water to a glass beaker. Weigh 36.3 g of Tris and transfer to the beaker. Mix and adjust pH with HCl (see Note 1). Make up to 200 mL with water and store at 4°C.
2. 20% CHAPS: Weigh 10 g of CHAPS and transfer to a 50-mL centrifuge tube. Make up to 50 mL with water and mix until dissolved.
3. Lysis buffer (8 M urea, 4% CHAPS, 20 mM Tris–HCl (pH 8.5)): Add approximately 10 mL of water to a 50-mL beaker. Weigh 12 g of urea and transfer to the beaker. Add 5 mL of 20% CHAPS and 333 µL of 1.5 M Tris–HCl buffer (pH 8.5). Mix and adjust volume to 25 mL with water. Make 1 mL aliquots and store at −20°C.
4. Complete protease inhibitors: Dissolve one tablet of Complete Mini, EDTA-free protease inhibitor cocktail (Roche, Indianapolis, IN, USA) in 1 mL of water. Store at −20°C.
5. Protein Phosphatase Inhibitor Set (Millipore, Billerica, MA, USA). This set contains inhibitors against serine/threonine phosphatases, tyrosine phosphatases, and acid and alkaline phosphatases. Store at −20°C.
6. Complete lysis buffer: Add 10 µL of complete protease inhibitors solution to 1 mL of lysis buffer. Add 1 µL of each Protein Phosphatase Inhibitor per 100 µL of lysis buffer. Store at −20°C.
7. Microcentrifuge pestles optimized for 1.5-mL tubes (USA Scientific, Ocala, FL, USA).

2.2. Posttranslational Modifications

2.2.1. Phosphorylation (see Note 2)

1. Calf intestinal alkaline phosphatase (NEB) at 10,000 U/mL.
2. 10×Reaction buffer (NEB): 50 mM Tris–HCl, 100 mM NaCl, 10 mM $MgCl_2$, and 1 mM dithiothreitol, pH 7.9 at 25°C.

2.2.2. Oxidation

AlexaFluor 488 C_5-aminooxyacetamide is light sensitive and very unstable while in solution. Prepare all stock solutions on ice and protect from light using aluminum foil or amber microcentrifuge tubes:

1. 1 mM AlexaFluor 488 C_5-aminooxyacetamide: Add 1.117 mL of water to 1 mg of AlexaFluor 488 C_5-aminooxyacetamide (Sigma, St. Louis, MO, USA). Make 7.5 µL aliquots in amber 500-µL microcentrifuge tubes. Lyophilize (or dry down) while protected from light (see Note 3). Store lyophilized aliquots at −80°C in a light safe freezer box (cardboard).

2. 1 M ascorbic acid stock: Weigh 1.761 g of ascorbic acid and make up to 10 mL with water. Store aliquots at −20°C.

2.2.3. Ubiquitination

1. 1 M Tris–HCl buffer (pH 7.4): Add approximately 100 mL of water to a glass beaker. Weigh 24.2 g of Tris and transfer to the beaker. Mix and adjust to pH 7.4 with HCl (see Note 1). Make up to 200 mL with water and store at 4°C.

2. 0.5 M $MgCl_2$: Add approximately 100 mL of water to a glass beaker. Weigh 9.52 g of $MgCl_2$ (anhydrous) and transfer to the beaker. Mix and adjust volume to 200 mL.

3. 588 µM ubiquitin aldehyde: Prepare fresh as required. Make a 1/100 dilution of the 58.8 mM stock supplied (Santa Cruz, Santa Cruz, CA, USA). Take 1 µL of stock and transfer to a 500-µL centrifuge tube. Add 99 µL of water and mix.

4. 188.7 mM AMP-PNP (adenylyl-imododiphosphate, tetra-lithium salt; Santa Cruz): Weigh 100 mg AMP-PNP into a 1.5-mL tube. Add 1 mL of water and mix. Store at −80°C.

5. 10 mM MG132 stock: Add 210 µL of DMSO per 1 mg of MG132 (Boston Biochem, Boston, MA, USA). Mix and store at −20°C.

6. 0.5 M ATP stock: Add 108.9 µL of water to a vial containing 30 mg of ATP (disodium hydrate; Sigma). Mix and store at −20°C.

7. 0.5 M creatine phosphate stock: Add 196 µL of water to a vial containing 25 mg of creatine phosphate (EMD Biochemicals, Gibbstown, NJ, USA). Mix and store at −20°C.

8. 2× Ubiquitin conjugation buffer (100 mM Tris–HCl, pH 7.8, 10 mM $MgCl_2$, 1 mM DTT, 150 µM MG132, 4 µM ubiquitin aldehyde, 5 mM AMP-PNP, 20 mM ATP, 20 mM creatine phosphate, 0.5 mg/mL creatine phosphokinase): Prepare fresh as required. Weigh 0.5 mg of creatine phosphokinase (~150 U) and transfer to a 1.5-mL centrifuge tube. Add 100 µL of 1 M Tris–HCl, pH 7.4, 20 µL of 0.5 M $MgCl_2$, 15 µL of 10 mM MG132 stock, 6.8 µL of 588 µM ubiquitin aldehyde, 40 µL of 0.5 M ATP, and 40 µL of 0.5 M creatine phosphate.

Weigh 1.54 mg of DTT into a separate 1.5-mL tube and dissolve in 100 μL of water. Add 10 μL to the tube containing the other buffer components. Make up to 1 mL with water and mix.

9. 1 M HEPES buffer: Add approximately 100 mL of water to a glass beaker. Weigh 52 g of HEPES and transfer to the beaker. Mix and adjust to pH 7.4 with HCl (see Note 1). Make up to 200 mL with water.

10. Isopeptidase-T: Concentrations and activity of commercially available stocks vary. Our current source (Enzo Life Sciences, Plymouth Meeting, PA, USA) is 25 μg per 12.5 μL, with >20 U/mg of protein.

11. UCH-L3: Concentrations and activity of commercially available stocks vary. Our current source (R&D Systems, Minneapolis, MN, USA) is 50 μg per 50 μL, with >1,000 pmol/min/μg.

12. Deubiquitinating buffer (25 mM HEPES, pH 7.4, 10 mM DTT, isopeptidase-T, and UCH-L3): Prepare fresh as required. Weigh 1.54 mg of DTT and transfer to a 1.5-mL tube. Add 25 μL of 1 M HEPES, 2.5 μL of isopeptidase-T, and 5 μL of UCH-L3. Make up to 1 mL with water and mix.

2.2.4. Palmitoylation

1. 1 M N-ethylmaleimide (NEM): Prepare fresh as required and store on ice. Weigh 125 mg of NEM into a 1.5-mL tube. Add 1 mL of ethanol and mix. Store at 4°C.

2. 1 M hydroxylamine hydrochloride: Prepare fresh as required and store at 4°C. Weigh 69.5 mg of hydroxylamine hydrochloride and transfer to a 1.5-mL centrifuge tube. Add 1 mL of complete lysis buffer and mix.

3. Cy-maleimide dyes: Lyophilized stocks are stored at −80°C, protected from light. Add 50 μL of anhydrous dimethylformamide (DMF) to 1 mg of Cy-maleimide dye. Mix thoroughly. Flush with nitrogen and protect from light. Store as 5–10 μL aliquots at −20°C.

2.3. Total Protein Labeling (Minimal Labeling)

1. NHS-Cy dyes (N-hydroxysuccinimide esters; GE Healthcare, Pittsburgh, PA, USA) are prepared according to the manufacturer's instructions. Lyophilized dyes are stored at −80°C, protected from light. A 5 nmol stock is dissolved in 5 μL of fresh DMF, aliquoted (2, 2, 1 μL), and stored at −80°C, protected from light. Prior to use, 1.5 times the volume of fresh DMF is added (i.e., 3, 3, 1.5 μL) and the aliquots are then stored at −20°C, protected from light.

2. 10 mM lysine: Weigh 146.19 mg of lysine and transfer to a 15-mL tube. Add water up to 10 mL. Aliquot into 1.5-mL tubes and store at −20°C.

2.4. Isoelectric Focusing (First Dimension)

1. 2× Sample buffer (2× SB; 8 M urea, 4% CHAPS, 130 mM DTT, 2% (v/v) IPG buffer): Prepare fresh as required. Weigh 480 mg of urea and 20 mg of DTT. Transfer to a graduated 1.5-mL centrifuge tube. Add 200 µL of 20% CHAPS (see Subheading 2.1, item 2), 20 µL of IPG buffer (pH 4–7 or 3–10) (see Note 4), and make up to 1 mL with water (see Note 5).

2. Rehydration buffer (8 M urea, 4% CHAPS, 13 mM DTT, 1% (v/v) IPG buffer): Prepare fresh as required. Weigh 480 mg of urea and 2 mg of DTT. Transfer to a graduated 1.5-mL centrifuge tube. Add 200 µL of 20% CHAPS, 10 µL of IPG buffer (pH 4–7 or 3–10) (see Note 4), and make up to 1 mL with water (see Note 5).

3. Cleaning solution for strip holder (see Note 6).

4. Cover fluid (see Note 7).

2.5. SDS-PAGE (Second Dimension)

1. 10× Running buffer (250 mM Tris, 1.92 M glycine, 1% SDS): Weigh 121.2 g of Tris base, 576.4 g of glycine, and 40 g of SDS and transfer to a 5-L beaker. Make up to 4 L with water. Do not adjust pH.

2. Equilibration buffer (for one IPG strip) (EB; 6 M urea, 50 mM Tris–HCl, pH 8.8, 30% glycerol, 2% SDS): Prepare fresh as required and approximately 30 min before use. Weigh 18.2 g of urea and 1 g of SDS. Transfer to a 100-mL beaker with a magnetic stir bar. Add 1.68 mL of 1.5 M Tris–HCl, pH 8.8, and 17.25 mL of 87% glycerol. Add water up to approximately 40 mL and mix to dissolve urea (see Note 8). Once dissolved, make up to 50 mL with water.

3. DTT-equilibration buffer (DTT-EB) for one IPG strip: Prepare fresh as required. Weigh 0.1 g of DTT and transfer to a 50-mL beaker. Add 20 mL of equilibration buffer prepared as described above and mix until DTT is fully dissolved (see Note 9).

4. Iodoacetamide-equilibration buffer (Iodo-EB) for one IPG strip: Prepared fresh as required. Weigh 0.9 g of iodoacetamide and transfer to a 50-mL beaker. Add 20 mL of equilibration buffer prepared as described above and mix until iodoacetamide is fully dissolved (see Note 9).

5. Saturated bromophenol blue solution: Add bromophenol blue to a quarter full in a 1.5-mL tube. Add 1 mL of 30 mM Tris–HCl, pH 8.5. Mix and let the undissolved bromophenol blue settle. When in use, take an aliquot from the top without disturbing the pellet.

6. 1% Agarose: Weigh 250 mg of low-melt agarose (Bio-Rad, Hercules, CA, USA) and add 25 mL of water. Melt the agarose by heating it on a low setting in a microwave. When fully dissolved, add 25 µL of saturated bromophenol blue solution. Aliquot into 1.5-mL tubes and store at room temperature.

3. Methods

3.1. Sample Preparation (see Note 10)

Samples should be kept on ice:

1. Add complete lysis buffer to the sample in the ratio of 500 μL per 100 mg of tissue or 200 μL/4–5 × 10^6 cells. Normally, in a 1.5-mL microcentrifuge tube.
2. Vortex for 5 min.
3. Grind or homogenize the sample. We use microcentrifuge pestles optimized for 1.5-mL tubes.
4. Vortex for 5 min.
5. Sonicate in ice water for 1 min, let the samples sit for 30 s, and sonicate for another minute (see Note 11).
6. Vortex again for 5 min, then centrifuge at 14,000 × g for 20 min at 4°C.
7. Collect the supernatant and store at −20°C; also store the pellet at −20°C.

3.2. PTM-Specific Procedures

3.2.1. Phosphorylation

Prepare samples as described above (see Subheading 3.1) in lysis buffer containing protease inhibitors but no phosphatase inhibitors. Store on ice and move to step 1 immediately:

1. Take half of each sample and add 1 μL of each protein phosphatase inhibitor per 100 μL of sample.
2. To the half without phosphatase inhibitors, add 10× reaction buffer to give a 1× final concentration (i.e., 1 μL of 10× buffer to 9 μL of sample). Add 0.5–1 U of alkaline phosphatase/μg of protein.
3. Incubate at 37°C for 1 h.
4. Clean the samples and resuspend them in complete lysis buffer (see Note 12).
5. Label both samples as described in Subheading 3.3 below.

3.2.2. Oxidation

1. Add 1 μL of 1 M ascorbic acid stock to 1 mL complete lysis buffer. Prepare samples as described above in Subheading 3.1 and store samples in lysis buffer with ascorbic acid.
2. To label oxidized proteins, add a maximum of 60 μg of protein to 7.5 μL of lysis buffer to a tube containing 7.5 nM aliquot of AlexaFluor 488 C$_5$-aminooxyacetamide (5).
3. Vortex and "quick-spin" to pool the sample in the bottom of the tube (see Note 13).
4. Incubate at room temperature, protected from light, for 2 h. Then transfer and keep at 4°C overnight.

5. Clean samples (see Note 12).

6. Label aliquots of the same samples to visualize total protein as described below in Subheading 3.3 (5).

3.2.3. Ubiquitination

Prepare samples as described above (see Subheading 3.1) in complete lysis buffer containing protease and phosphatase inhibitors. Divide the sample into two 20-μL aliquots (the amount of protein should be around 120 μg in each). Clean the sample aliquots as previously described (see Note 12), except resuspend the pellet as indicated below:

1. To one aliquot, add 20 μL of 2× ubiquitin conjugation buffer and 20 μL of water (see Note 14). Mix and incubate at 37°C for 1 h with shaking.

2. To the other aliquot, add 40 μL of deubiquitination buffer. Mix and incubate at room temperature for 1 h.

3. Add an equal volume of complete lysis buffer to the samples and incubate at 4°C overnight.

4. Label the samples with NHS-Cy3 or NHS-Cy5 dyes as described below in Subheading 3.3.

3.2.4. Palmitoylation

Prepare samples as described above in complete lysis buffer containing protease and phosphatase inhibitors but do not clean the samples:

1. Add 1 μL of 1 M NEM per 100 μL of sample to give a final concentration of 10 mM NEM. Incubate overnight at 4°C with rocking (see Note 15).

2. Clean the samples to remove NEM (see Note 12).

3. Resuspend the pellet in complete lysis buffer (50–100 μL) and divide the samples into two equal aliquots.

4. To one of the aliquots, add an equal volume of 1 M hydroxylamine hydrochloride. This is the depalmitoylated sample. To the other aliquot, add an equal volume of complete lysis buffer. Incubate for 1 h at room temperature on a rotating platform or vortex.

5. Clean the samples to remove the hydroxylamine hydrochloride (see Note 12). Resuspend the pellet in 25–50 μL of complete lysis buffer.

6. To label the depalmitoylated sample, add 1 μL of Cy-maleimide dye and mix by vortexing. Flush the tube with nitrogen gas and incubate the sample at room temperature for 2 h, protected from light, followed by overnight incubation at 4°C.

7. Clean the sample to remove the unbound Cy-maleimide dye (see Note 12).

8. Label the control samples, with an NHS-Cy dye of a different wavelength than the Cy-maleimide dye, as described in Subheading 3.3 below.

3.3. Total Protein Labeling (Minimal Labeling)

Strip holders should be cleaned before beginning the NHS-Cy dye labeling procedure (see Note 16). When running 24-cm strips, a 2-h prehydration step should also be included prior to labeling (see Note 17):

1. To label total proteins in a sample, add 1 µL of NHS-Cy dye (400 pmol) per 50 µg of sample protein and incubate on ice, protected from light, for 30 min.
2. Add 1 µL of 10 mM lysine and incubate on ice for 10 min.

3.4. Isoelectric Focusing (First Dimension)

Strip holders should be cleaned before beginning the NHS-Cy dye labeling procedure in order for the strip holders to have sufficient time to dry. Prepare fresh 2× SB and rehydration buffer as required:

1. Combine the labeled samples for each strip. For example, the oxidation (AlexaFluor 488 C_5-aminooxyacetamide) and Cy3-labeled aliquots of total sample plus the Cy5-labeled IC.
2. Add a volume of 2× SB equal to the total volume of the combined samples. Incubate on ice for 15 min.
3. Add rehydration buffer up to a final volume for the strip size used:

    7-cm strip—150 µL.

    13-cm strip—250 µL.

    24-cm strip—450 µL (see Note 17).

4. Load the sample into the strip holder being careful to avoid creating bubbles. Use a pipette tip to spread the sample onto the bottom of the strip holder, making sure there are no gaps between the walls and the bottom of the holder. Remove any bubbles before continuing.
5. Peel the protective layer from the strip while holding the non-gel end with forceps. Lay the strip gel-side down on top of the sample. Again, be careful to avoid creating bubbles (see Note 18). Align the line marked on the positive end of the strip with the electrode as shown in Fig. 2.
6. Apply cover fluid. Add 50–100 µL at a time. Use a maximum of 100 µL for 7-cm strips and up to 300 µL for 24-cm strips. Apply the cover fluid by dispensing slowly while dragging the tip gently along the back of the strip. Add the strip holder cover.
7. Place the strip holders on the IPG unit with the cathode and anode ends aligned as indicated on the apparatus. Close the lid carefully and protect samples from light.
8. Use the following settings for different strip sizes:

    7-cm strips:

    Rehydration: 14 h at 20°C.

    Step 1: 30 V for 30 min.

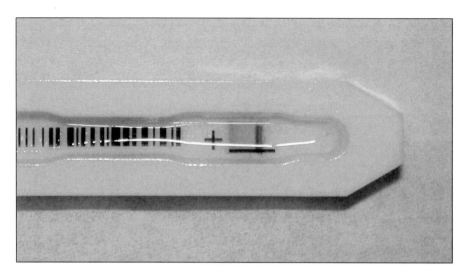

Fig. 2. Orientation and alignment of the IPG strip. Align the strip so that the line to the right of the positive sign overlaps the electrode as shown.

Step 2: step to 500 V for 30 min.
Step 3: step to 1,000 V for 30 min.
Step 4: step to 5,000 V for 2 h (see Note 19).

13-cm strips:
Step 1: 30 V for 12 h.
Step 2: step to 500 V for 1 h.
Step 3: step to 1,000 V for 1 h.
Step 4: step to 8,000 V up to 28,000 Vh (see Note 20).

24-cm strips:
Step 1: 30 V up to 450 Vh.
Step 2: step to 500 V up to 500 Vh.
Step 3: step to 1,000 V up to 1,000 Vh.
Step 4: step to 8,000 V up to 64,000 Vh.
Step 5: 500 V for 24 h (see Note 21).

3.5. SDS-PAGE (Second Dimension)

Use clean tubes when equilibrating IPG strips. For 7-cm strips, conical 15-mL tubes are useful; for other strips, alternative equilibration tubes are available (e.g., from GE Healthcare):

1. Prepare fresh equilibration buffer approximately 30 min before use, followed by fresh DTT-EB and Iodo-EB as required.
2. Prepare 1× running buffer for washing the IPG strips; usually 1–2 L, the remainder will be used when running the gel(s).
3. Remove the strip from the holder using forceps and wipe the back of the strip by dragging it across a folded Kimwipe. Rinse the strip in 1× running buffer.

4. Wash strips in DTT-EB. Use 10 mL of DTT-EB for 7-cm strips and 20 mL for 13- and 24-cm strips. Wrap tubes in aluminum foil and place them on an orbital shaker for 10 min.

5. Rinse strips in 1× running buffer, then transfer to fresh tubes with Iodo-EB. Again, use 10 mL for 7-cm strips and 20 mL for 13- and 24-cm strips. Wrap tubes in aluminum foil and place them on an orbital shaker for 10 min.

6. Rinse each strip in 1× running buffer and then load each strip onto a gel. Lay the top of the gel down onto a rack for 1.5-mL tubes with the long plate on the bottom. Place the nongel side of the strip on the long plate side of the gel that is facing up (Fig. 3). Carefully push the strip down between the glass plates,

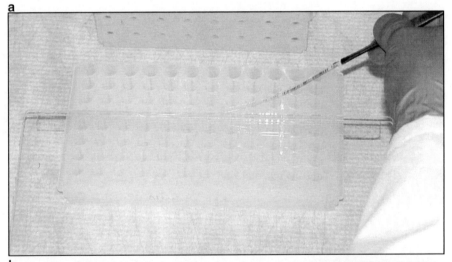

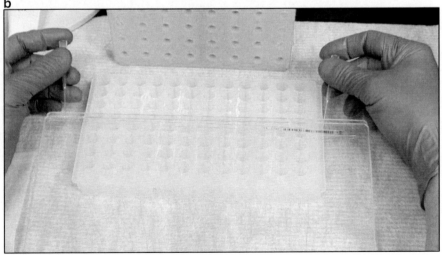

Fig. 3. Loading the strip onto the gel. (**a**) Lay the strip down on the long glass plate with the gel side facing up. (**b**) Using fine gel loading tips, push the strip down between the glass plates until it meets the acrylamide gel.

using fine gel loading tips, until it contacts the acrylamide gel. Be careful to avoid creating bubbles (see Note 22).

7. Melt an aliquot of 1% agarose containing bromophenol blue by microwaving for 5–10 s at a time. Cover the strip by slowly adding the agarose, cover the plate with aluminum foil, and allow the agarose to cool.

8. Load the plate(s) into the electrophoresis chamber as described by the manufacturer and add the appropriate amount of 1× running buffer. For large gels, connect the chamber to a circulating water cooler set at 14°C. If using fluorescent labels, cover the electrophoresis unit with aluminum foil.

9. Use the following settings for the different strip sizes and gels:

   7-cm strips/10-cm gels.

   100 mA for 1 h.

   13-cm strips/gels.

   600 V, 9 mA per gel, 1 W per gel for 16 h.

   24-cm strips/gels.

   600 V, 25 mA per gel, 1 W per gel for 16 h.

   At the end of the run, if the bromophenol blue has not reached the bottom of the gel, continue the electrophoresis until it does. Do not exceed 8 W per gel.

3.6. Gel Scanning and Data Analysis

Gels should be scanned with the appropriate imaging/scanning equipment for fluorescent dyes, being careful that the image intensity (PMT Volts) is kept just below saturation levels. Data analysis can be performed with various software packages. In our laboratory, we use a Typhoon Trio+ (GE Healthcare) imager and the DeCyder 2D 7.0 software. For additional statistical analysis, we have used Q-value (6) and SPSS.

Two examples of DIGE-based PTM analysis are shown in Fig. 4. Panel "a" shows postsynaptic densities isolated from rat hippocampus that were treated following the protocol described in Subheading 3.2.3. As indicated by the arrows, spots shown as green in the 2D gel correspond to ubiquitinated proteins, which can be removed from the polyacrylamide gels and identified by mass spectrometry. Panel "b" displays a portion of the 2D gel in which dephosphorylated proteins (see Subheading 3.2.1) were separated by gel electrophoresis. The absence of the phosphate group (or groups) induces pI shifts (lower portion of the figure) that can be used to determine the presence of the phosphate group in the untreated (phosphorylated) sample.

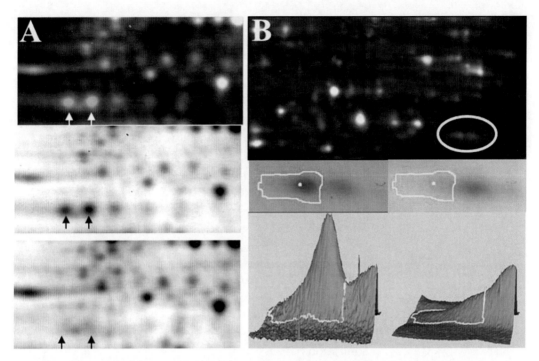

Fig. 4. Examples of PTMs analyzed by 2D DIGE. (**a**) Ubiquitination, deubiquitinated proteins are indicated by the *arrows*. (**b**) Phosphorylation, the *inserts* show how the dephosphorylated spots migrate differentially in the pI axis.

4. Notes

1. Concentrated HCl (12N) is first used to make large adjustments to the pH until close to the desired range. A lower concentration, such as 1N, is then used to make fine adjustments until the desired pH is reached.

2. In this example, we use calf intestinal alkaline phosphatase which is suggested by the manufacturer to remove phosphorylated serine, threonine, and tyrosine amino acid residues. However, not all sources/isoforms of alkaline phosphatase remove phosphates from these amino acids with equal efficiency. In addition, lambda alkaline phosphatase also removes phosphates from histidine residues. Therefore, it is possible to identify a residue-specific or complete phosphoproteome by using specific phosphatases or a cocktail of phosphatases covering all possible phosphorylated residues.

3. Here, we spin the samples in a speed vacuum dryer at 4,000 rpm, at 37°C, until dry. The lid of the speed vacuum dryer is transparent, so it is covered with aluminum foil.

4. The IPG buffer pH should match the pH of the strip. Use the highest grade of DTT available and store at 4°C.

5. Add water dropwise until almost 1 mL. Mix by inverting several times to dissolve the urea. Make up to final volume (1 mL) with water.
6. We use Strip Holder Cleaning Solution from GE Healthcare.
7. We use PlusOne DryStrip Cover Fluid from GE Healthcare.
8. The urea may dissolve slowly. If needed, heat the solution moderately (do not boil) and allow to cool before adding DTT or iodoacetamide.
9. Leave the weighed DDT and iodoacetamide in beakers at 4°C while EB is being mixed or cooled.
10. Due to the heterogeneous nature of biological samples, there are many varied protocols for preparing sample lysates. The procedure described here is used by our laboratory for most tissues samples, cell pellets, and subcellular organelles. We have also included a few references for preparing unique samples such as adipose tissue (7, 8), membrane proteins (9), and post-synaptic densities (10, 11).
11. DNase may be added if samples are still very viscous after sonication.
12. We use the 2D Clean-Up Kit from GE Healthcare. Alternatively, a chloroform-methanol precipitation method may be used (12).
13. With our microcentrifuges, a "quick-spin" consists of 5–10 s where the speed does not reach more than $1,699 \times g$.
14. If the pellet does not fully dissolve in 40 µL, add another 10 µL of 2× ubiquitin conjugation buffer and 10 µL of water until it does.
15. If necessary, samples may be incubated with NEM for 3 h or more on ice.
16. We use strip holder cleaning solution and a cleaning brush from GE Healthcare. Scrub the holder using the brush and rinse thoroughly with pure water.
17. Prehydrate 24-cm strips for 2 h in 200 µL of rehydration buffer (see Subheading 2.4, item 2). Account for this volume when preparing samples for isoelectric focusing (i.e., 450–200 µL = 250 µL rehydration buffer).
18. To remove bubbles once the strip has been applied, gently tap the back of the strip with a clean 200-µL tip. If bubbles remain, pull the strip up and try again.
19. If strips run overnight or finish before you are prepared to start the second dimension, a step 5 of 100 V for 24 h may be added at the end of the protocol as a "hold" step to keep the strip focused. This can be stopped once you are ready to proceed to the next stage.

20. A step 5 of 200 V for 24 h may be added as a "hold" step.

21. If step 5 is applied for more than 6 h, apply a further step of 8,000 V for 1 h before proceeding.

22. If bubbles are introduced between the strip and acrylamide gel, tapping the back of the strip (plastic) is usually enough to remove them. If not, lift the strip from one end and begin again, gently pushing the strip down at an angle so that bubbles are squeezed out toward the free end.

Acknowledgments

This work was supported by the UNC Systems-Proteomics Center and the Duke Neuroproteomics Center.

References

1. Yamagata A, Kristensen DB, Takeda Y, Miyamoto Y, Okada K, Inamatsu M, Yoshizato K (2002) Mapping of phosphorylated proteins on two-dimensional polyacrylamide gels using protein phosphatase. Proteomics **2**: 1267–1276.

2. Sun J, Morgan M, Shen R-F, Steenbergen C and Murphy E (2007) Preconditioning Results in S-Nitrosylation of Proteins Involved in Regulation of Mitochondrial Energetics and Calcium Transport. Circ Res **101**: 1155–1163.

3. Alban A, David SO, Bjorkesten L, Andersson C, Sloge E, Lewis S, and Currie I (2003) A novel experimental design for comparative two-dimensional gel analysis: two-dimensional difference gel electrophoresis incorporating a pooled internal standard. Proteomics **3**: 36–44.

4. Diez R, Herbstreith M, Osorio C, Alzate O (2009) 2-D Fluorescence Difference Gel Electrophoresis (DIGE) in neuroproteomics. In: Alzate O (ed) Neuroproteomics, CRC Press, Boca Raton.

5. DeKroon RM, Osorio C, Robinette JB, Mocanu M, Winnik WM, Alzate O (2011). Simultaneous detection of changes in protein expression and oxidative modification as a function of age and *APOE* genotype. J Proteome Res **10**: 1632–1644.

6. Storey JD and Tibshirani R (2003) Statistical significance for genomewide studies. PNAS **100**: 9440–9445.

7. Sanchez JC, Chiappe D, Converset V, Hoogland C, Binz PA, Paesano S, Appel RD, Wang S, Sennitt M, Nolan A, Cawthorne MA, Hochstrasser DF (2001) The mouse SWISS-2D PAGE database: a tool for proteomics study of diabetes and obesity. Proteomics **1**: 136–163.

8. Zhoa YM, Basu U, Dodson MV, Basarab JA, Guan LL (2010) Proteome difference associated with fat accumulation in bovine subcutaneous adipose tissues. Proteome Sci **8**: 1–14.

9. Babu GJ, Wheeler D, Alzate O, Periasamy M (2004) Solubilization of membrane proteins for two-dimensional gel electrophoresis: identification of sarcoplasmic reticulum membrane proteins. Anal Biochem **325**: 121–125.

10. Blackstone CD, Moss SJ, Martin LJ, Levey AI, Price DL, Huganir RL (1992) Biochemical characterization and localization of a non-N-methyl-D-aspartate glutamate receptor in rat brain. J Neurochem **58**: 1118–1126.

11. Lau LF, Mammen A, Ehlers MD, Kindler S, Chung WJ, Garner CC, Huganir RL (1996) Interaction of the N-methyl-D-aspartate receptor complex with a novel synapse-associated protein, SAP102. J Biol Chem **271**: 21622–21628.

12. Wessel D and Flugge UI (1983) A method for the quantitative recovery of protein in dilute solution in the presence of detergents and lipids. Anal Biochem **138**: 141–143.

Chapter 10

Comparative Analyses of Protein Complexes by Blue Native DIGE

Katrin Peters and Hans-Peter Braun

Abstract

Classically, DIGE is carried out on the basis of two-dimensional (2D) IEF/SDS PAGE. This allows comparative analyses of large protein sets. However, 2D IEF/SDS PAGE only poorly resolves hydrophobic proteins and is not compatible with native protein characterizations. Blue native PAGE represents a powerful alternative. Combined with CyDye labeling, blue native DIGE offers several useful applications like quantitative comparison of protein complexes of related protein fractions. Here we present a protocol for fluorophore labeling of native protein fractions for separation by blue native PAGE.

Key words: Blue native PAGE, DIGE labeling, Protein complexes, Membrane proteins, Mitochondria

Abbreviations

2D	Two-dimensional
BN PAGE	Blue native polyacrylamide gel electrophoresis
BN/SDS PAGE	Blue native/sodium dodecyl sulfate polyacrylamide gel electrophoresis
DIGE	Difference gel electrophoresis
DMF	Dimethylformamide
IEF/SDS PAGE	Isoelectric focusing/sodium dodecyl sulfate polyacrylamide gel electrophoresis
pI	Isoelectric point
PMSF	Phenylmethylsulfonyl fluoride

1. Introduction

Fluorophore-based difference gel electrophoresis (DIGE) represents an ideal system for comparative proteomics. Different protein fractions can be coelectrophoresed on a single gel, thereby eliminating gel-to-gel variations. If evaluated with special software tools, fluorophore detection allows exact relative quantification of proteins present in the protein fractions to be compared.

Classically, DIGE is carried out based on the two-dimensional (2D) isoelectric focusing/sodium dodecyl sulfate polyacrylamide gel electrophoresis (IEF/SDS PAGE) system (1). Indeed, using this experimental system, large protein sets are efficiently resolved. However, 2D IEF/SDS PAGE also has limitations. For instance, hydrophobic proteins are not well resolved during the IEF gel dimension, and proteins with very basic isoelectric points (pIs) often get lost. Additionally, 2D IEF/SDS PAGE is not compatible with native enzyme characterizations. 2D blue native (BN)/SDS PAGE represents an alternative gel system which allows overcoming these limitations. Hydrophobic as well as basic proteins are easily resolved. If applied as a one-dimensional system, blue native polyacrylamide gel electrophoresis (BN PAGE) is even compatible with in-gel enzyme activity stainings.

BN PAGE was first published by Hermann Schägger (2). The principle idea was to use Coomassie Blue not only for protein staining after gel electrophoresis but also for the introduction of negative charges into proteins by incubating protein fractions with this compound *before* gel electrophoresis. Coomassie Blue is an anionic molecule but, in contrast to SDS, does not denature proteins. Furthermore, protein complexes remain stable. If combined with low-percentage polyacrylamide gels, protein complexes can be resolved by BN PAGE. If sample preparation takes place in the presence of mild nonionic detergents, even hydrophobic membrane-bound protein complexes can easily be separated. As with IEF/SDS PAGE, strips of blue native gels can be horizontally transferred onto a second gel dimension which is carried out in the presence of SDS. Using this experimental setup, protein complexes are separated into their subunits which form vertical rows of spots on the resulting 2D gels.

BN PAGE and 2D blue native/sodium dodecyl sulfate polyacrylamide gel electrophoresis (BN/SDS PAGE) can be combined with the DIGE technology (3). BN DIGE allows systematic and quantitative comparison of protein complexes of related protein fractions, structural investigation of protein complexes, assignment of protein complexes to subcellular fractions like organelles, electrophoretic mapping of isoforms of subunits of protein complexes with respect to a larger proteome, and topological investigations (4). So far, BN DIGE has only been applied in a few studies (4–9). Indeed, the potential of BN DIGE was only very recently recognized. Here we present a BN DIGE protocol suitable for the comparative analyses of protein complexes of protein fractions.

2. Materials

Buffers, solutions, and reagents are given in the order of usage according to the methods protocol. Prepare all solutions freshly using analytical grade chemicals in combination with pure deionized water.

2.1. Preparation of a BN Gel

1. Acrylamide solution: 40%, acryl/bisacryl = 32/1 (AppliChem, Darmstadt, Germany).
2. BN gel buffer (6×): 1.5 M aminocaproic acid, 150 mM BisTris, pH 7.0 (adjust at 4°C).
3. N,N,N',N'-tetramethylethylenediamine (TEMED), 99% (Sigma, St. Louis, Missouri, USA).
4. Ammonium persulfate solution (APS): 10% (w/v) ammonium persulfate.
5. BN cathode buffer (5×): 250 mM tricine, 75 mM BisTris, 0.1% (w/v) Coomassie G 250, pH 7.0 (adjust at 4°C).
6. BN anode buffer (6×): 300 mM BisTris, pH 7.0 (adjust at 4°C).
7. Protean II gel unit (Bio-Rad, Richmond, Ca, USA).

2.2. Sample Preparation for BN DIGE

1. Solubilization buffer, pH 7.4, with digitonin: 30 mM HEPES, 150 mM potassium acetate, 10% (v/v) glycerol, pH 7.4, supplemented with 5% digitonin (see Note 1). Solution should be briefly boiled for dissolving digitonin. Buffer is stored at 4°C.
2. Solubilization buffer, pH 10: 30 mM HEPES, 150 mM potassium acetate, 10% (v/v) glycerol, pH 10 (see Note 2).
3. CyDye™ fluor minimal labeling reagents: Cy2™, Cy3™, and Cy5™ (GE Healthcare, Munich, Germany). The fluorophores (400 pmol) are diluted in dimethylformamide (DMF) according to the manufacturers' instructions. Diluted CyDyes are stored at −20°C and should be used within 3 months (see Note 3).
4. Lysine stock solution: 10 mM lysine, stored at 4°C.
5. Blue loading buffer: 750 mM aminocaproic acid, 5% (w/v) Coomassie 250 G, stored at 4°C.

2.3. SDS PAGE for Second Gel Dimension

1. Acrylamide solution: 40%, acryl/bisacryl = 32/1 (AppliChem).
2. N,N,N',N'-tetramethylethylenediamine (TEMED), 99% (Sigma).
3. APS: 10% (w/v) ammonium persulfate.
4. SDS gel buffer: 3 M Tris–HCl, 0.3% (w/v) SDS, pH 8.45.
5. BN gel buffer BN (6×): see Subheading 2.1.

6. SDS cathode buffer: 0.1 M Tris base, 0.1 M Tricine, 0.1% (w/v) SDS, pH 8.25.
7. SDS anode buffer: 0.2 M Tris–HCl, pH 8.9.
8. Overlay solution: 1 M Tris–HCl, 0.1% (w/v) SDS, pH 8.45.
9. Denaturation solution: 1.0% (w/v) SDS, 1.0% (v/v) β-mercaptoethanol.
10. SDS solution: 10% (w/v) SDS.
11. Protean II gel unit (Bio-Rad).

3. Methods

The following protocol is suitable for the comparison of protein complexes of two related protein fractions. It is based on the usage of the CyDye™ fluor labeling reagents (GE Healthcare, Munich, Germany). Both fractions should contain 100 µg protein and should be labeled with different fluorophores. Afterwards, the samples are combined and loaded onto a single BN gel. The BN gel should be prepared before sample separation.

3.1. Preparation of a BN Gel

Best resolution capacity of BN gels is achieved if the electrophoretic separation distance is greater than 12 cm. The following instructions refer to the Protean II electrophoresis unit (Bio-Rad, Richmond, CA, USA; gel dimensions 0.15 × 16 × 20 cm). However, units from other manufacturers are of comparable suitability for BN PAGE, e.g., the Hoefer SE-400 or SE-600 gel systems (GE Healthcare, Munich, Germany). Usage of gradient gels is recommended because the molecular masses of protein complexes can vary between 50 kDa and several 1,000 kDa (see Note 4):

1. Prepare a 4.5% separation gel solution by mixing 2.4 mL of acrylamide solution with 3.5 mL of BN gel buffer and 15.1 mL ddH$_2$O.
2. Prepare a 16% separation gel solution by mixing 7.4 mL of acrylamide solution with 3 mL of BN gel buffer, 4.6 mL ddH$_2$O, and 3.5 mL glycerol.
3. Transfer the two gel solutions into the two chambers of a gradient former and connect the gradient former via a hose and a needle with the space in-between two glass plates which are preassembled in a gel casting stand. Gradient gels can either be poured from the top (16% gel solution has to enter the gel sandwich first) or from the bottom (4.5% gel solution has to enter first). For pouring the gel from the bottom, an adjustable pump (e.g., Bio-Rad Econo Pump) is needed.

4. Add TEMED and APS to the two gel solutions (95 µL 10% APS/9.5 µL TEMED to the 4.5% gel solution and 61 µL APS/6.1 µL TEMED to the 16% gel solution).

5. Pour the gradient gel, leaving space for the stacking gel, and overlay with ddH$_2$O. The gel should polymerize in about 60 min.

6. Pour off the ddH$_2$O.

7. Prepare the stacking gel solution by mixing 1.5 mL of acrylamide solution, 2.5 mL of BN gel buffer, and 11 mL ddH$_2$O.

8. Add 65 µL APS and 6.5 µL TEMED and pour the stacking gel around an inserted comb. The stacking gel should polymerize within 30 min.

9. Prepare 1× BN anode buffer and 1× BN cathode buffer by diluting the corresponding stock solutions.

10. Once the polymerization of the stacking gel is finished, carefully remove the comb.

11. Add the BN cathode and anode buffers to the upper and lower chambers of the gel unit, respectively. Cool the unit down to 4°C.

3.2. Sample Preparation for BN DIGE

Starting material can either be whole cells or isolated organelles. Fractions should be treated under mild conditions in order to keep proteins in their native conformation (avoid high salt, ionic detergents, high temperatures, urea, etc.). All steps of the sample preparation should be carried out at 4°C. The BN gel should be prepared before the sample preparation is started (see Subheading 3.1):

1. Cell or organelle fractions, including 100 µg protein each, are centrifuged at full speed for 10 min at 4°C using an Eppendorf centrifuge.

2. Sedimented material is resuspended in 10 µL of solubilization buffer, pH 7.4, with digitonin and incubated for 20 min at 4°C (see Note 5).

3. Fractions are again centrifuged at full speed for 20 min at 4°C using an Eppendorf centrifuge to remove insoluble material.

4. The supernatants containing solubilized protein complexes are transferred into new 1.5-mL Eppendorf vessels and supplemented with 10 µL of solubilization buffer, pH 10, to adjust the pH value of the protein solutions to about 8.5 (see Note 6).

5. Labeling reaction: For CyDye™ labeling reactions, 100 µg protein of each sample is labeled with one CyDye by the addition of 1 µL of diluted CyDye solution to each protein fraction (see Note 7). Centrifuge the samples briefly and incubate for 30 min on ice in the dark.

6. Stop labeling reaction: Add 1 µL of lysine stock solution and incubate samples for 10 min in the dark.

7. Combine the two labeled protein fractions in one Eppendorf vessel.

8. Add 2 µL of blue loading buffer to the combined protein fraction.

3.3. BN PAGE for First Gel Dimension

1. Load the combined CyDye-labeled protein samples (see Subheading 3.2) into a well of a BN gel (see Subheading 3.1).

2. Connect the gel unit to a power supply. Start electrophoresis at a constant voltage (100 V for 45 min) and continue at a constant current (15 mA for about 11 h). Electrophoresis should be carried out at 4°C and in the dark in order to protect the dyes.

3. (Optionally): The BN cathode buffer, normally containing Coomassie Blue, may be replaced after half of the electrophoresis run against a BN cathode buffer without Coomassie Blue (see Note 8).

3.4. SDS PAGE for Second Gel Dimension

BN PAGE can be combined with a second gel dimension, which is carried out in the presence of SDS, to further separate the protein complexes into their subunits. All published protocols for SDS PAGE are suitable for combination with BN PAGE, e.g., the system published by Laemmli (10). However, the Tricine-SDS PAGE system developed by Schägger (11) generally gives the best resolution. The following instructions refer to this gel system carried out in the Protean II gel unit (Bio-Rad). As mentioned above, gel electrophoresis units of other manufacturers are of equal suitability:

1. Cut out a strip of a BN gel and incubate it in the denaturation solution for 30 min at room temperature.

2. Wash the strip with ddH$_2$O for 30–60 s. This step is important because β-mercaptoethanol inhibits polymerization of acrylamide.

3. Place the strip on a glass plate of an electrophoresis unit at the position of the teeth of a normal gel comb.

4. Assemble the gel electrophoresis unit by adding 1-mm spacers, the second glass plate and clamps, and transfer all into the gel casting stand. The reduced thickness of the gel of the second gel dimension (1 mm) in comparison to the strip of the first gel dimension (1.5 mm) avoids sliding down of the gel strip between the glass plates in vertical position.

5. Prepare a 16% separation gel solution by mixing 12.4 mL of acrylamide solution, 10 mL of SDS gel buffer, 7.6 mL ddH$_2$O, 100 µL APS, and 10 µL TEMED, and pour the solution into the space in-between the two glass plates below the BN gel strip. Leave space for the sample gel solution for embedding the BN gel strip. Overlay the separation gel with the overlay solution. The gel should polymerize in about 60 min (see Note 9).

6. Prepare a 10% sample gel solution by mixing 2.5 mL of acrylamide solution, 3.4 mL of BN gel buffer, 1 mL glycerol, 100 μL of SDS solution, 2.9 mL ddH$_2$O, 83 μL APS, and 8.3 μL TEMED.

7. Discard the overlay solution and pour the sample gel embedding the BN gel lane. The casting stand should be held slightly diagonally to avoid air bubbles to be captured underneath the BN gel lane (see Note 10).

8. Add the SDS cathode and anode buffers to the upper and lower chambers of the gel unit.

9. Connect the gel unit to a power supply. Carry out electrophoresis at 30 mA/gel over night. The gel run should take place in the dark to protect the fluorophores.

3.5. Visualization of CyDye™-Labeled Protein

After completion of gel electrophoresis, fluorophores can be detected in 1D BN or 2D BN/SDS gels using a fluorescence scanner (e.g., the Typhoon fluorescence scanner from GE Healthcare) (see Fig. 1). Keep gels in the dark before starting the scanning

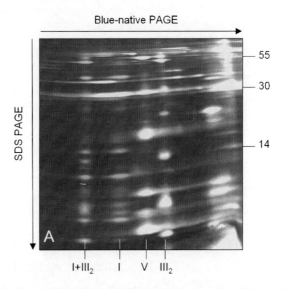

Fig. 1. Comparative analysis of the mitochondrial proteomes of a wild-type (Col-0) and a mutant (Col-0-ΔAt1g47260) cell line of the model plant *Arabidopsis thaliana*. Proteins of the Col-0 fraction were labeled with Cy3 and proteins of the Col-0-ΔAt1g47260 fraction with Cy5. Combined protein fractions were separated by 2D BN/SDS PAGE, and protein visualization was carried out by laser scanning at the respective wavelengths using the Typhoon laser scanner. On the resulting overlay image, Cy3-labeled proteins are indicated in *green* and Cy5-labeled proteins in *red*. Proteins of equal abundance in the two compared fractions are *yellow*. The molecular masses of standard proteins are given to the right of the figure; the identities of protein complexes are given below the gel. *I* complex I; *V* complex V; *III$_2$* dimeric complex III; *I + III$_2$* supercomplex containing complex I and dimeric complex III. The mutant line lacks a subunit of complex I. As a consequence, amounts of complex I and the I + III$_2$ supercomplex are drastically reduced (from Perales et al. (3), modified).

procedure. Using the CyDye™ fluorophores, gels should be scanned at 50–100 μm resolution at the appropriate excitation wavelengths (488 nm for Cy2™, 532 nm for Cy3™, and 633 nm for Cy5™). Digital gel images can be visualized using the ImageQuant analysis software (GE Healthcare). Quantification of relative differences of individual protein abundances can be carried out using specific software such as Delta 2D (Decodon, Greifswald, Germany) or DeCyder™ (GE Healthcare).

4. Notes

1. Digitonin is necessary for the solubilization of membrane-bound protein complexes of cellular or organellar fractions. Other nonionic detergents (e.g., dodecylmaltoside and Triton X-100) can be used but should be applied in the presence of modified buffer systems; see Wittig et al. (12).

2. The pH value of the buffer must be >9. Normally, it is at about pH 10 without adjustment. Therefore, the buffer can be directly used.

3. Always use CyDyes from the same reaction kit diluted with DMF of one batch to assure comparative labeling conditions. Consume diluted CyDyes within 3 months.

4. If very large protein complexes (>3 MDa) have to be resolved, the acrylamide gradient gel of the BN gel can be substituted by a 2.5% agarose gel prepared in BN gel buffer (13).

5. The resulting detergent-to-protein ratio is 5:1 (w/w).

6. A pH value of about 8.5 is a prerequisite for efficient labeling. The pH value of the protein solution can be easily controlled using pH test strips.

7. Additionally, a third fraction containing a mixture of both protein fractions (50 μg protein each) can be labeled with a third CyDye as internal standard.

8. Coomassie Blue might quench the fluorescence signal of the CyDyes. If the cathode buffer is substituted by the same buffer lacking Coomassie Blue after half of the electrophoresis run, the background of the resulting BN gel is much clearer. Alternatively, native gel electrophoresis can be performed following the "clear native" protocol of Wittig et al. (14).

9. With respect to the Tricine-SDS PAGE system for the second gel dimension, Schägger and von Jagow (11) originally proposed a two-step separation gel consisting of a large 16% phase and a smaller 10% phase (called "spacer gel"). The advantage of this slightly more complicated gel system is a better resolution

of large protein subunits. The two gel solutions are poured one upon the other, and glycerol is added to the 16% gel solution to avoid mixing with the 10% gel solution—for details see Schägger and von Jagow (11).

10. The transfer of the BN gel strip onto a second SDS gel dimension is recommended to be carried out by embedding the gel strip while casting a sample gel onto the second dimension gel. This procedure ensures optimal physical contact between the BN gel lane and the SDS gel of the second dimension. However, BN gel strips can also be fixed onto a prepoured SDS gel using agarose as usually carried out for 2D IEF/SDS PAGE. Physical contact of the gels might not be optimal, but the time period between the end of the first and the start of the second electrophoresis is shortened.

Acknowledgments

The authors would like to thank Christina Rode, Institute for Plant Genetics, Leibniz Universität Hannover, for critical reading of the manuscript.

References

1. O'Farrell P.Z. and Goodman H.M. (1976) Resolution of simian virus 40 proteins in whole cell extracts by two-dimensional electrophoresis: heterogeneity of the major capsid protein. *Cell.* 9: 289–298.
2. Schägger H. and von Jagow G. (1991) Blue native electrophoresis for isolation of membrane protein complexes in enzymatically active form. *Anal. Biochem.* 199: 223–231.
3. Perales M., Eubel H., Heinemeyer J., Colaneri A., Zabaleta E., Braun H.P. (2005) Disruption of a nuclear gene encoding a mitochondrial gamma carbonic anhydrase reduces complex I and supercomplex I + III$_2$ levels and alters mitochondrial physiology in Arabidopsis. *J Mol Biol* 350: 263–277.
4. Heinemeyer J., Scheibe B., Schmitz U.K., and Braun H.P. (2009) Blue native DIGE as a tool for comparative analyses of protein complexes. *J Prot* 72: 539–544.
5. Hedman E., Widén C., Asadi A., Dinnetz I., Schröder W.P., Gustafsson J.A., Wikström A.C. (2006) Proteomic identification of glucocorticoid receptor interacting proteins. *Proteomics* 6: 3114–3126.
6. Gillardon F., Rist W., Kussmaul L., Vogel J., Berg M., Danzer K., Kraut N., Hengerer B. (2007) Proteomic and functional alterations in brain mitochondria from Tg2576 mice occur before amyloid plaque deposition. *Proteomics* 7: 605–616.
7. Reisinger V. and Eichacker L.A. (2007) How to analyze protein complexes by 2D blue native SDS-PAGE. *Proteomics* 7: 6–16.
8. Dani D. and Dencher N.A. (2008) Native-DIGE: a new look at the mitochondrial membrane proteome. *Biotechnol J.* 3: 817–822.
9. Dani D. Shimokawa I., Komatsu T., Higami Y., Warnken U., Schokraie E., Schnölzer M., Krause F., Sugawa M.D., Dencher N.A. (2010) Modulation of oxidative phosphorylation machinery signifies a prime mode of anti-ageing mechanism of calorie restriction in male rat liver mitochondria. *Biogerontology* 11: 321–334.
10. Laemmli, U.K. (1970) Cleavage of structural proteins during the assembly of the head of bacteriophage T4. *Nature* 227: 680–685.
11. Schägger, H. and von Jagow, G. (1987) Tricine-sodium dodecyl sulfate-polyacrylamide gel electrophoresis for the separation of proteins in

the range from 1 to 100 kDa. *Anal. Biochem.* 166: 368–379.
12. Wittig I., Braun H.-P., Schägger, H. (2006) Blue-native PAGE. *Nature Protocols* 1: 418–428.
13. Henderson, N.S., Nijtmans, L.G., Lindsay, J.G., Lamantea, E., Zeviani, M. and Holt, I.J. (2000) Separation of intact pyruvate dehydrogenase complex using blue native agarose gel electrophoresis. *Electrophoresis* 21: 2925–2931.
14. Wittig I., Karas M., Schägger H. (2007) High resolution clear native electrophoresis for in-gel functional assays and fluorescence studies of membrane protein complexes. *Mol Cell Proteomics* 6: 1215–1225.

Chapter 11

2D DIGE Analysis of Protein Extracts from Muscle Tissue

Cecilia Gelfi and Sara De Palma

Abstract

2D DIGE, two-dimensional difference gel electrophoresis, is a technology used to study the protein expression on two-dimensional gels. Protein samples are labeled with different color fluorescent dyes designed not to affect the relative migration of proteins during electrophoresis. Here, we describe the practical procedures necessary to perform a 2D DIGE experiment for a muscle tissue protein extract followed by CyDye DIGE fluors minimal labeling and the analysis of 2D DIGE gels for the assessment of quantitative differences.

Key words: 2D DIGE, 2D electrophoresis, CyDye DIGE fluors, Differential analysis, Muscle proteome

1. Introduction

The DIGE technique was first described in 1997 (1); it is a modified form of 2DE, two-dimensional electrophoresis (2–4), developed to overcome the problem of the irreproducibility of gels and to increase the sensitivity of 2DE methodology by labeling samples with a different fluorescent dye prior to running them on the same gel (see Fig. 1).

Protein samples are labeled with up to three spectrally distinct, charge- and mass-matched fluorescent dyes (CyDye DIGE fluors) (1, 5, 6): Cy2, Cy3, and Cy5. This means that the same protein labeled with any of the CyDye DIGE fluors migrates to the same position on a 2D gel (7) eliminating intragel variation. Moreover, the dyes afford great sensitivity with a detection capability of 125 pg of a single protein and a linear response over at least five orders of magnitude. CyDyes DIGE fluors have an N-hydroxysuccinimide ester reactive group and are designed to covalently attach to the epsilon amino group of lysine via an amine linkage (see Fig. 2). The quantity of dye added to the sample limits the reaction; hence, this

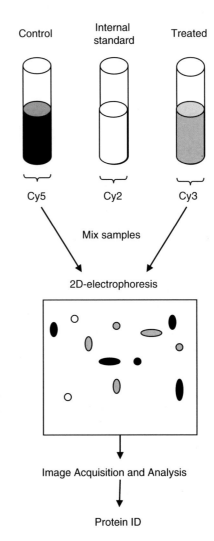

Fig. 1. Scheme of a DIGE experiment, using three fluors and a single gel.

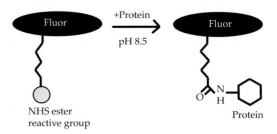

Fig. 2. Scheme of the DIGE labeling reaction.

method is referred to as "minimal labeling." This ensures that the stoichiometry of protein to fluor, results in only 1–2% of the total number of lysine residues being labeled.

After labeling, protein samples are mixed and separated on the same 2D gel. Different protein extracts labeled with different

CyDye fluors are then visualized separately using an imager containing appropriate laser wavelengths. Digital images of each sample will be analyzed by dedicated software. One of the most important advantages of 2D DIGE is the introduction of the internal standard (8, 9) created by pooling an aliquot of all biological samples and labeled with one of the CyDye DIGE fluors (usually Cy2 in a three-color experiment or Cy3 in a two-color experiment). This means that every protein from all samples will be represented in the internal standard. Linking every sample in-gel to a common internal standard gives some advantages compared to conventional 2DE; in particular, it allows an accurate quantification and the separation of experimental from biological variation, and statistical analysis of spots between gels can be applied (10).

2. Materials

All solutions have to be prepared using ultrapure water (18 MΩcm at 20°C) and analytical grade reagents. Filtering all solution is strongly recommended.

2.1. Sample Preparation and Quantitation (Compatible with CyDye DIGE Minimal Labeling)

1. Homogenizer for sample disruption (see Note 1).
2. Lysis buffer: 30 mM Tris, 7 M urea, 2 M thiourea, 4% (w/v) CHAPS, pH 8.5. For a final volume of 100 mL, dissolve 364 mg of Tris, 42 g of urea, 15.2 g of thiourea, and 4 g of CHAPS in about 50 mL of water. Adjust the pH with dilute HCl and make up to 100 mL with water. Prepare aliquots of 1 mL and store at −20°C. Prior to use, add 1 μL of PMSF 10 mM (see Note 2).
3. Sonicator (see Note 3).
4. Precipitation method to purify protein sample to remove non-protein impurities (see Note 4).
5. pH indicator strips (pH 4.5–10.0) to check the pH of the protein extract and sodium hydroxide (50 mM) to adjust the pH.
6. Quantitation method to determine sample concentration (see Note 5).

2.2. Protein Labeling

1. CyDye™ DIGE fluor Cy2, Cy3, and Cy5 minimal dye (GE Healthcare) (see Note 6).
2. 99.8% DMF (anhydrous dimethylformamide) for the reconstitution of CyDye, less than 3 months old from the day of opening (see Note 7).
3. 10 mM lysine solution to stop the labeling reaction. The solution can be stored at 4°C for several months.

2.3. Gel Electrophoresis

2.3.1. Isoelectric Focusing (IEF)

1. 2× sample buffer: 8 M urea, 130 mM DTT, 4% (w/v) CHAPS, and 2% (v/v) Pharmalyte™ 3–10 for isoelectric focusing (IEF). For a final volume of 100 mL, dissolve 48.05 g of urea, 2 g of DTT, and 4 g of CHAPS in 50 mL of water and add 1 mL of Pharmalyte™ 3–10. Add water to a volume of 100 mL. Aliquots can be stored at −20°C.

2. Immobiline™ DryStrip immobilized pH gradient (IPG) gel strips (GE Healthcare). Precast Immobiline DryStrips are available with different lengths (7, 11, 13, 18, and 24 cm) and different pH gradients (3.5–4.5, 4.5–5.5, 4–7 linear, 6–11 linear, 3–10 nonlinear and linear, 3–11 nonlinear and linear, etc.).

3. Rehydration stock solution to rehydrate IPG strips prior to IEF: 7 M urea, 2 M thiourea, and 2% (w/v) CHAPS. For a final volume of 100 mL, weigh out 42.04 g of urea, 15.22 g of thiourea, and 2 g of CHAPS and prepare a solution as in previous steps. Store at −20°C. Prior to use, add 65 mM DTT and 0.5% (v/v) IPG buffer (GE Healthcare).

4. Bromophenol blue as tracking dye.

5. IPG cover fluid (GE Healthcare) to minimize evaporation and urea crystallization.

6. IPGphor IEF units (GE Healthcare) or similar for first-dimension IEF of proteins.

7. IPGphor strip holder (GE Healthcare) having the same length of the IPG strips or, alternatively, the IPGphor Manifold (GE Healthcare) that allows protein separation on 7–24-cm-long Immobiline DryStrips.

2.3.2. IPG Strip Equilibration

1. Equilibration buffer: 6 M urea, 2% (w/v) SDS, 20% (v/v) glycerol, 375 mM Tris–HCl, pH 8.8. For a final volume of 100 mL, dissolve 36.036 g of urea and 2 g of SDS, add 20 mL of glycerol and 20 mL of Tris–HCl, pH 8.8, and prepare a solution as in previous steps. Store at −20°C. Prior to use, separate the equilibration buffer into two aliquots and add 65 mM DTT (100 mg/10 mL equilibration buffer) and 135 mM iodoacetamide (250 mg/10 mL equilibration buffer) respectively.

2.3.3. SDS-PAGE

1. Resolving gel buffer 5×: 1.875 M Tris–HCl, pH 8.8. Add about 200 mL of water to a graduated 1-L cylinder or a glass beaker. Weigh out 227.1 g of Tris and add water to a volume of 900 mL. Mix and adjust the pH with concentrated HCl. Make up to 1 L with water and store at 4°C in a glass bottle.

2. 40% acrylamide/bis solution (39:1 acrylamide/bis): To prepare 1 L of stock solution, weigh out 390 g of acrylamide monomer and 10 g of bis (cross-linker) and dissolve in a graduated cylinder with about 400 mL of water. Mix and make up to 1 L with water. Store at 4°C in a glass bottle (see Note 8).

3. 10% ammonium persulfate: Prepare aliquots of 1 mL and store at −20°C.

4. *N,N,N,N'*-tetramethylethylenediamine (TEMED). Store at 4°C.

5. SDS-PAGE running buffer (10×): 250 mM Tris, 1.92 M glycine, and 1% SDS. For a final volume of 1 L, weigh out 30.2 g of Tris and 144.2 g of glycine and dissolve in about 600 mL of water. Mix, filter the solution, add 10 g of SDS, and make up to 1 L with water. Store at 4°C in a glass bottle.

6. 30% (v/v) isopropanol. Store at room temperature in a dark glass bottle.

7. 0.5% agarose sealing solution: For a final volume of 100 mL, weigh out 0.5 g of agarose, add 100 mL of 1× SDS-PAGE running buffer, and melt in a heating block or boiling water (see Note 9).

8. Low-fluorescence glass plates (GE Healthcare).

9. Gel casting cassettes.

10. Electrophoretic unit.

2.4. Image Acquisition and Analysis

1. Typhoon variable mode imager (GE Healthcare) or similar.
2. DeCyder software (GE Healthcare) or similar.

3. Methods

3.1. Protein Extraction from Muscle Tissue

1. At least 15–20 mg of muscle tissue is necessary to perform a 2D DIGE experiment.
2. Grind muscle tissue in a frozen mortar.
3. After sample disruption, add lysis buffer and sonicate the sample on ice (20-s bursts with 60-s rest between bursts) until no particulate matter remains, or until no more of the particulate matter is dissolved.
4. Centrifuge the sample at $12,000 \times g$ for 5 min at 4°C to remove any insoluble material.
5. Precipitate proteins in order to separate them from nonprotein impurities (salts, nucleic acids, lipids, etc.).
6. Resuspend pellets with lysis buffer for DIGE.
7. Check the pH of the sample on ice. The sample pH must be pH 8.0–9.0. Adjust the pH using dilute NaOH, if required.
8. Accurately quantify protein samples.

3.2. Protein Labeling

1. Remove the CyDye from the −20°C-freezer and leave to warm for 5 min at room temperature.
2. After 5 min, add the specified volume of DMF to each vial of CyDye to prepare a CyDye stock solution (i.e., 5 μL of DMF

for the 5 nmol pack size). The stock solution is stable for 3 months at −20°C.

3. Vortex vigorously for 30 s.
4. Centrifuge the vial for 30 s at $12,000 \times g$ in a microcentrifuge.
5. Prepare the CyDye working solution adding 1 volume of CyDye stock solution to 1.5 volumes of high-grade DMF to make 400 µM CyDye solution. For example, take 2 µL CyDye stock solution and add 3 µL DMF to give 400 pmol of CyDye in 1 µL. The CyDye fluor working solution is stable for 1 week at −20°C.
6. Label 50 µg of each sample with 400 pmol of dye adding 1 µL of working solution.
7. Carry out the labeling reaction by incubating the sample on ice for 30 min in the dark.
8. Add 1 µL (for 50 µg of sample labeled) of 10 mM lysine solution to stop the reaction. Mix by pipetting and spin briefly in a microcentrifuge.
9. Leave for 10 min on ice in the dark.
10. After labeling, samples can be stored at −70°C for at least 3 months.

3.3. Loading Samples onto IPG Strips

1. Add to the CyDye-labeled samples an equal volume of 2× sample buffer and leave on ice for 10 min.
2. Combine the differentially labeled samples (two or three) into a single microfuge tube and mix. One of these samples should be the internal standard (see Note 10).
3. Add to the samples the appropriate volume of rehydration solution (see Note 11).
4. Pipette the appropriate volume of rehydration solution into each holder removing any larger bubbles (see Note 12).
5. Remove the protective cover from the IPG strip and position the strip with the gel side down. Be careful not to trap bubbles under the IPG strip.
6. Apply the appropriate volume of IPG cover fluid on the strip (see Note 13).
7. Allow the IPG strip to rehydrate for a minimum of 10 h (see Note 14).

3.4. First Dimension (IEF)

1. Run the appropriate IEF protocol (see Note 15). Figure 3 shows examples of 2D gels obtained separating labeled samples on different pH gradient. The IEF protocols used are also described.
2. If the IPG strips are not run immediately on the second dimension, they can be stored at −80°C in a sealed container. Do not equilibrate strips prior to storage.

IPG strip	Protocol (1st dimension conditions)	Images (2nd dimension: 12% polyacrylamide gels)
24 cm, pH 3-10 NL (Cy5 labelled Human skeletal muscle)	Rehydration 12h 200 V 2h 500 V 2h 1000 V 2h 2000 V 1.30h 3000 V 1.30h 3000-8000V 5h (Gradient) 8000 V 2h 50 µA/strip	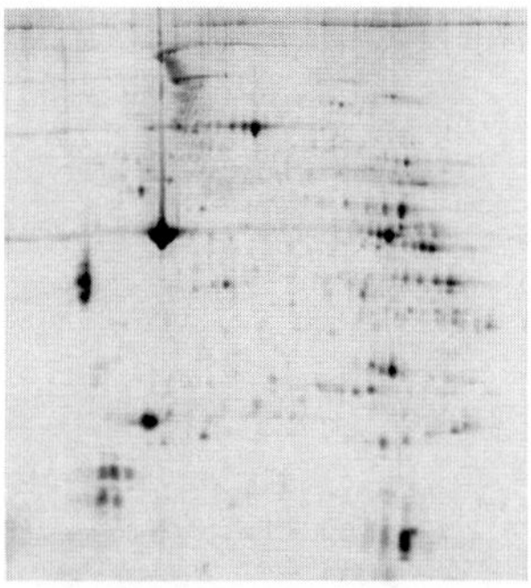
18 cm, pH 4.5-5.5 (Cy3 labelled Human skeletal muscle)	Rehydration 12h 200 V 2h 500 V 1h 1000 V 1h 2000 V 1h 3000 V 1h 3000-8000V 4h (Gradient) 8000 V 2 50 µA/strip	
18 cm, pH 4-7 (Cy5 labelled Human skeletal muscle)	Rehydration 12h 200 V 2h 500 V 2h 1000 V 2h 2000 V 1.30h 3000 V 1.30h 3000-8000V 5h (Gradient) 8000 V 2h 50 µA/strip	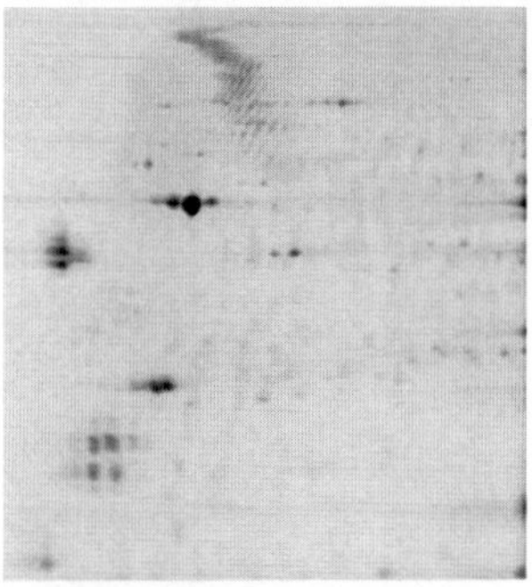
18 cm, pH 6-11 (Cy3 labelled Human skeletal muscle)	Rehydration 12h 300 V 2h 1000 V 1h 8000 V 2h (Gradient) 8000 V 1-2h 50 µA/strip	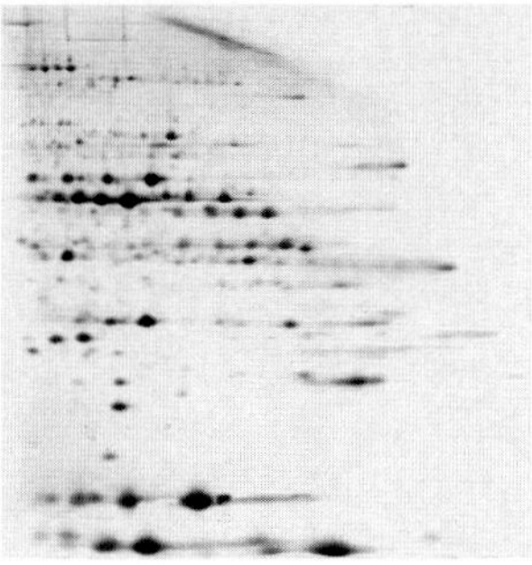

Fig. 3. IEF protocols and 2D DIGE images of human skeletal muscle protein lysates using different IPG strip pH gradients.

3.5. IPG Strip Equilibration

1. Separate the equilibration buffer into two aliquots and just prior to use add 65 mM DTT (100 mg/10 mL equilibration buffer) and 135 mM iodoacetamide (250 mg/10 mL equilibration buffer), respectively.
2. Equilibrate strips for 15 min in the first solution (DTT) followed by equilibration in the second solution (iodoacetamide) for 8 min.
3. After equilibration, place the strips on filter paper moistened with deionized water.

3.6. Second Dimension (SDS-PAGE)

A commonly used second dimension gel for 2DE is a homogeneous gel containing 12% total acrylamide (see Note 16):

1. Assemble the gel cassette.
2. Prepare a sufficient volume of gel solution without adding 10% (w/v) ammonium persulfate (APS) and TEMED.
3. Degas the gel solution for 10–15 min.
4. Add 10% APS solution and TEMED to the gel solution in order to make a 0.5% (v/v) APS/0.03% TEMED gel solution.
5. Pour the gel solution into the gel cassette and pipette 1–1.5 mL of 30% (v/v) isopropanol solution on top.
6. Allow gels to polymerize for at least 1–2 h.
7. Disassemble the gel cassette.
8. Prepare the appropriate volume of the 1× running buffer by diluting the 10× SDS-PAGE running buffer.
9. Melt the agarose sealing solution and, for each Immobiline DryStrip, slowly pipette the solution up to the top of the glass plate (see Note 17).
10. Carefully place the Immobiline DryStrip between the two glass plates holding one end of the strip with forceps. Using a thin plastic spacer, push the strip until it comes in contact with the surface of the gel avoiding the trapping of air bubbles (see Note 18).
11. Insert the glass plates into the electrophoresis apparatus and start the run (see Note 19).
12. At the end of the run, the gels are ready to be scanned.

3.7. Image Acquisition

After electrophoresis, take the images of the gel at the appropriate wavelength using a laser-based scanner such as a Typhoon variable scanner imager (GE Healthcare) (11) (see Note 20) or alternatively images can be taken using a CCD camera:

1. Rinse the gel sandwich with distilled water and wipe it dry with lint-free towels.
2. Place the gel sandwich on the scanner platen and take the images according to the manufacturer's instructions.

3.8. Image Analysis

Various fully automated DIGE image analysis software packages (DeCyder from GE Healthcare, Progenesis SameSpots from Nonlinear Dynamics, Melanie from Bio-Rad) developed for spot detection, matching, and differential protein expression analysis are commercially available.

The most important steps involved in image analysis are:

- Spot detection.
- Background subtraction.
- In-gel normalization.
- Artifact removal.
- Gel matching.
- Statistical analysis.

Here, we describe the principal steps of image analysis performed using the DeCyder software (5, 8, 11–13).

The software consists of three main modules:

DIA (differential in-gel analysis): to perform protein spot detection and quantification on a pair of images, from the same gel. A set of images is merged together, thereby incorporating all spot features in a single image. Spot detection and spot boundary definition is performed using pixel data from all the individual raw images and the merged image. An estimation of the number of spots present on the images must be entered—it is recommended that this value is overestimated to compensate for the detection of nonproteinaceous spots on the image. One can later use the *Exclude filter* in order to eliminate nonproteinaceous spots (based on their physical characteristics). Numerical data for individual spots are calculated (volume, area, peak height, and slope). Spot volumes are always expressed with the background subtracted. Background is subtracted by excluding the lowest tenth percentile pixel value on the spot boundary from all other pixel values within the spot boundary. The spot volume is the summation of these corrected values.

BVA (biological variation analysis): to match multiple images from different gels to provide statistical data on differential protein expression levels between multiple groups.

EDA module (extended data analysis): to elaborate a set of data. There are different calculations that can be performed in EDA:

- *Differential Expression Analysis.* This kind of calculation permits to investigate differential expression between two or more experimental groups. There are different subanalyses related to the experimental setup. Student's *t* test is used to evaluate the hypothesis of a difference between two groups. If more than two groups are included into the experimental design, ANOVA must be used. The simplest form of ANOVA is known as one-way ANOVA, and it is used to test for differences in standardized abundance among experimental groups; the test is accompanied by *Tukey's* test (post hoc test) to get an indication of which group is different. The two-way ANOVA is used to analyze two conditions in an experimental design in which two independent factors are taken into consideration. It calculates the significance of the difference between groups with the same condition 2 and different condition 1 (indicated as two-way ANOVA condition 1) and the other way around (two-way ANOVA condition 2). The two-way ANOVA analysis also calculates a significance value of the mutual effect of the two factors (two-way ANOVA interaction). Significantly changed proteins having a p value <0.01 are typically considered.

- *Principal Component Analysis* (*PCA*). This analysis is essentially a method for reducing the dimension of the variables in a multidimensional space getting a simpler view of the proteins and the spot maps in the data set. It is thus possible to detect outliers in the data set and to identify spot maps that have similar expression profiles.

- *Pattern Analysis.* This process consists of algorithms that can help to find the subsets of the data that show similar expression patterns. There are different types of pattern algorithms: one of the most important is *Hierarchical Clustering*, a method that combines or splits data pairwise and generates a tree-like structure called dendrogram. Protein and spot maps with similar expression profiles are grouped together.

4. Notes

1. The cells or tissue should be disrupted in such a way as to minimize proteolysis and at low temperature with a minimum of heat generation. There are different disruption methods, both mechanical and chemical (14, 15). Gentle lysis methods (osmotic lysis (16), freeze-thaw lysis (14, 17, 18), enzymatic lysis (19, 20)) are generally employed when the sample of interest consists of easily lysed cells. More vigorous lysis methods (sonication (21, 22), grinding (23–25), mechanical homogenization (17, 26), glass bead homogenization (27)) are employed when cells are less easily disrupted (cells in solid tissue or cells with tough cell walls).

2. Appropriate sample preparation is absolutely essential for good 2D DIGE results. The optimal procedure must be determined empirically for each sample type. Ideally, the process will result in the complete solubilization, disaggregation, denaturation, and reduction of the proteins. It is important to remember

that CyDye DIGE fluors are minimal dyes labeling primary amines. Exogenous sources of amines, such as DTT or ampholytes, must not be included in the buffer prior to labeling.

3. Sonicate samples in short bursts to avoid heating. Cool on ice between bursts.
   Sonication is complete when the solution appears significantly less cloudy than the starting solution.

4. Protein precipitation is employed to selectively separate proteins in the sample from contaminating substances. Current methods of protein precipitation (TCA precipitation, acetone precipitation, TCA combined with acetone precipitation, etc.) suffer from several disadvantages: incomplete precipitation, difficult protein resuspension, and introduction of ions that could interfere with IEF. Kits specifically designed for protein precipitation for 2DE are commercially available (e.g., 2-D Clean-Up Kit from GE Healthcare).

5. Electrophoretic analysis of protein samples requires accurate quantification of the sample to be analyzed. Use a standard protein quantitation method or, alternatively, use a kit designed specifically for the accurate determination of protein concentration in samples to be analyzed by 2DE (e.g., PlusOne™ 2-D Quant Kit from GE Healthcare). The recommended concentration of the protein lysate required for minimal labeling is between 5 and 10 mg/mL.

6. Store CyDyes in the dark at −20°C.

7. The quality of the DMF used in all experiment is critical to ensure a successful protein labeling. The DMF must be anhydrous, and it is necessary to avoid contamination with water. After opening, over a period of time, DMF will degrade producing amine compounds. Amines will react with CyDyes reducing fluors concentration available for protein labeling.

8. Acrylamide is a neurotoxin; it is important to wear gloves and use appropriate handling precautions.

9. Prepare the agarose sealing solution during equilibration.

10. Two kinds of experimental design can be performed: the two- or three-color experiment. In the two-color experiment, all samples are labeled with the same fluor (usually Cy5) and the internal standard with Cy3 (the internal standard is created by pooling an aliquot of all samples and labeled with one of the CyDye fluors). In the second case, all samples are labeled with two fluors (Cy3 and Cy5), enabling dye swapping to control any dye-specific effects that might result from preferential labeling or different fluorescence characteristics of the gel at the different excitation wavelengths. Here, the internal standard is labeled with Cy2. The major difference between the two

experimental designs is represented by the number of gels which is higher in a two-color experiment.

11. Just prior to use, add to the rehydration stock solution the appropriate amount of IPG buffer (0.5% v/v), 65 mM DTT, a trace of bromophenol blue, and the samples. The volume of the rehydration solution depends on the IPG strips' length (i.e., for 24-cm IPG strips, the rehydration solution volume required per strip is 450 μL). Make sure the IPG buffer matches the pH gradient of the strip used for IEF.

12. Select the strip holders corresponding to the IPG strip length. Wash each holder with detergent to remove residual protein and rinse with distilled water. Before use, the holders must be completely dry.

13. The IPG cover fluid volume to apply on the strip depends on the length of the holder used. Add the oil until the entire IPG strip is covered.

14. It is possible to rehydrate strips in the absence of protein samples. In this case, protein samples are loaded after rehydration, using a cup loading technique.

15. A typical IEF protocol generally proceeds through a series of voltage steps that begins at a relatively low value. Voltage is gradually increased to the final desired focusing voltage, which is held for up to several hours. A low initial voltage minimizes sample aggregation. It is important to remember that focusing parameters for different pH gradients and different protein loading need to be optimized.

16. Single percentage gels offer better resolution for a particular MW window. When a gradient gel is used, the separation interval is wider and spots are sharper because the decreasing pore size functions to minimize diffusion. Remember that stacking gels are not necessary for 2D gels.

17. Take care not to introduce bubbles and do not allow the agarose to cool or solidify before placing the strip.

18. By convention, the acidic end of the strip is placed on the left.

19. Recommended running condition:
    - 16 h overnight run, 15°C (2 W per gel).
    - 8 h duration, 15°C (4 W per gel).
    - 4 h duration, 15°C (8 W per gel).
    - The run is finished when the bromophenol blue dye front reaches the bottom of the gel.

20. The excitation and emission wavelength used for all three CyDyes are listed in Table 1.

Table 1
Excitation and emission characteristics for the three common DIGE fluors

Fluorescent dye	Laser excitation source (nm)	Emission filter (nm)
DIGE fluor Cy3 minimal dye	Green (532)	580 BP 30
DIGE fluor Cy5 minimal dye	Red (633)	670 BP 30
DIGE fluor Cy2 minimal dye	Blue (488)	520 BP 40

Acknowledgements

This work was supported by the Telethon foundation (grant N. GGP08107D to C.G.), EU community (grant BIO-NMD N. 241665 to C.G.), and Italian Ministry of University and Scientific Research (grant FIRB RBRN07BMCT to C.G.).

References

1. Unlu, M., Morgan, M. E., and Minden, J. S. (1997) Difference gel electrophoresis: a single gel method for detecting changes in protein extracts, *Electrophoresis 18*, 2071–2077.
2. Garrels, J. I. (1979) Two dimensional gel electrophoresis and computer analysis of proteins synthesized by clonal cell lines, *J Biol Chem 254*, 7961–7977.
3. Klose, J. (1975) Protein mapping by combined isoelectric focusing and electrophoresis of mouse tissues. A novel approach to testing for induced point mutations in mammals, *Humangenetik 26*, 231–243.
4. O'Farrell, P. H. (1975) High resolution two-dimensional electrophoresis of proteins, *J Biol Chem 250*, 4007–4021.
5. Gharbi, S., Gaffney, P., Yang, A., Zvelebil, M. J., Cramer, R., Waterfield, M. D., and Timms, J. F. (2002) Evaluation of two-dimensional differential gel electrophoresis for proteomic expression analysis of a model breast cancer cell system, *Mol Cell Proteomics 1*, 91–98.
6. Zhou, G., Li, H., DeCamp, D., Chen, S., Shu, H., Gong, Y., Flaig, M., Gillespie, J. W., Hu, N., Taylor, P. R., Emmert-Buck, M. R., Liotta, L. A., Petricoin, E. F., 3 rd, and Zhao, Y. (2002) 2D differential in-gel electrophoresis for the identification of esophageal scans cell cancer-specific protein markers, *Mol Cell Proteomics 1*, 117–124.
7. Swatton, J. E., Prabakaran, S., Karp, N. A., Lilley, K. S., and Bahn, S. (2004) Protein profiling of human postmortem brain using 2-dimensional fluorescence difference gel electrophoresis (2-D DIGE), *Mol Psychiatry 9*, 128–143.
8. Alban, A., David, S. O., Bjorkesten, L., Andersson, C., Sloge, E., Lewis, S., and Currie, I. (2003) A novel experimental design for comparative two-dimensional gel analysis: two-dimensional difference gel electrophoresis incorporating a pooled internal standard, *Proteomics 3*, 36–44.
9. Knowles, M. R., Cervino, S., Skynner, H. A., Hunt, S. P., de Felipe, C., Salim, K., Meneses-Lorente, G., McAllister, G., and Guest, P. C. (2003) Multiplex proteomic analysis by two-dimensional differential in-gel electrophoresis, *Proteomics 3*, 1162–1171.
10. Marouga, R., David, S., and Hawkins, E. (2005) The development of the DIGE system: 2D fluorescence difference gel analysis technology, *Anal Bioanal Chem 382*, 669–678.
11. Yan, J. X., Devenish, A. T., Wait, R., Stone, T., Lewis, S., and Fowler, S. (2002) Fluorescence two-dimensional difference gel electrophoresis

and mass spectrometry based proteomic analysis of Escherichia coli, *Proteomics* **2**, 1682–1698.
12. Friedman, D. B., Hill, S., Keller, J. W., Merchant, N. B., Levy, S. E., Coffey, R. J., and Caprioli, R. M. (2004) Proteome analysis of human colon cancer by two-dimensional difference gel electrophoresis and mass spectrometry, *Proteomics* **4**, 793–811.
13. Gade, D., Thiermann, J., Markowsky, D., and Rabus, R. (2003) Evaluation of two-dimensional difference gel electrophoresis for protein profiling. Soluble proteins of the marine bacterium Pirellula sp. strain 1, *J Mol Microbiol Biotechnol* **5**, 240–251.
14. Bollag, D., Edelstein, SJ. (1991) Protein extraction, in *Protein Methods* (Wiley-Liss, N., Ed.).
15. Scopes, R. (1987) Making an extract, in *Protein purification: Principles and Practice* (Springer Verlag, N., Ed.) 2nd ed.
16. Pennington, S. R., Wilkins, M. R., Hochstrasser, D. F., and Dunn, M. J. (1997) Proteome analysis: from protein characterization to biological function, *Trends Cell Biol* **7**, 168–173.
17. Lenstra, J. A., and Bloemendal, H. (1983) Topography of the total protein population from cultured cells upon fractionation by chemical extractions, *Eur J Biochem* **135**, 413–423.
18. Toda, T., Ishijima, Y., Matsushita, H., Yoshida, M., and Kimura, N. (1994) Detection of thymopoietin-responsive proteins in nude mouse spleen cells by two-dimensional polyacrylamide gel electrophoresis and image processing, *Electrophoresis* **15**, 984–987.
19. Cull, M., and McHenry, C. S. (1990) Preparation of extracts from prokaryotes, *Methods Enzymol* **182**, 147–153.
20. Jazwinski, S. M. (1990) Preparation of extracts from yeast, *Methods Enzymol* **182**, 154–174.
21. Kawaguchi, S., and Kuramitsu, S. (1995) Separation of heat-stable proteins from Thermus thermophilus HB8 by two-dimensional electrophoresis, *Electrophoresis* **16**, 1060–1066.
22. Teixeira-Gomes, A. P., Cloeckaert, A., Bezard, G., Dubray, G., and Zygmunt, M. S. (1997) Mapping and identification of Brucella melitensis proteins by two-dimensional electrophoresis and microsequencing, *Electrophoresis* **18**, 156–162.
23. Gorg, A., Boguth, G., Obermaier, C., Posch, A., and Weiss, W. (1995) Two-dimensional polyacrylamide gel electrophoresis with immobilized pH gradients in the first dimension (IPG-Dalt): the state of the art and the controversy of vertical versus horizontal systems, *Electrophoresis* **16**, 1079–1086.
24. Gorg, A., Postel, W., Domscheit, A., and Gunther, S. (1988) Two-dimensional electrophoresis with immobilized pH gradients of leaf proteins from barley (Hordeum vulgare): method, reproducibility and genetic aspects, *Electrophoresis* **9**, 681–692.
25. Gorg, A., Postel, W., and Gunther, S. (1988) The current state of two-dimensional electrophoresis with immobilized pH gradients, *Electrophoresis* **9**, 531–546.
26. Dignam, J. D. (1990) Preparation of extracts from higher eukaryotes, *Methods Enzymol* **182**, 194–203.
27. Blomberg, A., Blomberg, L., Norbeck, J., Fey, S. J., Larsen, P. M., Larsen, M., Roepstorff, P., Degand, H., Boutry, M., Posch, A., and et al. (1995) Interlaboratory reproducibility of yeast protein patterns analyzed by immobilized pH gradient two-dimensional gel electrophoresis, *Electrophoresis* **16**, 1935–1945.

Chapter 12

Combination of Highly Efficient Hexapeptide Ligand Library-Based Sample Preparation with 2D DIGE for the Analysis of the Hidden Human Serum/Plasma Proteome

Sonja Hartwig and Stefan Lehr

Abstract

Blood serum/plasma samples provide the major source to identify diagnostically relevant or treatment response-related proteins. But its complexity and the enormous dynamic range hinders comprehensive analysis. Therefore, more detailed analysis, in particular, of low-abundant peptides/proteins requires extensive pre-fractionation, which frequently influences the native composition and may lead to a loss of potentially important information. In order to overcome these barriers, we describe an innovative sample preparation tool utilizing combinatorial hexapeptide ligand libraries to efficiently enrich low-abundance proteins with a simple protocol. In contrast to the most common approaches based on the immunodepletion of antibody-targeted high-abundance proteins, this technology concentrates low-abundance proteins and concurrently reduces the high-abundance species. Thus, the dynamic range is compressed, and low-abundance proteins become more easily detectable. We show how this sample preparation technique can be easily combined with 2D DIGE analysis to enable more comprehensive and quantitative profiling of complex biological samples.

Key words: Serum profiling, Sample pre-fractionation, Sample preparation, Low-abundant proteins, Hexapeptide library beads, Combinatorial ligand libraries (CLL), Clinical proteomics, Biomarker discovery

1. Introduction

Obesity, type 2 diabetes, and the cardio-metabolic syndrome are epidemic diseases that represent a major threat to human health, meaning a serious burden for our society. To understand the factors causing such multifactorial diseases is one of the major driving forces of biomedical research. In this research field, quantitative proteomic technology gives the opportunity to identify novel biomarker enabling early detection, diagnosis, and monitoring of disease progression.

In this context, blood still serves as the most prominent human specimen for all kinds of analyses, even though it is characterized by an enormous diversity of proteins covering a concentration range of at least 10 orders of magnitude (1). The patients' good acceptance of phlebotomy and its nearly unlimited availability tempts one to underestimate the difficulties accompanied with its special composition.

Discovery and development of disease-related biomarkers, which are defined as a "characteristic that is objectively measured and evaluated as an indicator of normal biologic processes, pathogenic processes, or pharmacologic responses to a therapeutic intervention" (2), reveal a huge potential for substantial improvement in the early diagnosis and prognosis of diseases. Nevertheless, the complex and exhausting discovery and validation processes hinder the availability of novel biomarkers for the implementation into clinical practice, which is reflected by the very low number of FDA-approved biomarkers over the last couple of years. In view of these complex challenges, continuous improvements in study design, sample quality, reproducibility, as well as standardization of sample processing and harmonized data analysis are needed to provide the indispensable basis for driving biomarker usage into clinical practice. Especially in face of the complexity of the workflow and the enormous biological variability, it appears that an extraordinary diligence in each step of the analysis is of great importance. This begins with the first experimental step, i.e., sample collection and preparation, which is frequently underestimated. In this context, the close collaboration between the different scientific disciplines and the development of standard operation procedures is of particular importance.

To pave the way for the identification of novel disease-dependent marker proteins from blood, extensive fractionation procedures are often enforced. Several approaches have been developed (for an overview see ref. (3)), reducing sample complexity or the amount of high-abundant proteins in order to overcome suppression of the low-abundant ones, e.g., tissue leakage proteins. However, some methods dilute the sample, lack specificity, or remove proteins of interest in an uncontrolled way eventually leading to a loss of potentially important diagnostic information (4). Here we introduce a valuable alternative "single-step" sample preparation approach utilizing bead-based combinatorial hexapeptide ligand libraries, being compatible with a broad spectrum of profiling techniques, including all kinds of gel electrophoresis and also mass spectrometry-based approaches like SELDI or LC-MS. Hexapeptides or other short sequences of amino acids offer non-covalent binding sites for proteins due to the physicochemical properties being defined by the type and sequence of the amino acids (for an overview see ref. (5)). If a hexapeptide library is coupled to carrier beads and exposed to complex protein mixtures, sufficient diversity is given to bind potentially all proteins within the sample. As the amount of each specific hexapeptide is limited in a given amount of bead volume, high-abundant proteins saturating the beads will be reduced, whereas low-abundance proteins will be quantitatively trapped. Although the exact binding mechanism of complex samples to a hexapeptide library is not entirely understood (6), this technology provides a powerful tool to get access to the low-abundant protein fraction. Metaphorically speaking, this approach eases the problem of finding a needle in a haystack by removing most of the hay (high-abundant proteins) and attracting the needle (low-abundance protein) at the same time. In addition, this sample preparation method fulfills two important prerequisites for quantitative proteomics: it is highly reproducible (7), and the proportionality of the initial sample composition concerning the low-abundant proteins is preserved during the fractionation procedure (8). Therefore, this sample preparation approach can be combined in an ideal setting with 2D DIGE to allow valid and sensitive differential profiling of low-abundance proteins.

2. Materials

2.1. Serum Samples

Human blood samples were provided from healthy volunteers. The study was conducted in accordance with ethical guidelines of the Heinrich-Heine-University of Duesseldorf. Sample collection was exclusively performed in 2.7-mL S-Monovettes® (Sarstedt, Nuembrecht, Germany) according to a standard operation procedure (see Note 1).

2.2. Hexapeptide Library Sample Preparation

1. Prepacked spin-columns filled with hexapeptide-coupled beads for different starting protein amounts as well as bulk beads for customized applications are available under the commercial trademark ProteoMiner™ (Bio-Rad Laboratories, Hercules, CA, USA). In this study, prepacked columns containing 500 μL of bead slurry (20% beads, 20% (v/v) aqueous ethanol, 0.05% (w/v) sodium azide) were used (see Notes 2 and 3).

2. Deionized water.

3. Wash buffer: 150 mM NaCl, 10 mM NaH_2PO_4, pH 7.4.

4. 2D DIGE lysis buffer: 25 mM Tris-base, 4% (w/v) CHAPS, 7 M urea, and 2 M thiourea supplemented with 1% (v/v) phosphatase inhibitor 1 and 2 (Sigma Aldrich, Taufkirchen, Germany) and 1 tablet (per 50 mL) Complete™ protease inhibitor cocktail (Roche, Mannheim, Germany). Store at −20°C (see Note 4).

5. Advanced Protein Assay (Cytoskeleton, Distributor: TeBu-bio, Offenbach, Germany).

2.3. Protein Labeling (Minimal Labeling with CyDyes)

1. CyDyes: Cy2, Cy3, and Cy5; DIGE minimal labeling kit (GE Healthcare, Freiburg, Germany). These dyes are shipped lyophilized and should be stored under light protection at −20°C.
2. Anhydrous dimethylformamide (DMF; see Note 5).
3. 10 mM Lysine solution.

2.4. Isoelectric Focusing and Large-Format SDS PAGE

1. IPG strips (linear pH 4–7, GE Healthcare).
2. IPG buffer corresponding to the used pH range (pH 4–7, GE Healthcare).
3. Rehydration buffer: 2% (w/v) CHAPS, 0.4% (w/v) DTT, 7 M urea, 2 M thiourea, and 0.5% (v/v) corresponding IPG buffer (see Note 4).
4. 2D DIGE lysis buffer: 25 mM Tris, 4% (w/v) CHAPS, 7 M urea, and 2 M thiourea.
5. Equilibration buffer with DTT: 50 mM Tris–HCl (pH 8.8), 6 M urea, 30% (v/v) glycerol, 2% (w/v) SDS, and 0.5% (w/v) DTT (see Note 6).
6. Equilibration buffer with iodoacetamide: 50 mM Tris–HCl (pH 8.8), 6 M urea, 30% (v/v) glycerol, 2% (w/v) SDS, and 4.5% (w/v) iodoacetamide (see Note 6).
7. Polyacrylamide gels with corresponding buffer system (in our case large-format 12.5%-SDS-PAGE gels with 50 mM Tris/192 mM glycine buffer system; see Note 7).

2.5. Imaging and Image Analysis

1. Laser scanner with 488 nm-, 532 nm- and 633 nm-excitation wavelength and a resolution of at least 100 µm is essential. If 2D DIGE gels are scanned between the glass plates, a confocal optic is recommended for the scanner (specifications are fulfilled by the Typhoon 9400; GE Healthcare). Alternatively, CCD-based camera systems with sufficient resolution and multiplexing capabilities can be used.
2. Lint-free paper (see Note 8).
3. 2DE image analysis software package compatible with DIGE multi-channel images and the corresponding scanner file format (e.g., DeCyder, Proteomweaver).

3. Methods

Here we describe a protocol for reproducible and easy-to-use "one-step" sample preparation of blood samples utilizing hexapeptide ligand libraries. Due to its proven applicability for quantitative proteomic approaches, it is obvious to combine it as front-end sample preparation for DIGE profiling, which is widely accepted

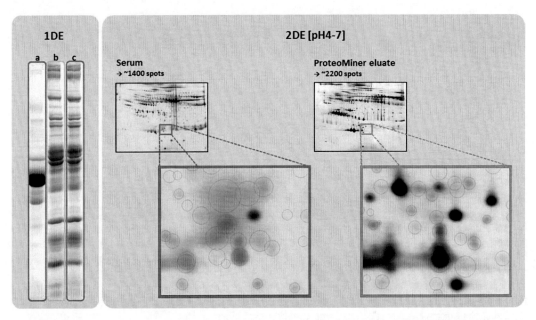

Fig. 1. Fractionation of human serum samples. 1DE (*left panel*): 10 μg of native human serum (*lane a*) and 10 μg of fractionated serum eluted either with acetic buffer (*lane b*) or with 2D DIGE lysis buffer (*lane c*) were separated on a 10%-SDS-PAGE gel. The proteins were stained with a ruthenium-based fluorescent dye, and protein patterns were visualized by laser scanning. In contrast to the native serum sample displaying more or less one albumin band after fractionation, a rich band pattern covering the whole separation range is detectable. No significant difference is detectable between the recommended acetic and 2D DIGE lysis buffer elution. 2DE (*right panel*): 150 μg of native human serum (*left*) and 150 μg of fractionated serum eluted with 2D DIGE lysis buffer (*right*) were separated in the first dimension by isoelectric focusing (IEF) (pH 4–7) and in the second dimension on a 12%-SDS-PAGE gel. The proteins were stained with a ruthenium-based fluorescent dye, and protein patterns were visualized by laser scanning. Image analysis was performed using Proteomweaver™ software. Fractionation of the serum samples resulted in a substantial increase of detectable protein spots. The spot count rose from approximately 1,400 to 2,200 which equates to a 60% increase.

for reproducible sensitive protein profiling. To minimize sample loss and experimental error between both methods, the 2D DIGE lysis buffer was selected for elution of captured proteins. This enables direct and effective analysis of protein populations that are usually masked by high-abundance proteins in the native sample (see Fig. 1). In addition, column-to-column reproducibility was tested to evaluate potential variability being introduced by the experimental procedure (see Fig. 2).

3.1. Serum Samples

1. Blood samples were exclusively collected in 2.7-mL S-Monovettes® (Sarstedt). After sample collection, the S-Monovettes® were incubated at room temperature (RT; 25°C) for 30 min.

2. Incubated samples were centrifuged at $1,400 \times g$ for 10 min and the supernatant was collected using a polyethylene transfer pipette.

3. Finally, samples were aliquoted (1.2 mL) and immediately transferred to −80°C for long-term storage.

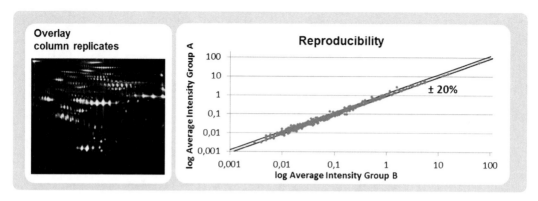

Fig. 2. DIGE analysis of fractionated serum samples to test column-to-column reproducibility. A reference serum sample was divided into three equal aliquots (1 mL), and each was separated on a different ProteoMiner™ column. Captured proteins were eluted using 2D DIGE lysis buffer; proteins were labeled with CyDyes and separated in the first dimension by IEF (pH 4–7) and in the second dimension on a 12%-SDS-PAGE gel. Protein patterns were visualized by laser scanning, and image analysis of the resulting nine images (three different columns crosswise labeled with Cy3 and Cy5 and internal standard labeled with Cy2) was performed using Proteomweaver™ software. Comparing the spot profiles from the three different columns reveals high reproducibility between the different replicates with CVs less than 20%. *Blue dots* in the scatter plot indicate intensity of detected protein spots in each group.

3.2. Hexapeptide Library Sample Preparation

1. Serum samples were thawed at RT (be sure that samples are fully thawed) and centrifuged for 10 min at $10,000 \times g$ at 4°C to remove sample particulates. Only the supernatant was used for the subsequent steps (see Note 9).

2. Top and bottom caps from the ProteoMiner™ columns were removed and columns centrifuged for 2 min at $1,000 \times g$. The flow-through was discarded. All subsequent centrifugations were for 2 min at $1,000 \times g$.

3. After addition of 1 mL of deionized water and replacement of the caps, the columns were rotated end-to-end for 5 min. Caps were removed, columns centrifuged again, and the flow-through was discarded. The centrifugation step was repeated once.

4. After addition of 1 mL of wash buffer and replacement of the caps, the columns were rotated end-to-end for 5 min. Caps were removed, columns centrifuged, and the flow-through was discarded. The centrifugation step was repeated once.

5. After addition of 1 mL of sample (serum) and replacement of the caps, the columns were rotated end-to-end for 2 h at room temperature. Caps were removed, columns centrifuged, and the flow-through was discarded. The centrifugation step was repeated once.

6. After addition of 1 mL of wash buffer and replacement of the caps, the columns were rotated end-to-end for 5 min. Caps were removed, columns centrifuged, and the flow-through was discarded. The centrifugation step was repeated once and the entire wash step was repeated 2–3 times.

7. After addition of 1 mL of deionized water and replacement of the caps, the columns were rotated end-to-end for 1 min. Caps were removed, columns centrifuged again, and the flow-through was discarded.

8. After addition of 100 μL of 2D DIGE lysis buffer (or rehydrated elution reagent; see Note 10), the columns were vortexed several times over a period of 15 min. The columns were placed in clean collection tubes and centrifuged for 2 min. The entire elution step was repeated twice and the eluates were pooled.

9. For protein quantification, the Cytoskeleton Advanced Protein Assay was used according to the manufacturer's instructions (see Note 11). Hexapeptide library-treated samples were processed immediately or stored at −20°C for later use (long-term storage at −80°C).

3.3. Protein Labeling (Minimal Labeling with CyDyes)

The protein labeling with cyanine dyes from GE Healthcare was performed according to the protocol provided by the manufacturer.

1. Briefly, the lyophilized dyes were resolved in DMF to a stock solution of 1 nM dye (each). These stock solutions can be stored up to 2 months at −20°C.

2. The stock solution was diluted with DMF to a final working concentration of 400 pM (see Note 12).

3. To eliminate dye-specific differences, a "dye swap" was applied, i.e., each sample was labeled crosswise with Cy3 and Cy5. In addition, for better statistical analysis, a technical replicate per sample was used.

4. 50 μg of Proteins/gel of each sample was labeled with 400 pmol of dye (Cy3 or Cy5). A pool of all samples (50 μg/gel) was labeled with 400 pmol of Cy2 to function as internal standard on each gel.

5. The label-reaction was accomplished for 30 min on ice and stopped with 1 μL of 10 mM lysine.

6. Labeled samples were directly used for isoelectric focusing (IEF) or stored up to 1 week at −20°C.

3.4. Isoelectric Focusing and Large-Format SDS PAGE

1. IPG strips were rehydrated in rehydration buffer in a rehydration tray overnight (see Note 4).

2. Labeled samples were combined (50 μg each labeled with Cy3, Cy5, and Cy2 per gel).

3. Prior to IEF, each labeled sample mix has to be made up to 100 μL with 2D DIGE lysis buffer, DTT (final concentration of 1% (w/v)), and the IPG buffer for the selected pH range (final concentration of 1% (v/v)).

4. We achieved good IEF results by using a Multiphor II electrophoresis unit and 24-cm IPG strips (pH 4–7, linear).

Samples were applied adjacent to the acidic end of the IPG strips by cup loading. IEF conditions were as follows: in a 1-h gradient up to 300 V, 1 h at 300 V, followed by a gradient of ~10 V/min up to 3,500 V, and finally 21 h and 26 min at 3,500 V (20°C). Total running time: ~28 h and 30 min (see Note 13).

5. Focused IPG strips can be stored at −20°C or directly applied to the second dimension.

6. Prior to second-dimension separation, IPG strips were equilibrated for 15 min in equilibration buffer.

7. Subsequently, IPG strips were re-equilibrated for 15 min in the same buffer containing 4.5% (w/v) iodoacetamide instead of DTT (see Note 6).

8. For protein separation by size in the second dimension, 12.5% polyacrylamide gels containing 5% glycerol combined with a Tris/glycine buffer system were used (thickness 1.0 mm). Electrophoresis of up to 12 large-format gels (24 × 18 cm) was conducted simultaneously at 3 W/gel and 20°C on an Ettan Dalt 12 system. The run was stopped when the blue dye front reached the end of the gels. Total run time: 16–19 h (see Note 7).

3.5. Imaging and Image Analysis

1. Closed glass plates were rinsed with double-distilled water, cleaned, and dried with lint-free paper. Gels were scanned directly between the glass plates using a Typhoon 9400 (GE Healthcare) laser scanner (see Note 14).

2. A resolution of 100 μm was sufficient for good quality results: photomultiplier tube settings of 500 V (see Note 15).

3. The Cy2 pattern was scanned at a wavelength of 488 nm (blue laser), Cy3 at 532 nm (green laser), and Cy5 at 633 nm (red laser).

4. Image analysis and evaluation of protein spot patterns, abundance, and statistics was accomplished automatically using Proteomweaver™ software (Bio-Rad). The gel images derived from identical sample types were grouped for comparison (see Note 16).

4. Notes

1. Sample quality is one important factor and significantly contributes to success or failure of biomarker discovery projects. In case of blood samples, special attention has to be paid to the collection tubes used (identical and appropriate tube types for the entire study), mix ratios (fill to the specified volume), and the stability of incubation conditions (use constant times,

temperature, and mixing/centrifugation conditions) in order to provide the required reproducibility.

2. In the experimental standard configuration, ProteoMiner™ columns (high-capacity kit) are designed to process 1 mL of sample. Although it might be challenging to gather these amounts of serum or plasma, we believe that at least this sample amount is needed to access low-abundance proteins for the following reason. Due to the extraordinary dynamic range (>10 orders of magnitude) of plasma/serum (1), it is almost impossible to discover potential biomarkers with concentrations lower than 10^3 pg/mL (e.g., tissue leakage proteins or cytokines) from unfractionated serum samples. Even if rather sensitive technologies are used at least 10 fmol of a single protein species is required. As a consequence, this protein amount has to be present in the starting volume. Typical concentrations of biomarkers discovered so far are in the range of 10^3 pg/mL (e.g., prostate specific antigen). Assuming a molecular weight of 25 kDa their detection requires at least 1 mL of whole plasma/serum to supply a sufficient concentration (1 ng). Despite this calculation, the hexapeptide material can easily be up- or downscaled to the available sample amount and concentration. For small sample amounts, a low-capacity kit for 200 μL serum/plasma is available and in addition dry bulk beads are offered for customized approaches. Recently, it has been shown that proteins of very low abundance could be identified after loading large volumes (>1,000 mL) of CSF sample (<0.5 mg/mL) to the hexapeptide library. The information revealed from this proteome mapping approach could be used to increase the quantity and quality of data revealed from small-scale experiments using as little as <1 mg CSF loaded to reduced volumes of hexapeptide beads (9).

3. Beside its use for protein enrichment, hexapeptide library beads can also be used for a final "polishing" step in the process of protein purification. Minor protein impurities are trapped by the beads, whereas the high-abundant proteins can be recaptured from the flow-through. In addition, this strategy allows better characterization of trace amounts of protein impurities by analyzing the protein fraction removed by the bead particles (10, 11).

4. To enhance efficiency and reproducibility, batches of 2D DIGE lysis and rehydration buffers should be produced and stored in appropriate aliquots (1–2 mL) at −80°C. Required amounts of DTT and IPG buffer have to be added just before use. To prevent decomposing, one should avoid heating any urea- or thiourea-containing buffer.

5. Always use the highest quality for anhydrous DMF to reconstitute the CyDyes. Ideally, include a 4-Å molecular sieve inside

the DMF bottle to trap any water and discard the DMF when the molecular sieve becomes viscous. Protect the dyes and the labeled samples from light exposure.

6. It is advantageous to store appropriate volumes of the equilibration buffer stock solution (50 mM Tris–HCl (pH 8.8), 6 M urea, 30% (v/v) glycerol, and 2% (w/v) SDS) at −20°C. Required volumes of iodoacetamide or DTT should be added just prior to use.

7. To maintain reproducibility especially for large-format 2D gels (20 × 26 cm), you have to pay increased attention to the casting conditions. For consistent results, we strongly recommend to develop a standard protocol defining all factors like polymerization time and temperature, high-quality chemicals, etc. Alternatively, precast gels are available from different suppliers (Gelcompany, GE Healthcare, etc.) including options with low-fluorescent plastic backing required for sensitive DIGE profiling.

8. When gels are scanned directly between glass plates, low-fluorescent glass plates should be used for gel casting to achieve highest sensitivity (see also Note 7). Best scanning results can be achieved with clean "gel packages"; rinse the plates with water and wipe them dry using lint-free paper. Reduced background will significantly improve and facilitate software-based pattern analysis.

9. For best results, samples should be free of precipitate as suspended components may coat the active bead surface. Therefore, in order to clarify the sample prior to ProteoMiner™ treatment, a centrifugation step ($10,000 \times g$ for 10 min) should be included, thus improving the capture efficiency of the beads as well as the reproducibility. Do not use hemolytic samples (red color). Please note that for the high-capacity columns the protein concentration has to be above 50 mg/mL which is usually the case for typical serum or plasma samples.

10. In most publications captured, proteins are eluted using acetic buffer conditions (recommended by the manufacturer). Although this elution procedure can easily be combined with other profiling techniques like SELDI, its use for 2D electrophoresis approaches like DIGE needs accurate pH adjustment or precipitation of the eluate prior to the IEF step. To circumvent this additional treatment step we established an elution protocol using 2D DIGE lysis buffer, which in our hands works as well as the acetic elution (see Fig. 1). Interestingly, our procedure levels the pH in almost all cases automatically to 8.5 which is exactly the pH needed for the DIGE labeling reaction making this protocol a very convenient alternative to the acetic elution. Nevertheless, we recommend checking the pH of

every diluted sample as little pH adjustment might become necessary in some rare cases.

11. This protein assay used tolerates many chaotropes and detergents including urea and CHAPS.

12. It is essential to reconstitute the CyDyes with anhydrous DMF (see Note 5) to a 1 nM-stock solution. If not used completely, aliquots (make sure that the vessels are sealed properly) can be stored at −20°C for several months without losing performance. In our hands, a working solution of 400 pM (prepare fresh before starting labeling) is sufficient for almost all cases. Tip: in the literature, some alternative dyes for protein labeling are described which may help to reduce the costs significantly (12, 13).

13. In our hands, we achieve best results for IEF with the classic Multiphor system (GE Healthcare) and a maximum voltage of 3,500 V instead of using state-of-the-art systems (e.g., IPGphor, protean IEF cell) with voltages of up to 10,000 V. We use an extended starting time at low voltages to reduce impurities that interfere with higher voltage focusing. As a consequence, our standard protocols for most pH ranges take 21–30 h, and usually, we prefer to run IEF overnight. Evaluation of optimized running conditions should be done for each different sample types (e.g., cell lysates, organelles, plasma).

14. If scanning between the glass plate assemblies is preferred, we recommend the use of a laser scanning system with confocal optics to minimize artificial background signals arising from the glass plate surface.

15. 500 V should be taken as initial value which has to be adjusted specifically for each wavelength to achieve reproducible results. Make sure that intensities in all three channels do not reach saturation.

16. Usually a DIGE approach is based on the comparison of at least two different samples. To minimize the impact of technical variation on experimental results, "dye swap" is recommended. Therefore, each sample has to be labeled with both dyes, i.e., Cy3 and Cy5. The differently labeled samples then have to be separated (together with a second sample and the internal standard) on different gels. In order to obtain statistically meaningful results, it is essential to analyze at least three different biological (not technical) replicates of each sample. Otherwise only experimental variability may be monitored and not the biological changes. Remember that the internal standard is composed of equal amounts of all the samples used in the experiment and has to be calculated from the number of samples and total number of gels to be run within the experiment. Tip: due to the immense biological variability, for profiling of

clinical human material (e.g., biopsies), careful calculation of the sample numbers needed (i.e., power calculation to determine the study scale) is essential to achieve statistically significant results.

References

1. Anderson NL, Anderson NG (2002) The human plasma proteome: history, character, and diagnostic prospects. Mol Cell Proteomics. 1(11):845–67.
2. Biomarkers Definitions Working Group (2001) Bimomarkers and surrogate endpoints: preferred definitions and conceptual framework. Clin. Pharmacol. Ther. 69, 89–95.
3. Bandow JE (2010) Comparison of protein enrichment strategies for proteome analysis of plasma Proteomics 10(7):1416–25.
4. Granger J, Siddiqui J, Copeland S et al (2005) Albumin depletion of human plasma also removes low abundance proteins including the cytokines. Proteomics 5(18):4713–8.
5. Guerrier L, Righetti PG, Boschetti E. (2008) Reduction of dynamic protein concentration range of biological extracts for the discovery of low-abundance proteins by means of hexapeptide ligand library. Nat Protoc. 3(5):883–90.
6. Fasoli E, Farinazzo A, Sun CJ et al (2010) Interaction among proteins and peptide libraries in proteome analysis: pH involvement for a larger capture of species. J Proteomics 73(4):733–42.
7. Dwivedi RC, Krokhin OV, Cortens JP et al. (2010) Assessment of the reproducibility of random hexapeptide peptide library-based protein normalization. J Proteome Res. 9(2):1144–9.
8. Hartwig S, Czibere A, Kotzka J et al (2009) Combinatorial hexapeptide ligand libraries (ProteoMiner): an innovative fractionation tool for differential quantitative clinical proteomics. Arch Physiol Biochem. 115(3):155–60.
9. Sjödin MO, Bergquist J, Wetterhall M (2010) Mining ventricular cerebrospinal fluid from patients with traumatic brain injury using hexapeptide ligand libraries to search for trauma biomarkers. J Chromatogr B Analyt Technol Biomed Life Sci. 878(22):2003–12.
10. Righetti PG, Boschetti E (2007) Sherlock Holmes and the proteome-a detective story. FEBS J. 274(4):897–905.
11. Antonioli P, Fortis F, Guerrier L et al (2007) Capturing and amplifying impurities from purified recombinant monoclonal antibodies via peptide library beads: a proteomic study. Proteomics. 7(10):1624–33.
12. Kaber G, Vormbrock I, Hartwig S et al (2009) Validation of self-made fluorescent cyanine dyes for 2-D gel-based multifluorescence protein analysis. *Arch. Physiol. Biochem.* 115:252–8.
13. Jung ME, Kim WJ. (2006). Practical syntheses of dyes for difference gel electrophoresis. *Bioorg Med Chem* 14:92–7.

Chapter 13

2D DIGE Analysis of Serum After Fractionation by ProteoMiner™ Beads

Cynthia Liang, Gek San Tan, and Maxey C.M. Chung

Abstract

Serum is a popular biofluid used for many protein biomarker discovery projects since the collection and processing of serum/plasma is relatively noninvasive and inexpensive. Unfortunately, the downstream analysis of serum/plasma is hampered severely by several high-abundant proteins which often interfere with the separation and detection of many of the proteins of lower abundance. Thus, a number of prefractionation methods have recently been developed with the view to reduce the dynamic range of these proteins. These include both dye- and immunoaffinity-based methods that are specifically designed to remove serum albumin. In this chapter, we describe an alternative method using ProteoMiner™ or Equalizer beads that is aimed at overcoming this problem in serum. This method uses a combinatorial library of hexapeptides bound to beads and works by binding proteins until saturation is reached. Thus, the high-abundant proteins will reach saturation quickly, while the lower-abundant proteins continue to bind. This results in a dramatic depletion of the most abundant proteins, with a concurrent concentration of the middle- to low-abundant proteins.

Key words: Serum, Equalizer beads, ProteoMiner™, 2D DIGE

1. Introduction

Serum or plasma is routinely used for disease diagnosis and for monitoring patient's health status. This is because blood circulates the whole body and any changes in protein expression in serum or plasma can reflect the pathophysiology of the body. As the collection and processing of serum/plasma is also relatively noninvasive and inexpensive, it is a popular biofluid used for many protein biomarker discovery projects. However, the downstream analysis of serum/plasma is hampered severely by several high-abundant proteins which often interfere with the separation and detection of many of the proteins of lower abundance. For example, serum or plasma is dominated by approximately 20 high-abundance proteins with concentrations as high as 20–50 mg/mL for serum albumin, as compared to concentrations as low as µg/mL or pg/mL for some known biomarkers (1). Presently, a number of prefractionation methods and devices have been developed with the view to reduce the abundance of these proteins. These include immunoaffinity-based methods that are specifically designed to remove serum albumin and in which the separation media are packed either in spin cartridges or liquid chromatography (LC) columns. However, there are several recent reports suggesting that potentially interesting protein candidates may be lost with the removal of albumin via this approach (2–7). ProteoMiner™ or Equalizer beads can be used as an alternative to address the problem of wide dynamic range of the proteins in serum and to overcome the loss of proteins by the removal of "total" albumin via immunodepletion methods. The ProteoMiner™ technology uses a combinatorial library of hexapeptides bound to beads. Due to the tremendous ligand diversity within the library, there is theoretically a ligand for every protein, antibody, and peptide present in the starting material (8, 9). The equalizer beads, as their name implies, work by binding proteins until saturation is reached. The high-abundant proteins will reach saturation quickly, while the lower-abundant proteins continue to bind. The excess unbound proteins will be washed away in the procedure. This results in a dramatic depletion of the most abundant proteins, with a concurrent concentration of the middle- to low-abundant proteins (8, 9).

In this chapter, we describe the preparation of ProteoMiner™-treated human serum for 2D DIGE analysis.

2. Materials

Prepare all solutions using ultrapure water and use analytical grade chemicals/reagents.

2.1. Equipment

1. Class II biohazard flow hood.
2. Rotary shaker.
3. Refrigerated centrifuge (with a centrifugation force of at least $12,000 \times g$).
4. Spectrophotometer.
5. IPGphor isoelectric focusing (IEF) unit (GE Healthcare, Princeton, NJ, USA).
6. Mini-PROTEAN 3 electrophoresis system (Bio-Rad, Hercules, CA, USA).

7. PROTEAN II xi electrophoresis unit (Bio-Rad).
8. Typhoon 9410 image scanner (GE Healthcare).
9. Orbital shaker/rocker.
10. Sonicator.

2.2. Sample Preparation

1. ProteoMiner™ small-capacity kit (Bio-Rad) (see Note 1).
2. 2D clean-up kit (GE Healthcare).
3. Coomassie plus protein assay reagent kit (Pierce, Rockford, IL, USA).

2.3. DIGE Labeling

1. CyDye™ DIGE fluors (minimal dyes): Cy2, Cy3, and Cy5 (GE Healthcare).
2. 99.8% Anhydrous *N,N*-dimethylformamide (DMF) (see Note 2).
3. pH indicator strips.
4. 10 mM L-lysine.
5. Cell lysis solution: 7 M urea, 2 M thiourea, 4% CHAPS, 30 mM Tris, pH being adjusted to 8.5 with HCl.
6. 1 M NaOH.

2.4. SDS Polyacrylamide Gel Electrophoresis (SDS PAGE)

1. 40% Acrylamide/bis solution (37.5:1 with 2.6% C) and *N,N,N′,N′*-tetramethylethylenediamine (TEMED) (Bio-Rad).
2. Separating buffer: 1.5 M Tris–HCl, pH 8.8. Store at 4°C.
3. Stacking buffer: 1 M Tris–HCl, pH 6.8. Store at 4°C.
4. 10% (w/v) SDS. Store at room temperature.
5. Ammonium persulfate (Bio-Rad): prepare a fresh 10% (w/v) solution in water.
6. Running buffer (10×): 25 mM Tris, 192 mM glycine, 0.1% (w/v) SDS. Store at room temperature.
7. Sample loading buffer (2×): 120 mM Tris–HCl pH 6.8, 20% (v/v) glycerol, 4% (w/v) SDS, 40 mM dithiothreitol (DTT) containing trace amount of bromophenol blue. To prepare this solution, add 10 μL of 2 M DTT to 490 μL of the buffer that had been prepared as a stock solution without the reducing agent just before use. After use, store the remaining aliquot at −20°C. The sample loading buffer without DTT can be stored in 490 μL aliquots at room temperature.
8. Water-saturated butanol (50:50 v/v).

2.5. First-Dimension IEF

1. Rehydration solution: 7 M urea, 2 M thiourea, 4% (w/v) CHAPS, with trace amount of bromophenol blue.
2. Immobilized pH gradient (IPG): 18-cm strips, 3–10 nonlinear (GE Healthcare).
3. IPG buffer: 3–10 nonlinear (GE Healthcare).

4. DTT: 2 M DTT in water. Store aliquots at –20°C.

5. Immobiline DryStrip cover fluid (GE Healthcare).

2.6. Second-Dimension SDS PAGE

1. Equilibration buffer: 6 M urea, 30% (v/v) glycerol, 50 mM Tris–HCl (pH 8.8), 2% (v/v) SDS, containing trace amount of bromophenol blue. Store at –20°C in aliquots of 10 mL.

2. 40% Acrylamide/bis solution (37.5:1 with 2.6% C) and TEMED (Bio-Rad).

3. 0.75% (w/v) agarose overlay: to make 100 mL of solution, use microwave oven to melt 0.75 g of low-melting agarose in 100 mL of 1× SDS-running buffer. Add trace amount of bromophenol blue and store in 2 mL aliquots at room temperature.

4. DTT.

5. Iodoacetamide (IAA).

2.7. Silver Staining

1. Fixing solution: 50% (v/v) methanol, 12% (v/v) acetic acid, and 0.05% (v/v) formalin.

2. Wash solution: 35% (v/v) ethanol in water.

3. Sensitizing solution: 0.02% (w/v) sodium thiosulfate in water.

4. Silver nitrate solution: 0.2% (w/v) silver nitrate and 0.076% (v/v) formalin.

5. Developing solution: 6% (w/v) sodium carbonate, 0.004% (w/v) sodium thiosulfate, and 0.05% (v/v) formalin.

6. Stop solution: 1.46% (w/v) sodium-EDTA in water.

2.8. Image Analysis

DeCyder™ 2D Differential Analysis Software.

---

3. Methods

All biological samples should be handled in a Class II biohazard hood. All waste generated should be decontaminated with 10% (v/v) bleach or autoclaved before discarding.

3.1. DIGE Experimental Design

Firstly, the experimental design is planned to determine the amount of sample required to be cleaned up for optimization and the actual DIGE experiment. Table 1 outlines the experimental design for labeling proteins for four sera samples. A total of 150 µg protein is loaded on each gel with a pooled sample as the internal standard. The internal standard in all the gels is composed of an equal aliquot of every sample in the entire DIGE experiment. The internal standard facilitates protein spots matching on different gels and allows relative protein abundance to be calculated. Dye swap is

Table 1
Experimental design for CyDye™ labeling of four serum samples with the incorporation of a pooled sample internal standard

	Cy2 (internal standard)	Cy3	Cy5
Gel 1	50 μg Pool	50 μg of Serum 1	50 μg of Serum 2
Gel 2	50 μg Pool	50 μg of Serum 3	50 μg of Serum 4
Gel 3	50 μg Pool	50 μg of Serum 2	50 μg of Serum 3
Gel 4	50 μg Pool	50 μg of Serum 4	50 μg of Serum 1
Gel 5	50 μg Pool	50 μg of Serum 1	50 μg of Serum 4
Gel 6	50 μg Pool	50 μg of Serum 3	50 μg of Serum 2
Total	300 μg (from a pool of 4×100 μg, each from serum 1 to 4)		

included to reduce bias in fluorescence or labeling from the different CyDye™. Triplicate gels are conducted to increase the statistical confidence of the experiment.

3.2. Fractionation of Sera Using ProteoMiner™

All steps in this section are conducted at room temperature. Incubation and washes on the rotary shaker are all conducted at 10 rpm.

3.2.1. Column Preparation

1. Remove the top cap of the spin column and snap off the bottom cap prior to use. Invert this bottom cap and subsequently use it as a plug for the remaining part of the experiment.
2. Place the spin column in a capless collection tube and centrifuge at $1,000 \times g$ for 1 min to remove the storage solution. Discard the storage solution.
3. Replace the bottom cap and add 200 μL of wash buffer to the beads. Replace the top cap and gently tap the column to mix.
4. Incubate the mixture in the spin column on a rotary shaker for 5 min. Place the spin column into a secondary tube that can fit into the rotary shaker if necessary.
5. Remove the top cap and then the bottom cap (see Note 3), place the spin column back to the same capless collection tube and replace the top cap. Centrifuge at $1,000 \times g$ for 1 min. Discard the solution.
6. Repeat steps 3–5.
7. Replace the bottom cap on spin column. The column now contains 20 μL of settled bead volume that is ready for sample binding.

3.2.2. Sample Binding

1. Centrifuge the serum sample at $10,000 \times g$ for 10 min if necessary. This is to ensure that the sample is free of precipitate or aggregate proteins and/or lipids, which will affect the binding of proteins to the beads.
2. Add 200 µL of serum to the column. Replace the top cap and tap the column to mix. Place the spin column on a rotary shaker and incubate for 2 h for sample binding to take effect (see Note 4).

3.2.3. Sample Wash

1. Place the spin column in a capless collection tube. Centrifuge at $1,000 \times g$ for 1 min. Discard the solution containing mainly the unbound proteins.
2. Replace the bottom cap and add 200 µL of wash buffer to the beads. Replace the top cap and tap the column to mix.
3. Wash the beads in the spin column by rotating on a rotary shaker for 5 min.
4. Place the spin column back to the capless collection tube. Centrifuge at $1,000 \times g$ for 1 min and discard the solution collected in the tube.
5. Repeat steps 2–4 twice.

3.2.4. Sample Elution

1. Replace the bottom cap on the spin column. Add 200 µL of deionized water. Replace the top cap and tap the column to mix.
2. Place on the rotary shaker to shake for 1 min.
3. Place the spin column in a capless collection tube. Centrifuge at $1,000 \times g$ for 1 min and discard the solution collected in the tube.
4. Repeat steps 1–3.
5. Replace and ensure that the bottom cap is tightly attached onto the spin column. Add 20 µL of rehydrated elution reagent. Replace the top cap and tap the column to mix.
6. Incubate the column on the rotary shaker for 15 min, with several rounds of light vortexing.
7. Place the spin column in a clean collection tube. Centrifuge at $1,000 \times g$ for 1 min. Retain the eluate.
8. Repeat steps 5–7 twice.
9. Repeat steps 5–7 with a final elution volume of 60 µL of rehydrated elution reagent.
10. Eluates may be pooled or analyzed individually (see Fig. 1). Store sample at −80°C or proceed with sample precipitation using 2D Clean-up kit.

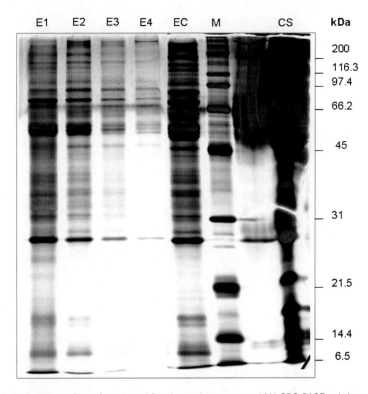

Fig. 1. Protein profiles of crude and fractionated serum on a 13% SDS-PAGE gel. *Lane M* is molecular weight marker. *Lanes E1–E4* are fractions derived from sequential elution of bound proteins from the ProteoMiner™ column using 8 M urea, 2% (w/v) CHAPS, and 5% acetic acid. *Lane EC* is a combined pool of E1–E4. *Lane CS* is loaded with 1 μL of unfractionated serum, while *lanes E1–E3* are loaded with 0.2 μL, and *lane E4* is loaded with 0.6 μL of the sequential eluates.

3.3. Sample Precipitation Using 2D Clean-Up Kit

All steps should be carried out on ice (4–5°C) at all times unless otherwise stated according to the manufacturer's manual.

1. Prechill the wash buffer prior to use for at least 1 h at –20°C.
2. Transfer the ProteoMiner™ fractionated sample into a 1.5-mL microcentrifuge tube (see Note 5). Ensure that the tube capacity is at least 8 times greater than the sample volume.
3. For each sample, add 3 volumes of precipitant to 1 volume of sample. Vortex and incubate on ice for 15 min.
4. Then, add 3 volumes of coprecipitant to the mixture and vortex briefly to mix.
5. Centrifuge at a maximum centrifugal force of at least $12,000 \times g$ for 5 min.
6. Remove the tube immediately and discard as much of the supernatant as possible using a pipette. Avoid disturbing the pellet.
7. Next, carefully reposition the tube in the centrifuge as before and pulse-spin it to bring down the remaining liquid to the bottom of the tube. Remove all the liquid from the tube.

8. Without disturbing the pellet, layer the coprecipitant onto the pellet. The volume of coprecipitant used is 3–4 times the volume of the pellet. Leave the tube on ice for 5 min.

9. Carefully place the tube in the centrifuge as before and centrifuge at the maximum centrifugal force for 5 min.

10. Using a pipette, remove and discard the wash solution without disturbing the pellet.

11. Next, add sufficient volume of deionized water to cover the pellet. Vortex to mix and ensure the pellet is dispersed well.

12. Following this, add 1 mL of the prechilled wash buffer and 5 µL of the wash additive. Vortex (hard) until the pellet is fully dispersed.

13. Incubate the tubes at −20°C for 2 h. Vortex for 30 s once every 15 min during this step of incubation (see Note 6).

14. Centrifuge the tube at a maximum speed for 5 min.

15. Carefully remove and discard the supernatant. A white pellet should now be visible. Allow the pellet to air-dry briefly (for no more than 5 min). Note that overdrying the pellet can result in difficulty in resuspension of the pellet.

16. Finally, resuspend the pellet in 30 µL of 2DE rehydration solution (see Subheading 2.5) but without bromophenol blue. If the resolubilization of the pellet is too slow (perhaps due to a large pellet or a pellet being overdried), sonication can help to speed up the resolubilization. Store the reconstituted sample at −80°C.

3.4. Protein Quantification

Protein concentrations are determined using the Coomassie plus protein assay kit based on the manufacturer's instructions with some modifications. Perform the assay at least in duplicates (see Note 7). Finally, measure the absorbance at 595 nm and determine the protein concentration from a standard curve.

3.5. CyDye™-Labeling Efficiency Check Prior to 2D DIGE

Always perform a preliminary 2D gel run for each sample. This will facilitate the optimization of the separation parameters for the 2D DIGE gel. It can also be used to check if there is any problem with the sample. If the sample amount is limited, conduct a 1D PAGE run to check that the calculated amount of protein to be used for the 2D DIGE gel for a given pair of sample is similar based on the staining intensity of the protein bands on the 1D gel. This is highly recommended before proceeding with the CyDye™ labeling optimization check as detailed in the next section.

3.5.1. pH Adjustment of Sample for CyDye™ Labeling

For successful labeling, the pH of the sample needs to be in the range of pH 8.0–9.0.

1. Check the pH of the sample by spotting 0.1–0.5 µL on a pH indicator strip (see Note 8).

2. Add cell lysis solution or sodium hydroxide (1 M) to adjust the pH of acidic samples to around pH 8.5 (see Note 9).

3. Recalculate the protein concentration of the pH-adjusted sample, if required.

3.5.2. Reconstitution of CyDye™

1. Transfer a small volume of DMF with a syringe and needle into a microfuge tube.

2. Take the vial of CyDye™ DIGE fluors from a −20°C freezer and leave it to warm up at room temperature for 5 min.

3. Reconstitute CyDye™ DIGE fluors with DMF to give a stock concentration of 1 nmol/μL. For example, add 25 μL of DMF to 25 nmol of fluor.

4. Replace the cap of the microfuge tube containing the dye and vortex vigorously.

5. Centrifuge briefly to bring the reconstituted CyDye™ contents down to the bottom of the tube. Store the remaining stock of CyDye™ at −20°C and protect from light as soon as possible.

6. Add 1 volume of CyDye™ stock solution to 1.5 volumes of DMF to make a diluted CyDye™ working solution (400 pmol/μL).

3.5.3. Labeling a Small Amount of Protein to Test the Labeling Efficiency

1. Prepare a 1-mm thick, 12.5% SDS-PAGE gel by mixing 1.6 mL of 40% acrylamide/bis solution with 1.25 mL of 1.5 M Tris–HCl pH 8.8 separating buffer, 50 μL of 10% SDS, 50 μL of 10% APS, 2.05 mL of water, and 2.5 μL of TEMED. Pour the gel solution into the mini-PROTEAN 3 gel-casting assembly and leave sufficient space for the stacking gel. Overlay with 200 μL of water-saturated butanol. Allow the gel to polymerize for at least 30 min.

2. Pour off the water-saturated butanol.

3. Prepare the stacking gel by mixing 200 μL of 40% acrylamide/bis solution with 252 μL of 1 M Tris–HCl-stacking buffer pH 6.8, 20 μL of 10% SDS, 20 μL of 10% APS, 1.51 mL of water, and 2 μL of TEMED.

4. Pipette in the stacking gel and insert the comb. Allow the gel to polymerize for 30 min.

5. Prepare the running buffer by diluting 100 mL of 10× running buffer with 900 mL of ultrapure cold water. Mix the content and leave the running buffer at 4°C.

6. Transfer 5 μg of sample into a microfuge tube.

7. Add 0.1 μL of diluted CyDye™ working solution to the tube (i.e., 5 μg of protein is labeled with 40 pmol of CyDye™).

8. Mix and centrifuge briefly. Leave on ice for 30 min in the dark.

9. Add 0.1 μL of 10 mM lysine to quench the reaction. Mix and centrifuge briefly.

10. Leave the sample on ice for another 10 min in the dark.
11. At least two different sample amounts were used for assessment of the labeling efficiency in SDS PAGE. We routinely use 3 and 1 µg of protein. Add an equal volume of 2× SDS sample loading buffer to the labeled protein and heat at 95°C for 5 min.
12. Run the sample on the 12.5% SDS-PAGE gel.
13. Scan the gel with the Typhoon imager.
14. Quantify the labeling of each sample using the ImageQuant software.
15. Generate a volume report to check that labeling efficiency is similar across all the samples with the same dilution and load. If labeling is comparable, proceed with the labeling of all samples for the first-dimension IEF (see Note 10).
16. If the labeling is not comparable, poststaining of the gel using a total protein stain such as silver or fluorescent stain is carried out to determine if it is due to differences in labeling efficiency or sample load inconsistency.

3.6. CyDye™ Labeling for 2D DIGE

1. Aliquot 50 µg of each sample into properly labeled microfuge tubes. To prepare the internal standard, an equal amount of all the samples in the experiment are mixed together to give a pooled sample (see Table 1) (see Note 11).
2. Incubate the samples with 1 µL of 400 pmol/µL CyDye™ on ice in the dark for 30 min.
3. Quench the reaction by adding 1 µL L-lysine with incubation on ice in the dark for 10 min.

3.7. First-Dimension Isoelectric Focusing

1. Mix 50 µg of Cy3-labeled sample and 50 µg of Cy5-labeled sample with 50 µg of Cy2-labeled internal standard to give a total of 150 µg protein load/gel (see Table 1).
2. Add rehydration buffer containing 7 M urea, 2 M thiourea, 4% w/v CHAPS, 20 mM DTT, and 0.5% (v/v) pH 3–10 NL IPG buffer, with trace amount of bromophenol blue, to a total volume of 350 µL.
3. The IPG strips are subjected to passive rehydration (remember—gel face down!) on the Immobiline DryStrip Reswelling Tray for 12–16 h at room temperature in the dark.
4. Transfer the rehydrated strips to the Ettan IPGphor ceramic cup loading strip holder (gel face up) for IEF on the IPGphor unit.
5. Place electrode pads moisted with deionized water on each end of the IPG strips for better contact between the gel and electrodes.
6. Cover the IPG strips with 4 mL of mineral oil.

7. IEF is then carried out to a total of 55,100 Vh with the following parameters: (a) 200 V, 200 Vh; (b) 500 V, 250 Vh; (c) 1,000 V, 500 Vh; (d) 1,000–8,000 V, 4,500 Vh; and (e) 8,000 V, 40,000 Vh. Voltage increases are on a stepwise basis for all steps except for step (d) which is on a linear gradient mode. IEF separation is conducted in the dark by covering the IPGphor unit with the black cover provided with the unit. The temperature is maintained at 20°C.

8. The strips can be used for the second dimensional separation immediately or they can be stored in capped glass tubes at −80°C until use.

3.8. Second-Dimension SDS PAGE

1. Prepare a 1-mm thick, 13% gel (18×20 cm) by mixing 12.35 mL of 40% acrylamide/bis solution with 9.5 mL of 1.5 M Tris–HCl pH 8.8 separating buffer, 380 μL of 10% SDS, 380 μL of 10% APS, 15.39 mL of water, and 19 μL of TEMED. Pour the gel solution into the PROTEAN II xi gel-casting assembly, leaving space for the insertion of the IPG strip (~0.5 cm from the top of the short glass plate). Overlay with 500 μL of water-saturated butanol. Allow the gel to polymerize for at least 6 h.

2. Equilibrate each of the IPG strips with 10 mL of equilibration buffer containing 1% (w/v) DTT for 15 min on a rocker.

3. Discard the DTT-containing equilibration buffer and equilibrate each of the strips with 10 mL of equilibration buffer containing 2.5% (w/v) of IAA for 15 min.

4. Briefly rinse the equilibrated strips with ultrapure water.

5. Place the IPG strip on the top of the 13% gel and seal with 0.75% agarose (see Note 12).

6. Separate the proteins in the second dimension using 15 mA/gel for 20 min, followed by 30 mA/gel at 10°C until the bromophenol blue dye front comes close to the bottom of the gel.

7. After electrophoresis, wrap the gels together with the glass plates undetached in plastic wraps to keep them moist and keep them in the dark.

3.9. Image Acquisition and Analysis

1. Gels with glass plate intact are rinsed with ultrapure water and wiped dry before placing onto the platen of the Typhoon 9410 scanner for image acquisition (see Note 13).

2. Optimal excitation/emission wavelengths for detection are Cy2: 488/520 nm, Cy3: 532/580 nm, and Cy5: 633/680 nm, respectively.

3. A preliminary scan at low resolution of 1,000 μm and the Photomultiplier tube (PMT) voltage being set at 500 V on each channel (Cy2, Cy3, and Cy5) are conducted to assess the overall profile of the 2D gel.

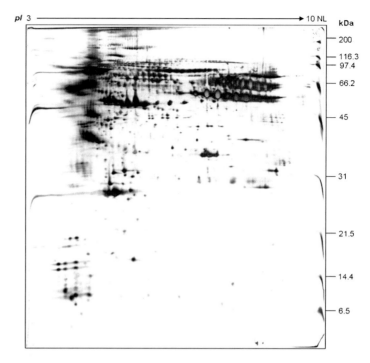

Fig. 2. Representative 2D image of ProteoMiner™ fractionated serum.

4. The PMT voltage for each channel is adjusted to ensure that the maximum pixel values of the three channels only differ within 5–10% and that the PMT voltages used do not saturate all spots of interest in the 2D gel (see Note 14).

5. Two gels can be scanned simultaneously once the PMT voltages are optimized. The images should be acquired at high resolution of 100 μm for image analysis.

6. The images files are kept with .gel extension.

7. After the gels are removed from the glass plates, they can be placed in fixing solution until use.

8. Scanned images are analyzed using the DeCyder™ 2D Differential Analysis Software version 6.5.

9. An example of a 2DE gel image for a ProteoMiner™-treated serum is shown in Fig. 2.

4. Notes

1. The ProteoMiner™ kit provides reagents enough for processing ten reactions per samples. The lyophilized elution reagent after reconstitution can only be stored up to 1 week at −20°C. It is recommended to use freshly prepared buffer when conducting

the experiment. Therefore, the following are the recipe for preparing the solutions:

Wash buffer (PBS buffer): 150 mM NaCl, 10 mM NaH_2PO_4, and pH 7.4.

Elution reagent: 8 M Urea, 2% (w/v) CHAPS.

Rehydration reagent: 5% (v/v) acetic acid.

2. DMF must be of high quality and anhydrous (specification: ≤0.005% H_2O, ≥99.8% pure). Once opened, DMF will start to degrade, producing amine compounds. Amines will react with the NHS ester fluor and reduce the concentration of dye available for protein labeling. Therefore, it is not recommended to use DMF that is open for more than 3 months. Store the opened DMF in a desiccator.

3. Removing the top cap prior to the removal of the bottom cap before placing the column onto the capless tube is necessary to relieve the pressure in the column. If the top cap is not removed before the bottom cap, the solution in the column may leak out immediately. This sequence in removal and replacing the column cap need to be observed throughout the ProteoMiner™ fractionation step to avoid sample loss.

4. For plasma samples, there may be clumping after 1 h of binding. Heparinized plasma is not compatible with ProteoMiner™.

5. Precipitate twice the amount of sample required for the DIGE experiment as there will be sample loss in the process.

6. The samples can be incubated from 30 min to 1 week at −20°C with minimal protein degradation or modification.

7. The BSA standard used in the assay has to be diluted in the same buffer as the sample being assayed. In this case, a diluted rehydration buffer was used to ensure buffer components are within the compatible concentration range as the protein assay kit.

8. Cut the pH indicator into narrower strips so that less volume of sample is consumed for pH testing.

9. Depending on the sample, usually an addition of 0.1–0.2 μL of 1 M NaOH is sufficient to adjust the pH of the sample to pH 8.5.

10. It is recommended that CyDye™ labeling of samples is conducted on the same day as testing the labeling efficiency. In our hands, we have experienced drifts in the pH of adjusted samples after a day of storage.

11. Label slightly more protein for the Cy2 internal standard than according to the intended number of gels (e.g., 10% more) to allow for variation in pipetting. This is to ensure that there is sufficient Cy2-labeled protein for all the replicate gels.

12. Trim off a small portion of the IPG strips at both ends to provide space for the addition of molecular weight markers if required. The molecular weight markers can be pipetted onto a small piece of filter paper and allowed to air-dry prior to introducing it onto the second dimension gel.
13. Two gels can be scanned simultaneously in a Typhoon variable mode imager at approximately 10 min/channel using 100-μm scan resolution. The remaining gels have to be protected from light and kept moist before image acquisition can be performed.
14. Optimization of PMT voltage can be done by scanning a smaller area at high resolution or the final resolution intended. In this way, we can save time and the pixel output is more representative of the final value than pixel output obtained from 1,000-μm scan resolution.

References

1. States DJ, Omenn GS, Blackwell TW *et al* (2006) Challenges in deriving high-confidence protein identifications from data gathered by a HUPO plasma proteome collaborative study. *Nat. Biotechnol.* 24:333–338.
2. Lowenthal MS, Mehta AI, Frogale *et al* (2005) Analysis of albumin-associated peptides and proteins from ovarian cancer patients. *Clin. Chem.* 51:1933–1945.
3. Zhou M, Lucas DA, Chan KC *et al* (2004) An investigation into the human serum "interactome". *Electrophoresis* 25:1289–1298.
4. Gundry R, Fu Q, Jelinek C (2007) Investigation of an albumin-enriched fraction of human serum and its albuminome. *Proteomics Clin. Appl.* 1:73–88.
5. Mehta AI, Ross S, Lowenthal MS, Fusaro V *et al* (2003) Biomarker amplification by serum carrier protein binding. *Dis. Markers* 19:1–10.
6. Gundry RL, Cotter RJ (2007) From Diagnosis to Therapy. In: Dunn, MJ, Van Eyk, JE (Eds.) Clinical Proteomic, Wiley, London p. 263.
7. Gundry RL, Fu Q, Jelinek CA *et al* (2007) Investigation of an albumin-enriched fraction of human serum and its albuminome. *Proteomics Clin. Appl.* 1:73–88.
8. Thulasiraman V, Lin S, Gheorghiu L *et al* (2005) Reduction of the concentration difference of proteins in biological liquids using a library of combinatorial ligands. *Electrophoresis* 26: 3561–3571.
9. Marrocco C, Rinalducci S, Mohamadkhan A *et al* (2010) Plasma gelsolin protein: a candidate biomarker for hepatitis B-associated liver cirrhosis identified by proteomic approach. *Blood Transfus* 8: Suppl 3:s105–112.

Chapter 14

Study Design in DIGE-Based Biomarker Discovery

Alexandra Graf and Rudolf Oehler

Abstract

The DIGE technology allows the detection of small differences in the expression level of abundant proteins. Many diseases are associated with quantitative deviations of proteins which might represent useful biomarkers for diagnosis or prognosis. DIGE is therefore a highly convenient method for the characterization of disease-related expression changes. This chapter focuses on the study design in DIGE-based biomarker discovery. It introduces the statistical implications of testing thousands of proteins in parallel and discusses the solutions proposed by the literature. The outline provided in the method section tries to guide the researcher through the different statistical considerations, which have to be taken into account in biomarker detection. Special emphasis is given to the use of sample sizes of sufficient statistical power and to the statistical evaluation of the results.

Key words: Sample size calculation, Power calculation, Clinical proteomics, Biomarker research, Study design, DIGE

1. Introduction

Clinical proteomics tries to get a "holistic" view on the pathobiochemistry of a disease. The hope is that this approach increases the probability to find those proteins which are indicative for the pathogenic process and can be used as biomarkers. Biomarkers are classified as antecedent biomarkers (identifying the risk of developing an illness), screening biomarkers (screening for subclinical disease), diagnostic biomarkers (recognizing overt disease), staging biomarkers (categorizing disease severity), or prognostic biomarkers (predicting future disease course, including recurrence and response to therapy, and monitoring efficacy of therapy). Biomarkers may also serve as surrogate end points. Although there is limited consensus on this issue, a surrogate end point is one that can be used as an outcome in clinical trials to evaluate safety and effectiveness of therapies in lieu of measurement of the true outcome of interest. The exact yardstick for evaluating the performance of a biomarker varies on the basis of the intended use. Standards have been proposed for designing and reporting the results of studies evaluating the performance of biomarkers for diagnosis (1) and for prognosis (2). In general, the performance of biomarkers is seldom as good in a second sample as in the sample in which they were initially assessed. Consequently, it has been proposed that biomarker discovery studies should be organized in two experiments: (a) an initial discovery experiment (in our case a DIGE analysis) which identifies potential biomarkers in a training set of patient samples and (b) a validation experiment which determines the performance of these markers in an independent validation set. Such an approach results then in a list of validated biomarkers. These guidelines have been generally accepted in the intervening years and potential biomarkers without validation are regarded today as insufficiently characterized for publication (3). Here we try to provide an outline for the study design of a DIGE-based biomarker discovery study including both the discovery experiment and the validation experiment.

The discovery experiment can be considered a sophisticated comparative analysis of many (approximately 500–1,500) parameters in parallel. However, the task of selecting useful biomarkers from a very large number of candidate proteins is a difficult exercise. Generally, statistical tests are applied for each of the candidate proteins, and the significant proteins are considered as potential biomarkers. However, if a large number of statistical tests are performed simultaneously, the problem of multiple testing arises. Consider first as an example a clinical study, where expression values of only one single protein are compared between two groups (e.g., between responder and nonresponder to a specific therapy), i.e., only one statistical hypothesis is tested. In this case, we may calculate a two-sided two-sample t test at a prespecified level α for the type 1 error (usually, this so-called significance level is set to 0.05; see also Table 1 for a list of statistical terms). Thus, we consider the expression values to be differentially expressed between the two groups if the standard p value of the t test is smaller than the significance level. Here α is the probability of identifying the protein to be differentially expressed if, in truth, the protein is not related with the outcome (and we therefore make a type 1 error, i.e., a false-positive decision).

Multiple testing refers to testing of more than one hypothesis (protein) at the same time. In proteomic studies, hundreds of proteins can be investigated simultaneously. Now, using for each of the proteins a standard t test with significance level 0.05, the probability of declaring at least one protein as differentially expressed, which in truth, is not related with the clinical outcome, will greatly exceed the prespecified level 0.05. The expected number of false-positive decisions is increasing with an increasing number of investigated

Table 1
List of statistical terms

Term	Symbol	Definition
In the single-hypothesis setting		
Type 1 error	α	Probability that a protein is identified, which in truth, is not related to the clinical outcome
Type 2 error	β	Probability that a protein is not identified, which in truth, is related to the clinical outcome
Power	$1-\beta$	Probability that a protein is identified, which in truth, is related to the clinical outcome
In the multiple-hypotheses setting		
Family-wise type 1 error rate	FWER	Probability that at least one of the proteins, which are in truth not related to the clinical outcome, is identified
False discovery rate	FDR	Expected proportion of identified proteins, which are in truth not related to the clinical outcome, among all identified proteins
Power	$1-\beta$	Expected proportion of identified proteins, which are in truth, related to the clinical outcome
Sample size	n	Sample size per group
Standard deviation	σ	Common standard deviation (assumed to be equal in both groups)
Effect size	θ	Difference in group means times the common standard deviation
	m	Number of tested hypotheses (proteins)
	m_1	Number of effective proteins, in truth, related with the clinical outcome
	π_1	Proportion of proteins, which are in truth related to the clinical outcome, among all investigated proteins
	Δ	Difference of group means of the expression values

proteins. Testing about 20 proteins simultaneously, each with a 0.05 level, we already expect 1 protein to be a false-positive (if, in truth, all 20 proteins are not related with the clinical outcome). If about 3,000 proteins are tested and assuming none of them to be, in truth, related with the clinical outcome, again using a 0.05 level for each test would lead to 150 false-positive decisions.

To avoid these pitfalls, a number of multiplicity adjustment procedures are available (see refs. (4, 5)). These adjustment procedures generally result in so-called adjusted p values (instead of the standard p values) that are to be compared to the significance level 0.05. Literature is commonly discussing methods for gene expression studies; however, for proteomic studies, these methods can generally be applied. Two approaches are widely used to adjust for multiplicity:

1. Control of the family-wise error rate (FWER), i.e., the probability to identify at least one protein that is not related with the clinical outcome.

2. Control of the false discovery rate (FDR), i.e., the expected proportion of the identified proteins not related with the clinical outcome among all identified proteins (see ref. (6)). If an FDR of 0.05 is controlled, we can expect in the long run a proportion of 5% false-positives among the significant proteins.

In this chapter, a protein is called "identified" if the adjusted p value is smaller than the prespecified significance level for the type 1 error. Controlling the FDR generally relaxes the multiple testing criteria compared with controlling the FWER and may increase the number of truly identified proteins (methods to control the FDR and further investigations are, e.g., given in refs. (6–9)). The FDR algorithm has been implemented in the DeCyder software for analysis of DIGE experiments as described below.

Irrespective of the method used to deal with the problem of multiple testing, mostly a large number of parallel analyses (sample size) per group are needed to get statistically interpretable results. When testing only a single protein, the sample size problem in case of a two-sample t test is formulated as the number of gels needed to ensure a specific power (probability of identifying a protein that is, in truth, related with the clinical outcome, called "effective" protein in this chapter) to detect a standardized effect size θ (difference of group means in terms of the common standard deviation) with a prespecified type 1 error. In clinical studies, the power is generally set to 0.80. For the example of a two-sided two-sample t test with α-level 0.05, we would need a sample size of about 17 gels/group to detect a difference between group means of size of 1 standard deviation with 80% power. Increasing the effect size to 1.5 or 2 would reduce the sample size of each group to 9 or 6, respectively. However, an effect size of 1.5 or more is regarded as a very large effect in clinical studies. The main problem for the sample size calculation is that the true effect size is unknown. Sample size calculations are therefore often based on the minimum clinical relevant effect size an experimenter wants to detect with their study. On the other hand, before starting a clinical trial, frequently pilot studies are performed to get an intention of the effect size. In the proteomic setting, where thousands of proteins are tested simultaneously, things again get more complicated. The first problem is that, as for the type 1 error rate, also the concept of power can be generalized in various ways in the context of multiple hypothesis testing. Usually, the power is defined as the proportion of truly identified effective proteins among all investigated effective proteins. There have been several publications on sample size estimation in microarray studies. For example, Pawitan et al. (10) investigated the relationships between FDR, power, and sample size. Although their approach may be useful in deciding on the sample size, no direct algorithm was provided. Jung (11) gives an algorithm to calculate the sample size for a prespecific power based on Storey's asymptotic results (see also refs. (8, 9)). The power was defined as discussed above.

As in the single-hypothesis case, for sample size calculations in proteomic studies, assumptions on the effect sizes of the effective proteins have to be made. In clinical practice, it may be useful to specify the minimum effect size the experimenter wants to detect and therefore assuming in the sample size calculations that all effective proteins have the same effect size. For this assumption, sample size calculations become easier. In the sample size calculations for proteomic studies (and thus in the context of multiple testing), one also needs to make assumptions on the proportion of effective proteins among all investigated proteins. Note that using Storey's approach, this (unknown) proportion is important for the sample size calculations and not the total number of investigated proteins. The existing methods for sample size calculations generally assume independence of test statistics. This may not be a realistic assumption; however, when accounting for correlation in the sample size calculations, one also has to specify the size and type of correlation between the test statistics, which may be difficult in praxis.

In the outline provided below, we try to guide the researcher through the above indicated statistical considerations and explain them with practical examples.

2. Materials

2.1. Sample Size Calculations

The sample size calculations can be performed using the graphs shown in Fig. 1 and the data in Table 2. They were calculated using the software R (Version 2.11.1; http://www.r-project.org).

2.2. Statistical Evaluation of the DIGE Gels

The statistical evaluation can be done using the DeCyder Software package (Versions 6.5; GE Healthcare, Uppsala, Sweden).

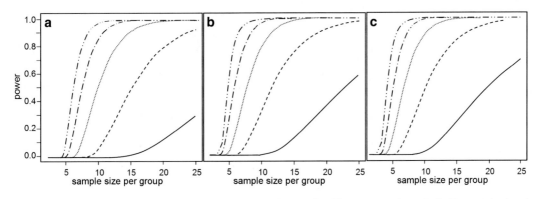

Fig. 1. Statistical power as a function of the sample size of each group for different proportions of effective proteins (graph (**a**): $\pi_1 = 0.01$, graph (**b**): $\pi_1 = 0.05$, and graph (**c**): $\pi_1 = 0.1$) and different effect sizes (*solid line*: $\theta = 1$, *dashed line*: $\theta = 1.5$, *dotted line*: $\theta = 2$, *dot-dashed line*: $\theta = 2.5$, and *dot-dot-dashed line*: $\theta = 3$). The calculations are based on two-sided two-sample *t* tests and an FDR of 0.05.

Table 2
Sample size of each group to achieve a power of 0.8 or 0.9 for different effect sizes and proportions of effective proteins

Sample size of each group

Effect size	Power = 0.80			Power = 0.90		
	$\pi_1 = 0.01$	$\pi_1 = 0.05$	$\pi_1 = 0.10$	$\pi_1 = 0.01$	$\pi_1 = 0.05$	$\pi_1 = 0.10$
1.5	21	16	15	24	19	17
2.0	13	11	9	15	12	11
2.5	10	8	7	11	9	8
3.0	8	6	6	9	7	6

3. Methods

As described in the introduction, it is advisable to subdivide a biomarker discovery study into two experiments: a "discovery experiment" for detection of biomarker candidates followed by a "validation experiment" for their confirmation.

3.1. Discovery Experiment

3.1.1. Sample Size Calculation

To make an adequate sample size calculation (i.e., to estimate how many samples have to be collected for the discovery experiment), the discussion of the following parameters is required: false discovery rate, statistical power, proportion of effective proteins, and minimal effect size.

Fix a Level for the False Discovery Rate

The FDR value is usually set to 0.05 (see Note 1).

Fix a Level for the Statistical Power

The minimal level for the statistical power which is required for a discovery experiment is usually 0.8. The statistical power is 1 minus the type 2 error (β, i.e., a power of 0.8 corresponds to a type 2 error of 0.2). The type 2 error is the proportion of false-negative decisions among the assumed number of effective proteins (m_1), i.e., the proportion of effective proteins not identified using a statistical test. A power of 0.8 is sufficient for most genuine biomarker studies. However, it has to be stressed that such a level may be too low to draw mechanistic conclusions (e.g., on pathways) from the results of the discovery experiment (see Note 2).

Make an Assumption on the Proportion of Effective Proteins (π_1)

π_1 is the ratio of the number of effective proteins (m_1, i.e., those proteins which, in truth, differ between the two experimental groups) to the total number of proteins on the gel (m). The absolute values of both parameters (m_1 and m) are unknown prior to the discovery experiment. But for the calculation of the sample size of the discovery set, it is necessary to make an assumption on their ratio ($\pi_1 = m_1/m$), i.e., to make an educated guess of the percentage of all proteins which expression might be affected by the treatment or the disease. The π_1 value is usually in the range from 0.01 to 0.10 (12) in dependence on the sample material and on the biological question of the study (see Note 3).

Make an Assumption on the Minimal Effect Size

The effect size (θ) is the ratio of the difference between group means (Δ) of the expression levels of a protein for the two experimental groups to the common standard deviation of this protein (σ). For simplicity, the standard deviation is assumed to be equal in both groups. The effect size can be either estimated from the result of a pilot DIGE experiment (see Note 4) or has to be assumed. Therefore, it may be set to the minimal clinical relevant effect size, one wants to detect in the experiment. An effect size of more than four would indicate that there is almost no overlap in the expression level of the respective protein between the two groups (see Note 5). But such an extraordinary high effect size of a biomarker is rare in clinical proteomics. It is very likely to find no such biomarker in a study. Most biomarkers have a much lower effect size showing considerable overlap of their distribution with the control group. However, the larger the overlap, the lower may be the clinical relevance of the biomarker. In the proteomic setting, a minimal effect size smaller than 1.0 often does not seem to be worthwhile.

Calculate the Adequate Sample Size

The sample size per group can be calculated on the basis of the parameters defined above. The diagrams in Fig. 1 show the relationship between the sample size per group and the statistical power for different effect sizes θ and different proportions of effective proteins π_1 (0.01 in a, 0.05 in b, or 0.10 in c). The curves were calculated using Storey's approach (see ref. (8)) using an FDR of 0.05 to correct for multiple testing. The effect size (θ) was assumed to range from 1.0 to 3.0. The informative value of the graphs can be explained with the help of an example: A study where we assume that 1% of the investigated proteins are related with the clinical outcome ($\pi_1 = 0.01$; graph a) with a common effect size of $\theta = 1.5$ (dashed line) would require 21 samples/group to reach a statistical power of 0.80. A power of 0.8 (corresponding to a type 2 error of 0.2) indicates that in the long run, 80% of all effective proteins are identified. In contrast, assuming $\pi_1 = 0.1$ 15 samples/group would be sufficient (graph c) to reach the same power. Table 2 shows the required sample sizes (for different θ and π_1) to achieve a statistical power of 0.80 or 0.90. Note again that sample size calculations are based on two-sided two-sample t tests and an FDR of 0.05.

3.1.2. Perform the Discovery Experiment

Collect the required number of protein samples in correspondence to the calculations shown above. Then split each sample into two parts. One part will be mixed with aliquots from all other samples and then used as an internal standard on each gel. The remaining part has to be analyzed individually in the DIGE experiment. On each gel, two samples (ideally one per experimental group) and one internal standard are applied. From the statistical point of view, it is disadvantageous to pool the samples within the experimental groups for diminution of the number of gels (leaving the overall patient-sample size unchanged). Such pooling strategy does not take into account the biological variation within the experimental groups. It leads to a dramatic reduction of the power. After the gel run, the gel images are loaded into the DeCyder software package. Then spots are detected in the differential in-gel analysis (DIA) module and matched in the biological variation analysis (BVA) module.

3.1.3. Statistical Evaluation of the Results from the DIGE Experiment

A Student's t test-based statistical evaluation of the DIGE experiment can be done in the BVA module or in the extended data analysis (EDA) module of the DeCyder software package (see Note 6). The software calculates a list with the FDR-adjusted p values and the *average ratio* for each spot. The *average ratio* is a measure for the ratio of the two mean standardized abundances of the two experimental groups (see Note 7). To determine those spots which can be regarded as potential biomarkers, first, sort the spot list according to the p value of the t test. Then, exclude all spots with a p value above the fixed FDR level (usually 0.05). The expression levels of all remaining spots differ significantly between the experimental groups. However, most of them probably show a very small difference in expression. To exclude such spots, sort the list of the significant spots according to the *average ratio*. Then eliminate all spots with a small *average ratio* (e.g., between −1.30 and +1.30; see Note 8). The remaining spots may be regarded as potential biomarkers. The list of potential biomarkers might be considered too long to include them all in the validation experiment. Then the researcher can apply more stringent selection criteria increasing the *average ratio* (i.e., ±1.5 or ±2.0) or alternatively select the proteins for the validation experiment due to biological reasons. For practical reasons, usually not more than 10–15 potential biomarkers are selected for validation.

3.2. Validation Experiment

3.2.1. Type of Assay

A DIGE experiment is very time intensive and not suited for routine diagnostic analysis. It is therefore advisable to change the analytical method before starting the validation experiment. Ideally, the analytical method used in the validation experiment should allow high-throughput analysis at low costs. Such methods include ELISA, Western blotting, and protein biochip. But in principle, a validation experiment may also be performed using a DIGE platform. However, it should be taken into account that the validation experiment is only applicable to confirm biomarkers found in the discovery experiment and it is not eligible to find and validate new biomarkers.

In case of a change of the analytical platform, all samples of the discovery experiment should be reanalyzed using the validation platform. This is especially necessary for immunological methods to prove the ability of the antibodies to quantify the potential biomarker.

3.2.2. Sample Size of the Validation Experiment

To validate the potential biomarkers, it is advisable to collect at least the same number of new samples as in the discovery experiment. Many researchers use the validation experiment to estimate a threshold value for each biomarker for future studies. An expression level above or below such a threshold could then be used in future to allocate an unknown sample to an experimental group (e.g., disease or control). In this context, a much larger number of samples should be included in the validation experiment than in the discovery experiment (4–10 times can be found in literature, clearly the larger the better).

3.2.3. Statistical Evaluation of the Results of the Validation Experiment

The number of potential biomarkers tested in the validation experiment is much smaller than the number of spots investigated in the discovery set. Nevertheless, the resulting p values of the corresponding t test should be adjusted for multiple testing. Often it is suggested to apply the control of the FWER rather the FDR (see Note 9). Those proteins which show also in the validation experiments a significant difference in their expression level between the experimental groups can be considered as validated biomarkers.

4. Notes

1. A level of 0.05 is generally accepted in clinical routine diagnostics. Therefore, it is reasonable to use the same significance level in a proteomic discovery experiment. Using a smaller level (e.g., 0.025 or 0.01) decreases the power of the statistical test, and several biomarkers might not be found.

2. In a biomarker study, the researcher wants to find one or more protein spots, whose expressions are associated, for example, with the incidence of a disease. Due to the type 2 error, some of the proteins declared as statistically not significant (adjusted p value of >0.05) might in reality correlate very well with the disease. However, the needs of the researcher are satisfied as long as a sufficient number of true-positive proteins have been found. Thus, fixing a moderate level β for the type 2 error (e.g., 0.2 which is also applied in conventionally clinical studies) in the sample size calculations is acceptable. Note that applying a larger level (and therefore a lower statistical power) increases the false-negative rate so far that only a few (or maybe no) true biomarkers might be found in the analysis.

3. For example, in a DIGE-based comparison of plasma samples from cancer patients with samples from healthy controls, it has to be expected that only a few protein spots differ between the experimental groups. The tumor itself and the surrounding tissue might secrete a number of different proteins which are normally not present in the blood flow of healthy subjects. However, due to the huge dilution of these proteins by the total plasma volume, only a few of them might be detectable in the peripheral blood. Therefore, the π_1 is often set to 0.01 for such experiments (indicating that about 1% of all investigated spots might show disease-related expression changes). In contrast, in a proteomic comparison of tumor tissue with surrounding normal tissue, it has to be expected that the portion of spots showing disease-related expression changes is much higher ($\pi_1 = 0.05$ or even 0.10).

4. The true effect size can be calculated from an experimental data set as follows:

   (a) $\theta = \Delta / \sigma$

   θ ... effect size, Δ ... difference in group means of the expression levels, σ ... common standard deviation.

   In a comparison of the expression pattern of two experimental groups (control and group-1), Δ is calculated as follows:

   (b) $\Delta = \text{meanSA}_1 - \text{meanSA}_c$

   meanSA_1 or $_c$ is mean standardized abundance of group-1 and control, respectively.

   The SA values of each spot are calculated according to the DIGE algorithm dividing the spot volume of the sample by the spot volume of the internal standard on the same gel (13). The DeCyder software shows the SA values of the spot of interest on the different gels in the graph view of the BVA module. The values can be exported using the XML toolbox.

5. The relationship between Δ and θ is illustrated in Fig. 2 with the help of two examples.

6. For analysis in the BVA module, open the workspace and select the Protein Table (PT). Then open the Protein Statistics dialog (in the Process menu). Select the type of statistical test (independent or dependent) according to your biological question. Activate the Average Ratio and the Student's t test option and define the two experimental groups which should be compared. Finally, select the FDR option (this is essential to avoid the problem of multiple testing).

   For analysis in the EDA module, open the workspace and select the Differential Expression Analysis dialog (in the Calculations window). Select the type of statistical test

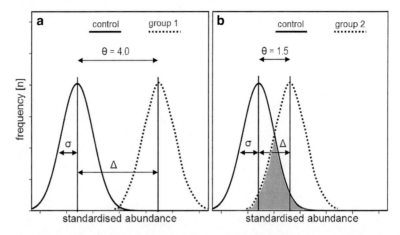

Fig. 2. Two examples for the relationship between the difference in mean standardized abundance (Δ) and the effect size (θ). The effect size of control vs. group 1 is 4.0 and of control vs. group 2 is 1.5. The graphs show the different degree of overlap (indicated in *gray*) of the distributions of expression values of the control and the respective group.

(independent or dependent) according to your study design. Activate the Average Ratio and the Student's t test option and define the two experimental groups which should be compared. Finally, select the FDR option.

7. The DeCyder software package expresses the degree of disparity in spot mean abundance between two experimental groups as "average ratio" or "avR." Other software packages and many publications use the term "fold-change" for the same value. avR derives from the ratio of the mean standardized abundances (here termed R'):

   (a) $R' = \mathrm{meanSA}_1 / \mathrm{meanSA}_c$.
   (b) If $R' > 1$, then $\mathrm{av}R = R'$.
   (c) If $R' < 1$, then $\mathrm{av}R = -1 / R'$ mean SA_1 or $_c$ mean standardized abundance of group-1 and control, respectively.

   Note that the "average ratio" (or "fold-change") should not be confused with the effect size (θ). Although both measures are used to describe the disparity in expression of a protein, only the effect size considers the standard deviation (and hence the biological variation). Accordingly, the "average ratio" cannot directly be converted to effect size.

8. According to the guidelines of the European (EMA) and the American (FDA) regulatory agencies, the maximal allowed technical variation of routine diagnostic methods (e.g., ELISA) is 20% (14, 15). It seems therefore reasonable to include only spots with an average ratio above ±1.2 in the validation experiment.

9. If several statistical tests are being performed simultaneously, controlling the FWER to adjust for multiple testing is more

conservative than controlling the FDR. In order to avoid a lot of false-positives, the significance level for each p value is reduced to account for the number of comparisons being performed. Using the Bonferroni correction to control the FWER, the p values of the m investigated proteins are compared with a significance level of $0.05/m$ instead of 0.05 to control an overall level of the type 1 error rate of 0.05. For example, in a validation experiment of five potential biomarkers, the p values of the individual t tests have to be lower than $0.05/5 = 0.01$. However, if m is very large (as in the discovery set), the Bonferroni correction is too stringent (i.e., the power of the investigated test may be rather small), and therefore, the control of the FDR should be applied.

Acknowledgments

We thank Sonja Zehetmayer for helpful comments.

References

1. Bossuyt, P.M., et al. Towards complete and accurate reporting of studies of diagnostic accuracy: the STARD initiative. Standards for Reporting of Diagnostic Accuracy. *Clin Chem* **49**, 1–6 (2003).
2. McShane, L.M., et al. Reporting recommendations for tumor marker prognostic studies. *J Clin Oncol* **23**, 9067–9072 (2005).
3. Mischak, H., et al. Recommendations for biomarker identification and qualification in clinical proteomics. *Sci Transl Med* **2**, 46ps42 (2010).
4. Dudoit, S., Shaffer, J. & Boldrick, J. Multiple hypothesis testing in microarray experiments. *Statistical Science* **18**, 71–103 (2003).
5. Shaffer, J.P. Multiple Hypothesis Testing. *Annual Review of Psychology* **46**, 561–584 (1995).
6. Benjamini, Y. & Hochberg, Y. Controlling the false discovery rate—A practical and powerful approach to multiple testing. *Journal of the Royal Statistical Society, Series B* **57**, 289–300 (1995).
7. Benjamini, Y. & Yekutieli, D. The control of the false discovery rate in multiple testing under dependency. *The Annals of Statistics* **29**, 1165–1188 (2001).
8. Storey, J.D. A direct approach to false discovery rate. *Journal of the Royal Statistical Society, Series B* **64**, 479–498 (2002).
9. Storey, J.D., Taylor, J.E. & Siegmund, D. Strong control, conservative point estimation and simultaneous conservative consistency of false discovery rates: a unified approach. *Journal of the Royal Statistical Society, Series B* **66**, 187–205 (2004).
10. Pawitan Y., Michiels, S., Koscielny, S., Gusnanto, A. & Ploner, A. False discovery rate, sensitivity and sample size for microarray studies. *Bioinformatics* **21**, 3017–3024 (2005).
11. Jung, S.H. Sample size for FDR-control in microarray data analysis. *Bioinformatics* **21**, 3079–3104 (2005).
12. Petrak, J., et al. Deja vu in proteomics. A hit parade of repeatedly identified differentially expressed proteins. *Proteomics* **8**, 1744–1749 (2008).
13. Unlu, M., Morgan, M.E. & Minden, J.S. Difference gel electrophoresis: a single gel method for detecting changes in protein extracts. *Electrophoresis* **18**, 2071–2077 (1997).
14. Directorate-General-Health-and-Consumer-Protection. Guidance for generating and reporting methods of analysis in support of pre-registration data requirements. Vol. Annex II (part A, Section 4) and Annex III (part A, Section 5) of Directive 91/414. (ed. EUROPEAN-COMMISSION) (Brussels, BE, 2000).
15. Biopharmaceutics-Coordinating-Committee. Guidance for Industry: Bioanalytical Method Validation. (ed. U.S. Department of Health and Human Services, F.a.D.A., Center for Drug Evaluation and Research (CDER), Center for Veterinary Medicine (CVM)) (Rockville MD 2057, USA, 2001).

Chapter 15

Comparative 2D DIGE Analysis of the Depleted Serum Proteome for Biomarker Discovery

Megan Penno, Matthias Ernst, and Peter Hoffmann

Abstract

Serum is unarguably the most used diagnostic fluid. As it circulates throughout the body, leakage peptides/proteins from damaged and dying cells, host-response proteins including inflammatory mediators, and aberrant secretions from tumors and diseased tissues are released into serum, potentially providing a rich source of disease biomarkers. Here, a method for extending access to the serum proteome by removing highly abundant proteins prior to comparative two-dimensional difference gel electrophoresis (2D DIGE) and subsequent protein digestion for identification by mass spectrometry is described.

Key words: Biomarkers, Serum, Depletion, Multiple affinity removal system, Tryptic digestion

1. Introduction

The constituents of the serum proteome can be divided into eight functional groups: (1) the classical serum proteins produced by solid organs such as the liver; (2) the immunoglobulins; (3) long-distance receptor ligands including the peptide and protein hormones; (4) local receptor ligands such as cytokines and other short-distance mediators of cellular functions; (5) temporary passenger proteins that may navigate through the serum on their way to the site of primary function; (6) foreign proteins derived from bacterial, viral, parasitic, or proteinaceous agents within the circulating blood volume; (7) tissue leakage products that normally function within cells but are released into the serum as a result of cell damage or apoptosis; and (8) aberrant secretions released from tumors and other diseased tissues (1). Accordingly, serum is the single most informative sample that can be collected from an individual that describes their current state of health (2).

One of the most significant challenges associated with analyzing the serum proteome is the broad protein dynamic range spanning

10–15 orders of magnitude (3, 4). In order to visualize the low-abundance proteins originating from damaged tissues and tumors, it is necessary to remove the high-abundance classical proteins and immunoglobulins. Numerous prefractionation methods have been developed based on dye-based affinity chromatography (e.g., Cibacron Blue), peptide affinity (e.g., refs. 5–7), and immunoaffinity depletion strategies. The immunoaffinity systems have shown the most promise due to their high selectivity, specificity, and reproducibility. Several devices using liquid chromatography (LC) columns and spin filters incorporating antibodies from different species are commercially available. One of these is the multiple affinity removal system (MARS) developed by Agilent Technologies (Santa Clara, CA, USA), which has been shown to significantly improve coverage of the human serum proteome (8).

Here, the depletion of mouse serum using a MARS-M3 LC column is outlined, and a subsequent processing strategy for labeling the resulting fractions with difference gel electrophoresis (DIGE) fluor minimal CyDyes (GE Healthcare, Uppsala, Sweden) for comparative 2D DIGE is described. When separated by two-dimensional electrophoresis (2DE) using large-format precast gels, these samples produce highly reproducible, well-resolved spot patterns.

2. Materials

The following protocol for high-abundance protein depletion and subsequent 2DDIGE analysis is compatible with either serum or plasma. Great care should be taken during sample collection and storage as these factors can dramatically affect proteomic analysis (9). Species specificity is dependent on the MARS column used. Presently, five different MARS columns are available for humans that remove albumin, albumin/immunoglobulin G and the top 6, 7, and 14 most abundant proteins, and a single column for mice that depletes the top three most abundant proteins. The columns come in three dimensions: 4.6×50, 4.6×100, and 10×100 mm. Described here is the depletion of mouse serum using a 4.6×100 mm MARS-M3 column on a Hewlett Packard 1090 HPLC system. Volumes and instrument-specific parameters should be modified accordingly for alternative configurations. All solutions should be prepared using high-purity water (ultrapure water: resistivity ≥ 18.2 MΩ cm, total organic content ≤ 1 ppb, at 25°C). Reagents are analytical grade unless specified otherwise. Products/solutions should be maintained at room temperature unless otherwise specified. Prior to use, ensure all buffers are thawed completely and mixed well by vortexing.

2.1. Depletion of High-Abundance Proteins

1. MARS-M3 column (4.6×100 mm; Agilent Technologies, Santa Clara, CA, USA). Store at 4°C and maintain at room temperature when in use.
2. Spin filters, 0.22-μm pore size cellulose acetate (Agilent Technologies).
3. MARS buffer A and buffer B (Agilent Technologies).
4. HP 1090 HPLC system (Hewlett Packard, Palo Alto, CA, USA).
5. 15- and 50-mL Falcon polypropylene tubes (BD Biosciences, San Jose, CA, USA).
6. 100% acetone, stored at −20°C in a solvent-compatible freezer.

2.2. Sample Preparation for DIGE

1. 80% acetone, stored at −20°C in a solvent-compatible freezer.
2. TUC4 buffer: 7 M urea, 2 M thiourea, 30 mM Tris–HCl, 4% (w/v) 3-[(3-cholamidopropyl)dimethylammonio]-1-propane-sulfonate (CHAPS), and pH 8.5. Stored at −20°C in 1 mL aliquots.
3. 10 kDa MWCO Vivaspin 500 centrifugal devices (GE Healthcare).
4. Twin Cond B-173 conductivity meter (Horiba, Kyoto, Japan).
5. Gel-loading pipette tips.
6. EZQ protein quantitation assay (Invitrogen, Carlsbad, CA, USA).

2.3. DIGE Labeling

1. Cy2, Cy3, and Cy5 DIGE fluor CyDyes (GE Healthcare): working solutions of 200 pmol CyDye per 1 μL of anhydrous 99.8% N,N-dimethylformamide. Store 1 μL aliquots at −80°C (see Note 1). Thaw on ice when required, protected from light.
2. 10 mM L-lysine. Store in aliquots at −20°C.
3. 1 M dithiothreitol (DTT). Store aliquots at −80°C and thaw at room temperature when required.
4. IPG buffer pH 4–7 (GE Healthcare). Store at 4°C.

2.4. Two-Dimensional Electrophoresis (2DE)

1. Rehydration buffer: 7 M urea, 2 M thiourea, 30 mM Tris–HCl, 2% (w/v) CHAPS, 0.5% (v/v) IPG buffer, 1.2% (v/v) DeStreak reagent (GE Healthcare; store at 4°C), and 0.05% (v/v) saturated bromophenol blue solution (prepared in water). IPG buffer, DeStreak, and bromophenol blue can be added freshly to 1 mL aliquots of TUC2 buffer (7 M urea, 2 M thiourea, 30 mM Tris–HCl, 2% (w/v) CHAPS), which can be stored at −20°C.
2. 24-cm IPG strips, pH 4–7 (GE Healthcare). Store at −20°C.
3. PlusOne dry strip cover fluid (GE Healthcare).
4. Ettan IPGphor II (GE Healthcare).

5. Reswelling tray (GE Healthcare).
6. Paper wicks (GE Healthcare).
7. Loading cups (GE Healthcare).
8. Ceramics IPG strip tray (GE Healthcare).
9. DTT equilibration solution: dissolve 0.05 g of DTT and 1.8 g urea in 5 mL of proprietary IPG equilibration buffer (Serva, Heidelberg, Germany). This recipe is designed for two strips. Adjust the amount and volume according to the number of strips.
10. Iodoacetamide (IAA) equilibration solution: dissolve 0.125 g IAA and 1.8 g of urea in 5 mL IPG equilibration buffer (Serva). Add a sufficient volume of saturated bromophenol blue solution until the IAA equilibration solution is a deep blue color and no longer transparent. This ensures that the dye front will be visible during electrophoretic separation. This recipe is designed for two strips. Adjust the amount and volume according to the number of strips.
11. Anode buffer: dissolve the content of the anode buffer bag (Serva) in 7.5 L of water.
12. Cathode buffer: dissolve the powder in the cathode buffer bag (Serva) in 5 L of water.
13. Flat cassettes 2D (Gel Company).
14. 2DGel DALT 12.5% NF (Serva). Store at 4°C until required.
15. ECL Plex rainbow fluorescent markers (GE Healthcare). Store at −20°C.
16. Reference markers.
17. Low-melt agarose: weigh 1 g of low-melt agarose (Quantum Scientific, Queensland, Australia) and dissolve in 100 mL of cathode buffer (see Subheading 2.4, item 12) on a heated magnetic stirrer (70°C). Once dissolved, store at room temperature.
18. IPG strip pusher (Gel Company).
19. Ettan DALT12 electrophoresis system (GE Healthcare).
20. Ettan DIGE Imager (GE Healthcare).
21. DeCyder 2D (v 7.0; GE Healthcare).

2.5. Proteolytic Digestion for Protein Identification

1. Ettan spot cutting robot (GE Healthcare).
2. Polypropylene 96-well plates with silicone sealing mats (Axygen, Union City, CA, USA).
3. 100 mM NH_4HCO_3 stock solution, prepared in water on the same day as required.

4. Porcine modified trypsin (Promega, Madison, WI, USA). Create a stock solution by resuspending the lyophilized material at 100 ng/μL in 5 mM NH_4HCO_3 and store in 10 μL aliquots at −80°C. The 10 ng/μL working solution is made with the addition of 90 μL of 5 mM NH_4HCO_3 to the 10 μL aliquot, which should be thawed and maintained on ice. This is done immediately before use.

5. Acetonitrile.

6. Formic acid.

3. Methods

3.1. Depletion of High-Abundance Proteins

1. Dilute 20 μL of mouse serum in 90 μL of MARS buffer A. Add to the concentrator pocked of the 0.22-μm centrifugal filtration device and spin for 1 min at $16,000 \times g$.

2. Program the HPLC system with the following method: 0–10 min: 100% MARS buffer A at 0.5 mL/min; 10.01–17 min: 100% MARS buffer B at 1 mL/min; and 17.01–28 min: 100% MARS buffer A at 1 mL/min. The maximum operating pressure is 120 bar.

3. Prime the HPLC system (see Note 2).

4. Inject 100 μL of the diluted serum at 0.5 mL/min. Collect the "flow-through" peak, which contains the low-abundance proteins, into a 15-mL Falcon polypropylene tube. This elutes from the column between 1.5 and 5.5 min after commencing the run and has a volume of ~2 mL. Immediately, add 8 mL of ice-cold acetone to the flow-through peak and store on ice while the "bound" peak is collected.

5. Collect the bound peak containing the high-abundance proteins (albumin, transferrin, and IgG for the MARS-M3 column) into a 50-mL Falcon polypropylene tube. This elutes between 12 and 16 min following the commencement of the run and has a volume of ~4 mL. Add 16 mL of ice-cold acetone to the bound fraction, then immediately store both fractions at −20°C for at least 16 h to allow proteins to precipitate. The samples should be processed within 5 days of their collection or transferred to −80°C.

3.2. Sample Preparation for DIGE

1. Recover the protein pellets by centrifugation at $5,000 \times g$ for 10 min at 10°C. Remove the supernatants, ensuring the protein pellets remain intact. Invert the tubes onto a lint-free paper towel (e.g., Kimwipe, Kimberly-Clark, Dallas, TX, USA) for 5 min to ensure maximum liquid removal and transfer into a fresh tube.

2. Wash pellets with the addition of 10 mL of ice-cold 80% acetone. Incubate on ice for 30 min, vortexing for 20–30 s every 10 min (see Note 3). Centrifuge at $5,000 \times g$ for 10 min at 10°C. Remove the supernatants as described in Subheading 3.2, step 1.

3. Solubilize the pellets in 500 µL of TUC4 buffer. Mix by thorough pipetting for 20 s, then place sample in a sonicating water bath for 5 min. Incubate at room temperature for 20 min with occasional vortexing. Centrifuge at $1,000 \times g$ for 1 min to bring the material to the bottom of the tube.

4. Transfer the samples to labeled microfuge tubes and centrifuge at $8,000 \times g$ for 10 min at room temperature. Collect the supernatants in fresh, labeled microfuge tubes and discard the insoluble pellets. At this point, the samples can be stored at −20°C for up to 1 week or at −80°C for long-term storage. If stored prior to further processing, thaw in the refrigerator or on ice (see Note 4).

5. To further desalt the samples, thus ensuring their compatibility with isoelectric focusing (IEF), add the entire 500 µL volume to the concentrator pocket of a Vivaspin concentration device. Centrifuge at $14,000 \times g$ for 30 min at room temperature (see Note 5). The volume should be reduced to ~20 µL. If not, continue centrifugation until the volume is sufficiently reduced.

6. Discard the liquid in the bottom collection tube. Add 180 µL of TUC4 to the concentrator pocket using a gel-loading pipette tip and mix by thorough pipetting (Buffer Exchange 1). Centrifuge at $14,000 \times g$ for 45 min at room temperature. The volume should be reduced to ~20 µL. If not, continue centrifugation.

7. Discard the liquid in the bottom collection tube. Add 180 µL of TUC4 to the concentrator pocket and mix by thorough pipetting (Buffer Exchange 2). Centrifuge at $14,000 \times g$ for 60 min at room temperature. The volumes should be reduced to ~20 µL. If not, continue centrifugation.

8. Discard the liquid in the bottom collection tube. Add 180 µL of TUC4 to the concentrator pockets and mix by thorough pipetting (Buffer Exchange 3). The final volume should be approximately 200 µL.

9. Check conductivity of the protein solutions using a conductivity meter according to the manufacturer's instructions. If conductivity is <300 µS, remove the sample from the Vivaspin using a gel-loading pipette tip and transfer to a fresh, labeled microfuge tube. If the conductivity of a sample is >300 µS, perform an additional buffer exchange step and reevaluate.

10. Determine the protein concentrations of the samples using the EZQ protein quantitation assay as per the manufacturer's recommendations.

3.3. DIGE Labeling

1. Prepare an experimental design whereby the biological replicates within an experimental group are labeled equally with either Cy3 or Cy5 (i.e., for an experiment with six biological replicates per group, three should be labeled with Cy3 and three with Cy5) (10). This removes a potential bias associated with a specific dye. The internal standard, which consists of equal amounts of protein from all samples, should be labeled with Cy2.

2. To prepare the internal standard, pool an appropriate volume of each sample containing 50 μg of protein based on the results of the protein assay. Add sufficient Cy2 working solution such that the dye/protein ratio is 200 pmol/100 μg (see Note 6).

3. For the individual-labeling reactions, take the required volume of each sample containing 100 μg of protein and add directly to a thawed aliquot of either Cy3 and Cy5 (dye/protein ratio of 200 pmol/100 μg).

4. Perform the labeling reactions on ice for 30 min in darkness. Quench with the addition of 1 μL of 10 mM lysine per 200 pmol of dye, incubating on ice for 10 min in darkness.

5. Combine the appropriate Cy3 and Cy5 samples with 100 μg of the Cy2-labeled internal standard and add a sufficient amount of TUC4 to bring the samples to equivalent volumes. Add the required amounts of DTT solution, carrier ampholytes (IPG buffer), and bromophenol blue such that the final sample consists of 10 mM DTT, 0.5% (v/v) ampholytes, and trace amounts of bromophenol blue for color, along with 300 μg protein, 7 M urea, 2 M thiourea, 30 mM Tris–HCl, and 4% CHAPS.

6. If sample volumes exceed the maximum allowable volume for cup loading using the IPGphor II (150 μL), they may be concentrated by centrifugation using Vivaspin devices at $14,000 \times g$ at room temperature.

3.4. Two-Dimensional Electrophoresis (2DE)

1. Prior to IEF, rehydrate the 24-cm IPG strips by evenly pipetting 450 μL of rehydration solution into the groove of the reswelling tray. Care should be taken to avoid introducing air bubbles. Remove the plastic backing that protects the IPG gel, then, without touching the gel, lay the strip down such that the gel is facing the solution. Ensure the buffer is evenly distributed under the strip. Add 1 mL of cover fluid on top of each strip to prevent evaporation of the buffer. Incubate strips overnight at room temperature.

2. Position the ceramic IPG strip tray on the IPGphor II. Align the rehydrated IPG strips along the groove inside the manifold with gel side facing up. If the experiment contains fewer than 12 strips, use the innermost wells first leaving the outside wells empty.

3. Soak two paper wicks per IPG strip in 150 μL of reverse osmosis water (see Note 7). Place the wet paper wick over each end of the IPG strip such that the wick covers approximately one-third of the end of the gel, with two-thirds extending over the plastic backing.

4. Position the loading cup immediately below the paper wick at the anodic end of the strip. Pour 100 mL of cover fluid into the manifold ensuring the strips are covered but do not pour into the cups. Observe that the cups are correctly sealed against the IPG strips and do not fill with cover fluid. If the cups are not leaking, gently pipette the samples into the cups taking care not to introduce any bubbles. Cover the samples with 50 μL of cover fluid to prevent drying during the run.

5. Place the electrodes on the paper wicks such that two-thirds of the wick is positioned in front of the electrode wire and one-third behind.

6. Perform IEF according to the program outlined in Table 1.

7. Prepare the DTT and IAA equilibration solutions. Place a maximum of two focused IPG strips into a single, long, rod-shaped, plastic tube (e.g., the tubes that sample cups are provided in from GE Healthcare). Incubate the IPG strips in the DTT equilibration solution for 15 min in a flat position with gentle agitation. During this incubation time, prepare the anode and cathode buffers.

8. Decant the DTT equilibration solution and replace with IAA equilibration solution. Cover the tubes with aluminum foil and incubate for 15 min at room temperature with gentle agitation. Remove the IAA equilibration solution and add 5 mL of cathode buffer per tube to rinse off the equilibration solution.

Table 1
Isoelectric focusing program on the IPGphor II

Step	Type	Voltage (V)	Duration (h)	Accum. (Vh)
1	Step	300	2	600
2	Step	500	2	1,600
3	Step	1,000	2	3,600
4	Gradient	8,000	5	28,600
5	Step	8,000	5	68,600
6	Step	500	10	–

The purpose of step 6 is to maintain the electric field overnight and prevent protein diffusion

9. Clean the glass surface of the flat cassettes thoroughly with ethanol and water using lint-free wipes. Apply a small volume of ultrapure water onto the glass plate along the closure. Remove but do not discard the plastic coversheet of a 2DGel DALT 12.5% NF precast gel (see Note 8). With the gel facing downwards, gently lay the gel on the glass plate first at the closure region then moving towards the hinge. Use a roller to remove any air bubbles and excess liquid between the gel and the glass plate. With the cassette lying horizontally on the bench, close the flap and press the edge of closure tightly to ensure the gel is firmly in place.

10. If molecular weight markers are required, pipette 20 μL of the ECL Plex marker onto a small piece of filter paper (approximately 0.5×1 cm). Allow to dry at room temperature.

11. Melt the solidified agarose in microwave oven. Stand the loaded cassettes vertically using a rack. Pipette 5 mL of molten agarose on top of the gel (between the glass plate and the plastic backing) using a 10-mL syringe fitted with a needle. Lay the IPG strip into the molten agarose orientated such that the lower pH end of the strip will be situated on the left-hand side of the gel. Gently push the strip against the gel surface using an IPG strip pusher. Remove any trapped air bubbles in between the gel and IPG strip by pressing the strip towards the gel, but do not damage the gel edge. Place the piece of filter paper loaded with the molecular weight marker on the left-hand side of the gel next to the acidic end of the strip. Allow the agarose to set at room temperature.

12. Set up the Ettan DALT12 system. Pour 7.5 L of anode buffer into the separation unit, turn the pump on, and ensure the system is in circulation. Spray the silicone rubber seals and cassettes with 0.1% SDS in water to facilitate loading. If running fewer than 12 gels, insert the gel cassettes first, then place the blank cassettes at the front and rear of the unit, and, if necessary, in between each gel. Ensure the seals are not bowed by gently pulling up on the cassettes then pushing them down again.

13. Pour a portion of cathode buffer into the upper buffer chamber. Check for leaks between the upper and lower chambers. If the system is not leaking, pour the remaining cathode buffer in the upper chamber. Close the lid and run the apparatus according to the program outlined in Table 2.

14. Cease electrophoretic separation when the dye front reaches the bottom of the gel (approximately 16 h). Open the lid and begin disassembling the tank by firstly removing the cassette/blank at the rear or the apparatus. Rinse the cassettes thoroughly with deionized water. To remove the gels, lay the glass side of the cassette on the bench and open the flap carefully.

Table 2
Second-dimension electrophoresis program on the Ettan DALT12

Phase	Current (mA)/gel	Power (W)/gel	Voltage (V)	Duration (h)
1	5	0.5	50	2
2	10	0.5	100	1
3	30	2.5	250	14

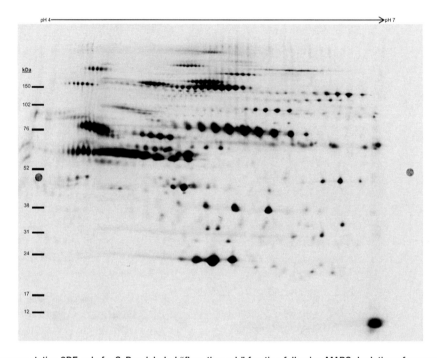

Fig. 1. A representative 2DE gel of a CyDye-labeled "flow-through" fraction following MARS depletion of mouse serum.

The gel should still attach to the glass plate. Hold the edges of the plastic backing and start pulling the gel away from the glass plate. Cover the gel with the plastic coversheet to reduce exposure to environmental contaminants (see Note 8). Attach fluorescent references markers to the central left and right sides of the film support, approximately 1 cm inside each margin.

15. Scanning should be performed immediately using an Ettan DIGE Imager (GE Healthcare) at 100 μm resolution. Appropriate exposure settings for the precast gels are 0.4 s for the Cy2 channel, 0.07 s for the Cy3 channel, and 0.1 s for the Cy5 channel. Protect those gels not currently being scanned from light. Once scanned, replace the plastic coversheet, place the gel in an individual ziplock bag, then store at −80°C (see Note 9). Fixation of the gels is not required unless performing a subsequent staining procedure. An example gel is presented in Fig. 1.

3.5. Comparative Image Analysis

1. Rotate and crop the gel images using the Ettan DIGE Imager software to an appropriate field of interest that includes the reference markers with the lower pH end of the strip on the left.

2. Perform comparative image analysis using DeCyder 2D. Each gel should be processed separately in the differential in-gel analysis (DIA) module of DeCyder prior to export to the biological variation analysis (BVA) module. In DIA, spot detection is performed based on an estimated 10,000 spots with an exclusion filter rejecting spots with volumes <30,000. Background subtraction and normalization are carried out automatically. Use the reference marker detection tool to ensure that the left and right reference markers are highlighted correctly.

3. Import the DIA workspaces in BVA for spot matching to a designated master gel. Given that serum proteins contain a large number of posttranslational modifications that produce long clusters of spots on the 2D gels (see Fig. 1), it may be necessary to undertake spot matching manually.

4. Statistical analysis can be performed using the numerous tests offered by DeCyder in both the BVA and extended data analysis (EDA) modules. Alternatively, normalized volumes of matched spots can be exported into a spreadsheet (e.g., Excel, Microsoft, Redmond, CA, USA) using DeCyder's XML Toolbox function. Standardized abundance values can be calculated by dividing the spot volumes of the Cy3 and Cy5 channels by the Cy2 channel for each gel. The resulting values are identical to those displayed in DeCyder. Alternative software packages can then be used to analyze the data.

5. Once spots of interest have been determined, specify these as "pick" spots in the BVA experiment. Create pick lists for a minimum of three analytical gels. Each spot of interest should be picked from three gels to provide sufficient material for MS analysis.

3.6. Proteolytic Digestion for Protein Identification

1. Load the generated pick lists into the Ettan spot cutting robot software ensuring the appropriate gel is placed on the spot cutting tray. Align the reference markers with the white grid lines of the spot cutting tray to facilitate automatic recognition of the reference markers by the software.

2. Pick the gel plugs into a 96-well plate. Multiple gels can be successfully picked into the same plate provided that great care is taken to ensure matched spots are dispensed into the correct wells. It may be necessary to remove the dispensing solution (20% ethanol) between each round of picking. Once complete, cover the plate with a silicone sealing mat and store at 4°C until proceeding with the tryptic digest.

3. Remove the 20% ethanol dispensing solution from the plate wells. Add 250 µL of 50 mM NH_4HCO_3 to each well containing

gel plugs. Replace the silicone mat to prevent dust contamination of the samples during incubation steps and stand for 5 min at room temperature. Remove the solution and repeat.

4. Add 100 μL of 5 mM NH_4HCO_3 and incubate for 10 min at room temperature. Remove and replace with 200 μL of acetonitrile. Incubate for 15 min at room temperature.

5. Remove the acetonitrile into an appropriate waste disposal container. Allow the plugs to air-dry for 10 min at room temperature then rehydrate with 10 μL of trypsin working solution.

6. Allow the gel pieces to rehydrate for 15 min at room temperature. Add 10 μL of 5 mM NH_4HCO_3 in 20% acetonitrile to make up the total volume of each sample well to 20 μL containing 100 ng of trypsin, 5 mM NH_4HCO_3, and 10% acetonitrile (see Note 10). Replace the silicon sealing mat and incubate the plate overnight at 37°C.

7. Following the overnight digestion, add 20 μL of 1% formic acid to each sample well to inhibit the reaction. Place the plate in a sonicating water bath such that only the bottom half of the plate is submerged for 15 min.

8. Remove the liquid from each well and transfer to the Final Sample Tube. If protein identification is being performed by LC-MS, the Final Sample Tube can be a labeled autosampler vial.

9. Add 50 μL of 50% acetonitrile containing 1% formic acid to each well. Incubate the sample in a sonicating water bath for 15 min, remove the liquid, and transfer to the appropriate Final Sample Tube.

10. Add 100 μL of 100% acetonitrile. Incubate the sample in a sonicating water bath for 15 min, remove the liquid, and transfer to the appropriate Final Sample Tube.

11. Dry down the peptide-containing extracts in the Final Sample Tubes using a vacuum centrifuge until they are ~1 μL in volume. Reconstitute the samples to an appropriate volume for LC-MS (between 5 and 8 μL, depending on the LC injection volume) in 3% acetonitrile and 0.1% formic acid. Incubate the sample in a sonicating water bath for 15 min to ensure the peptides are fully solubilized. Store the samples at −20°C until required. Centrifuge briefly prior to LC-MS analysis.

4. Notes

1. Resuspend the lyophilized 25 nmol DIGE fluor minimal Cy2, Cy3, and Cy5 CyDyes in 125 μL of anhydrous *N,N*-dimethylformamide to produce 200 pmol/μL stock solutions. Aliquot the stocks into clearly labeled screw-top microfuge tubes (1 μL), which should be flushed with argon before capping.

The aliquots can be stored at −80°C for >12 months without appreciable loss of sensitivity. When required, simply add the sample directly to the tube of dye.

2. Set up buffer A and buffer B as the mobile phases. Purge both lines at a flow rate of 1 mL/min for 10 min with the column off-line. Set up the chromatography timetable (see Subheading 3.1, step 2) and perform two blank runs with 100-μL injections of buffer A with the column off-line. Attach the column and equilibrate with buffer A for 4 min at 1 mL/min. Place the MARS column online and perform a blank injection of 100 μL of buffer A at 0.5 mL/min.

3. Additional acetone washing does not reduce the conductivity of the samples and thus is unnecessary, potentially resulting in sample loss.

4. If samples are frozen at this stage, all subsequent processes (i.e., buffer exchange, DIGE labeling, and 2DE) should be performed without refreezing the sample. Repeat freeze-thaw cycles should be avoided.

5. It is important not to centrifuge at colder temperatures, otherwise the urea will precipitate and concentrate in the Vivaspin device.

6. This is not the standard dye/protein ratio recommended by the manufacturer of the CyDyes (GE Healthcare). In our experience, however, the efficiency of labeling at 200 pmol/100 μg of protein is comparable to the recommended 400 pmol/50 μg. The economic benefits of reducing the dye requirement are significant. Furthermore, we have found that a total 300 μg (100 μg contributed from each labeling reaction) is the maximum amount of protein that can be run on a large-format gel without affecting spot resolution, yet still provides enough material for subsequent MS analysis. This negates the requirement for a preparative gel(s) in the experiment as spots are excised directly from the analytical gels.

7. It is important to use reverse osmosis rather than ultrapure water for soaking the paper wicks as ultrapure water is not sufficiently conductive.

8. Keep the coversheet that is used to protect the precast gels prior to electrophoresis. It is useful when it comes to storing the gels in the freezer.

9. It has been found that the gels can be safely thawed at room temperature and rescanned after several months of storage at −80°C (possibly indefinitely) without significant diffusion of the protein spots or damage to the gel or support film.

10. The inclusion of 10% acetonitrile in the digest solution ensures the trypsin has maximum activity and helps to maintain the solubility of hydrophobic peptides.

References

1. Anderson NL, Anderson NG (2002) The human plasma proteome: history, character, and diagnostic prospects. Mol Cell Proteomics. 1(11):845–67.
2. Jacobs JM, Adkins JN, Qian WJ, Liu T, Shen Y, Camp DG 2nd, Smith RD (2005) Utilizing human blood plasma for proteomic biomarker discovery. J Proteome Res. 4(4):1073–85
3. Anderson L (2005) Candidate-based proteomics in the search for biomarkers of cardiovascular disease. J Physiol. 563(Pt 1):23–60.
4. Thadikkaran L, Siegenthaler MA, Crettaz D, Queloz PA, Schneider P, Tissot JD (2005) Recent advances in blood-related proteomics. Proteomics. 5(12):3019–34.
5. Lollo BA, Harvey S, Liao J, Stevens AC, Wagenknecht R, Sayen R, Whaley J, Sajjadi FG (1999) Improved two-dimensional gel electrophoresis representation of serum proteins by using ProtoClear. Electrophoresis. 20(4–5):854–9.
6. Sato AK, Sexton DJ, Morganelli LA, Cohen EH, Wu QL, Conley GP, Streltsova Z, Lee SW, Devlin M, DeOliveira DB, Enright J, Kent RB, Wescott CR, Ransohoff TC, Ley AC, Ladner RC (2002) Development of mammalian serum albumin affinity purification media by peptide phage display. Biotechnol Prog. 18(2):182–92.
7. Boschetti E, Righetti PG (2008) The ProteoMiner in the proteomic arena: a non-depleting tool for discovering low-abundance species. J Proteomics. 71(3):255–64.
8. Gong Y, Li X, Yang B, Ying W, Li D, Zhang Y, Dai S, Cai Y, Wang J, He F, Qian X (2006) Different immunoaffinity fractionation strategies to characterize the human plasma proteome. J Proteome Res. 5(6):1379–87.
9. Rai AJ, Vitzthum F (2006) Effects of preanalytical variables on peptide and protein measurements in human serum and plasma: implications for clinical proteomics. Expert Rev Proteomics. 3(4):409–26.
10. Alban A, David SO, Bjorkesten L, Andersson C, Sloge E, Lewis S, Currie I (2003) A novel experimental design for comparative two-dimensional gel analysis: two-dimensional difference gel electrophoresis incorporating a pooled internal standard. Proteomics. 3(1):36–44.

Part III

Applications in Clinical Proteomics

Chapter 16

Differential Gel-Based Proteomic Approach for Cancer Biomarker Discovery Using Human Plasma

Keun Na, Min-Jung Lee, Hye-Jin Jeong, Hoguen Kim, and Young-Ki Paik

Abstract

Two-dimensional fluorescence difference gel electrophoresis (2D DIGE) has become a general platform for analysis of various clinical samples such as biofluids and tissues. In comparison to conventional 2-D polyacrylamide gel electrophoresis (2D PAGE), 2D DIGE offers several advantages, such as accuracy and reproducibility between experiments, which facilitate spot-to-spot comparisons. Although whole plasma can be easily obtained, the complexity of plasma samples makes it challenging to analyze samples with good reproducibility. Here, we describe a method for decreasing protein complexity in plasma samples within a narrow pH range by depleting high-abundance plasma proteins. In combination with analysis of differentially expressed spots, trypsin digestion, identification of protein by mass spectrometry, and standard 2D PAGE and DIGE, this method has been optimized for comparison of plasma samples from healthy donors and patients diagnosed with hepatocellular carcinoma.

Key words: Two-dimensional fluorescence difference gel electrophoresis, Narrow pH range, Plasma proteomics, Hepatocellular carcinoma, Biomarker

1. Introduction

Human plasma is one of the most readily available clinical samples for discovery of disease biomarkers because it is commonly collected in the clinic and provides noninvasive, rapid analysis for any type of disease (1). Most human plasma proteins are synthesized in the liver, with the exception of γ-globulin.

Separation of plasma proteins by electrophoresis offers a valuable diagnostic tool, as well as a way to monitor clinical progress (2). However, plasma is known to contain a very complex proteome with a dynamic range of more than ten orders and proteins secreted by metabolic trauma from various organs in the human body. For example, approximately 51–71% of plasma protein is albumin, which is a major contributor to osmotic plasma pressure, and assists in the transport of fatty acids and steroid hormones (3). Immunoglobulins make up 8–26% of the plasma protein and play a role in the transport of ions, hormones, and lipids through the circulation system. Approximately 4% is fibrinogen, which can be converted into insoluble fibrin and is essential for the clotting of blood. Regulatory proteins, which make up less than 1% of plasma protein, include cytokines, enzymes, proenzymes, and hormones. Current research regarding plasma protein is centered on performing proteomic analysis of serum/plasma samples to identify disease biomarkers. Gel-based proteomic approaches rely on reducing the complexity of whole plasma by depleting high-abundance proteins with affinity chromatography (4) and/or by using premade IPG strips within a narrow pH range.

Hepatocellular carcinoma (HCC) is a common cancer and accounts for nearly 40% of all cancers and approximately 90% of primary liver cancers in Southeast Asia (5). HCC usually develops in cirrhotic livers that are infected with chronic hepatitis B virus (HBV), hepatitis C virus (HCV), or coinfected with human immunodeficiency virus (HIV) and HBV or HCV (6). Although HCC has been the subject of considerable research interest, the associated prognosis and death rates have remained nearly constant, which has been attributed to inefficient diagnosis. Current techniques for diagnosing HCC involve screening for the presence of one or more biomarkers including alpha-fetoprotein (AFP), des-gamma-carboxyprothrombin (DCP), glypican-3 (GPC3), alpha-L-fucosidase (AFU), and transforming growth factor (TGF)-beta1 (7, 8). Although these biomarkers have proven useful for detecting HCC, they generally suffer from limited sensitivity and/or specificity (9). Thus, the development of a new class of biomarkers for the diagnosis of HCC is an urgent research priority (10–12).

In our laboratory, we have previously used various proteomic techniques, such as two-dimensional electrophoresis (2DE), 2-D liquid chromatography (LC) coupled to the ProteomeLab Protein Fractionation System (PF2D), and isotope labeling, to identify differences in protein expression between clinical plasma and liver tissue samples (13, 14). These proteomic studies suggest that the characterization of proteins with posttranslational modifications (PTMs) and selection of the optimal proteomic methods are the key factors that drive the discovery of novel biomarkers (15–17).

Although 2D PAGE is the most powerful gel-based method to separate and visualize proteins, the recognized problems with this approach are inconsistent gel-to-gel reproducibility and limited dynamic range due to low sensitivity. An improved method is two-dimensional fluorescence difference gel electrophoresis (2D DIGE), in which samples are labeled individually with fluorescent cyanine dye (Cy2, Cy3, and Cy5) and then pooled before separation and scanning in a single gel. This approach overcomes the limitations of 2D PAGE by increasing the quantitative accuracy of detecting spot-to-spot differences (10, 18). To accelerate the discovery of fundamental biomarker candidates in clinical samples, this chapter describes a processing method for plasma samples that facilitates the comparison of healthy donor and HCC patient plasma proteomes using 2D DIGE, narrow pH strips, and nanoLC tandem mass spectrometry (nanoLC-MS/MS) (17).

2. Materials

2.1. Preparation and Pretreatment of Clinical Samples

1. Blood collection tube: K_2-EDTA 7.2 mg BD Vacutainer® (BD Bioscience, San Diego, CA, USA).
2. HPLC system, e.g., HP1100 LC system (Agilent Technologies, Palo Alto, CA).
3. Multiple Affinity Removal System (MARS): LC column (Agilent Technologies; 5185–5984), Buffer A (Agilent Technologies; 5185–5987), Buffer B (Agilent Technologies; 5185–5988).
4. Protease inhibitor (Complete Protease Inhibitor Cocktail, Roche, 11 697 498 001, 20 tablets): Dissolve one tablet containing protease inhibitors (antipain, bestatin, chymostatin, leupeptin, pepstatin, aprotinin, phosphoramidon, and EDTA) in 2 mL of distilled water.
5. Amicon Ultra-15 (5-kDa molecular weight cutoff; Millipore, Barcelona, Spain).
6. 50% (w/v) or 6N trichloroacetic acid (TCA).
7. Lysis buffer: 7 M urea, 2 M thiourea, 4% CHAPS, 30 mM Tris–HCl (pH 8.5) (see Note 1).
8. pH indicator strip.
9. Protein assay: 2-D Quant Kit (GE Healthcare, Piscataway, NJ, USA) or similar assay.

2.2. Components for 2D DIGE and 2D PAGE

1. CyDye reagent: CyDye DIGE Fluor minimal dye (GE Healthcare). Dissolve each dye to 400 pmol/µL in dimethylformamide. Store as 1-µL aliquots in individual tubes at −85°C until use.
2. IPG strip: Immobiline Dry Strip, pH 3.5–4.5, pH 4.0–5.0, pH 4.5–5.5, pH 5.0–6.0, pH 5.5–6.7, pH 7–11, pH 3–10 NL, pH 4–7, 24 cm long, 0.5 mm thick (GE Healthcare).
3. Quenching solution: 10 mM lysine.
4. Sample buffer: 7 M urea, 2 M thiourea, 4% (w/v) 3-((3-cholamidopropyl)dimethylammonio)-1-propanesulfonate (CHAPS), 65 mM DTT, 30 mM Tris–HCl (pH 8.5), trace bromophenol blue (BPB).

5. Sample buffer (2×): 7 M urea, 2 M thiourea, 4% CHAPS, 130 mM DTT, 30 mM Tris–HCl (pH 8.5), trace BPB.
6. Reswelling tray for 24-cm strip.
7. Multiphor™ II and Immobiline Dry Strip cover fluid (GE Healthcare).
8. Power supply: EPS 3501 XL power supply (GE Healthcare).
9. Thermostatic circulator: MultiTemp III thermostatic circulator (GE Healthcare).
10. Carrier ampholyte mixtures: IPG buffer for pH 3.5–4.5, pH 4.0–5.0, pH 4.5–5.5, pH 5.0–6.0, pH 5.5–6.7, pH 7–11, pH 3–10 NL, pH 4–7 (GE Healthcare).
11. Gradient former: Model 395 (Bio-Rad, Milan, Italy).
12. SDS PAGE gel cast: Ettan DALTtwelve Electrophoresis System (GE Healthcare).
13. Ettan DALT low fluorescence (LF) glass plate set (26×20 cm) (GE Healthcare).
14. Ettan DALT glass plate set (26×20 cm) (GE Healthcare).
15. Tris–HCl buffer (5×): 227 g Tris in 1 L of distilled water (adjusted to pH 8.8 with concentrated HCl).
16. SDS buffer (5×): 15 g Tris, 72 g glycine, and 5 g sodium dodecyl sulfate (SDS) in 1 L of distilled water (pH 8.8).
17. Acrylamide stock solution: Acrylamide/Bis-acrylamide 37:5.1, 40% (w/v) solution (AMRESCO, Solon, OH, USA).
18. Equilibration buffer: 180 g urea, 10 g SDS, 100 mL of 5× Tris–HCl buffer, 200 mL of 50% (v/v) glycerol, 31.25 mL of acrylamide stock solution, 5 mM tributylphosphine (TBP) (see Note 2).
19. Gel solution for making 14 gels (26×20 cm, 1-mm spacer, 9–16% gradient): Heavy solution (93.4 mL of 5× Tris–HCl buffer, 199 mL of 40% acrylamide stock solution, 175 mL of 50% glycerol, 1 mL of 10% (w/v) ammonium persulfate (APS), and 100 μL TEMED); light solution (93.4 mL of 5× Tris–HCl buffer, 105 mL of 40% acrylamide stock solution, 1 mL of 10% (w/v) APS, 100 μL TEMED, and 269 mL distilled water).
20. Fixing solution: 40% (v/v) methanol and 5% (v/v) phosphoric acid in distilled water.
21. Coomassie Brilliant Blue G-250 staining solution: 17% (w/v) ammonium sulfate, 3% (v/v) phosphoric acid, 34% (v/v) methanol, and 0.1% (w/v) Coomassie Brilliant Blue G-250 in distilled water.
22. Preparative gel scanner: GS710 model (Bio-Rad), 100-μm high-resolution unit.
23. 2D DIGE gel scanner: Typhoon 9400 imager (GE Healthcare).
24. Image preprocessor: ImageQuant V2005 (GE Healthcare).

25. Intra-gel spot analysis: DeCyder v6.5.11 (GE Healthcare).
26. Evaporator: Speed vacuum (Heto, Copenhagen, Denmark).
27. In-gel digestion buffer: 50 mM NH_4HCO_3 (pH 7.8).
28. Trypsin stock solution: Sequencing grade modified trypsin (Promega, Madison, WI, USA), V5111, 5 vials (20 μg each), 18,100 U/mg. Dissolve 20 μg of one vial in 1 mL of 50 mM NH_4HCO_3.
29. Spot destaining buffer: 40% (v/v) 50 mM NH_4HCO_3 in acetonitrile.

2.3. Analysis by NanoLC-MS/MS

1. NanoLC-MS/MS system (Agilent).
2. LTQ mass spectrometer (Thermo Fisher Scientific, San Jose, CA, USA).
3. Capillary column: 150 × 0.075 mm (Proxeon/Thermo Fisher Scientific).
4. Slurry matrix: 5 μm, 100-Å pore-size Magic C18 stationary phase (Michrom Bioresources, Auburn, CA).
5. Mobile phase A: 0.1% formic acid in distilled water.
6. Mobile phase B: 0.1% formic acid in acetonitrile.
7. Peak list generation: Xcalibur 2.1 (Thermo Fisher Scientific).
8. Peptide data searching: Mascot 2.1.03. (Matrix Science, London, UK) using the NCBInr 06/08/2010 database.
9. MS/MS raw data conversion: BioWorks software (version 3.2, Thermo Fisher Scientific).

3. Methods

3.1. Collection and Preparation of Clinical Samples

1. According to the standard protocol for reference plasma sample collection recommended by the Human Proteome Organization (HUPO) (1), collect the blood of healthy donors and HCC patients into K_2 EDTA tubes, and leave at room temperature for 30 min. Then centrifuge the tubes at 2,400 × g for 15 min to remove red blood cells and cellular particles. Transfer the upper liquid phase (plasma) into cryovials and store at −85°C until use (see Note 3).
2. Dilute 500 μL of human plasma with 2 mL of MARS Buffer A, and add 100 μL of protease inhibitor cocktail solution. Inject 100 μL of the diluted plasma into the Agilent HP1100 LC system equipped with a MARS affinity column at a flow rate of 0.25 mL/min. Collect flow-through fractions, precipitate by addition of 50% TCA solution, and then store the pellet at −20°C overnight (see Note 4).

3. Thaw the pellet at room temperature and resuspend it as small particles in 700 μL of 100% ice-cold acetone using the end part of a 200-μL tip or long-nose tip and repeated aspiration and dispensing. Centrifuge at 20,000×g for 10 min, and discard the supernatant. Resuspend again in 700 μL of 100% ice-cold acetone, centrifuge at 20,000×g for 10 min, and discard the supernatant. Move the pellet against the tube side to easily dissolve it in the lysis buffer using the end part of a tip and dry the pellet at room temperature for 5 min. Add an adequate lysis buffer volume (usually 100–150 μL), vortex gently to prevent the creation of any bubble for 5 min, detach the non-dissolved pellet from the tube wall using a tip, and then vortex again as described above. Centrifuge at 20,000×g for 20 min at 4°C, recover the supernatant, and adjust the protein solution to pH 8.0–9.0 with 1N NaOH, as assessed with a pH indicator strip. Measure the protein concentration and adjust 1,000 μg of each sample to 5 μg/μL concentration for CyDye labeling (see Note 5).

3.2. CyDye Minimal Labeling and Protein Separation by 2DE

1. Prepare the pooled standard (25 μg each of normal and HCC, pooled into one 50 μg total sample), normal (50 μg), and HCC (50 μg) samples as shown in Table 1 (see Note 6). Add 400 pmol of the appropriate dye (Cy2, Cy3, or Cy5) to each sample and vortex, then incubate on ice in the dark for 30 min

Table 1
Experimental design for 2D DIGE using reciprocal labeling, two replicates, and a pooled internal standard

Gel no.	pH range	Cy2	Cy3	Cy5
1	3.5–4.5	Pooled standard (normal+HCC)	Normal	HCC
2	4.0–5.0	Pooled standard (normal+HCC)	Normal	HCC
3	4.5–5.5	Pooled standard (normal+HCC)	Normal	HCC
4	5.0–6.0	Pooled standard (normal+HCC)	Normal	HCC
5	5.5–6.7	Pooled standard (normal+HCC)	Normal	HCC
6	7.0–11.0	Pooled standard (normal+HCC)	Normal	HCC
7	3.5–4.5	Pooled standard (normal+HCC)	HCC	Normal
8	4.0–5.0	Pooled standard (normal+HCC)	HCC	Normal
9	4.5–5.5	Pooled standard (normal+HCC)	HCC	Normal
10	5.0–6.0	Pooled standard (normal+HCC)	HCC	Normal
11	5.5–6.7	Pooled standard (normal+HCC)	HCC	Normal
12	7.0–11.0	Pooled standard (normal+HCC)	HCC	Normal

(see Note 7). Quench by adding 1 μL of 10 mM lysine and incubate on ice for 10 min. Mix the three samples (150 μg) together, and add an equal volume of 2× sample buffer to a final volume of 450 μL. For each preparative gel, mix 1 mg of unlabeled pooled standard proteins and sample buffer to a final volume of 450 μL.

2. Mix 9 μL of IPG buffer for each pH range into 450 μL of the protein solution and incubate for 30 min at room temperature. Rehydrate Immobiline 24-cm Dry Strips of the six pH ranges with protein solution in the strip holder for 16 h at room temperature. Perform first-dimension isoelectric focusing (IEF) using the MultiPhor II electrophoresis system at 20°C with the following conditions: step 1: 100 V for 4 h, step 2: 300 V for 2 h, step 3: 600 V for 1 h, step 4: 1,000 V for 1 h, step 5: 2,000 V for 1 h, step 6: 3,500 V for 29 h (see Note 8).

3. Before the end of the IEF process, prepare all 9–16% 2-D gels, using 12 LF glass plates for the 2D DIGE gels and six general glass plates for the preparative gels.

4. After IEF, transfer the strips into capped glass tubes and soak the strip gels in equilibration buffer containing 5 mM TBP for 25 min (see Note 9). Apply the strips onto the precast 9–16% 2-D gels. Perform electrophoresis with an Ettan DALTtwelve electrophoresis system using the following electrophoresis conditions at 20°C: step 1: 2.5W/gel for 30 min, step 2: 10W/gel for 3 h, step 3: 16W/gel for 4 h.

5. Scan the gels containing the DIGE-labeled proteins using a Typhoon 9400 Imager® set for the excitation/emission wavelengths of each DIGE fluor; Cy2 (488/520 nm), Cy3 (532/580 nm), and Cy5 (633/670 nm) (see Note 10). Crop and save the area of interest using ImageQuant V2005 software.

6. Fix each preparative gel in fixing solution for 2 h. Stain with Coomassie Brilliant Blue G-250 staining solution for 6 h, and destain by washing with distilled water at least three times. Scan each gel, and then pack each one in a clean vinyl bag with water, and store at 4°C.

Figure 1 shows typical 2-D gel spot patterns of whole plasma (a, b) and plasma depleted of High-abundance protein (HAPs) (c, d), respectively. In the image of whole plasma, over 90% of spots contain mainly albumin, IgG heavy and light chain, alpha-1-antitrypsin, IgA, transferrin, haptoglobin, fibrinogen, apolipoprotein A-1, and alpha-1-acid glycoprotein as HAPs. Therefore, differently expressed targets may be included in less than 10% of all spots detected and are likely masked by HAPs. The tools for HAP depletion are commercially available (e.g., Qproteome Albumin/IgG Depletion Kit, QIAGEN; MARS, Agilent Technologies; Seppro® MIXED12-LC20 column, GenWay

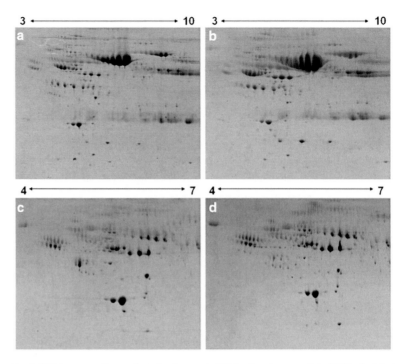

Fig. 1. 2DE image patterns of whole plasma and high-abundance protein (HAP)–depleted plasma by MARS. One milligram of whole plasma (**a, b**) and HAP-depleted plasma (**c, d**) for "normal" (**a, c**) and "HCC" (**b, d**).

Biotech; ProteoPrep® 20 Plasma Immunodepletion Kit, Sigma-Aldrich; etc.). We used MARS (Agilent) for depletion of six HAPs (albumin, IgG heavy and light chain, alpha-1-antitrypsin, IgA, transferrin, haptoglobin), and the recovery of low-abundance proteins was about 10%. The HAP depletion of C (normal) and D (HCC) shows clearer spot images than those of A and B, but many spots appear to be clustered. To solve these problems, we applied narrow-pH-range strips (single pI, 1.0) and run the 2D DIGE to minimize spot intensity variations. In Fig. 2, the protein spots shown in a wide-pH-range strip were separated well, and many spots appeared to be differentially expressed. Some of the 43 target spots identified by MALDI-TOF MS turned out to be the same protein with different pI on the 2-D gel (Table 2), indicating that these are modified (e.g., by glycosylation or phosphorylation).

3.3. Image Analysis and In-Gel Tryptic Digestion

1. Load the DIGE images of the gels into the DeCyder program. Group the images as "Standard," "Normal," or "HCC" in accordance with Table 1. Set the estimated number of spots for each codetection procedure to "2500" and select "Student's t test" as the test for statistical confidence of the analysis. Perform intra-gel analysis and spot matching using the difference in-gel

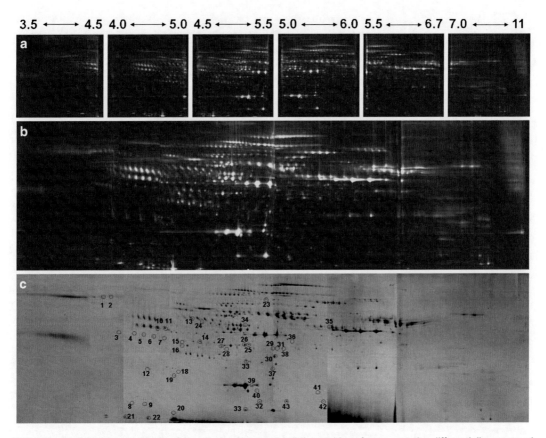

Fig. 2. The 2D DIGE image patterns of six narrow pH ranges and the position of representative differentially expressed spots with a fold change of >±2. Six images (**a**) were combined into one image. (**b**) The "normal" sample was labeled with Cy3 (*green*) and the "HCC" sample with Cy5 (*red*). Forty-three spots were identified from preparative gels by nanoLC-MS/MS (**c**).

analysis (DIA) and biological variation analysis (BVA) mode (see Note 11).

2. Using a master gel, match and merge accurately the spots of the other gels, if necessary. Accept statistically significant spots ($p < 0.05$), and filter over the average volume ratio of ±2. Select and check for accuracy across filtered spots of the 2D DIGE and preparative gels (Fig. 3) (see Note 12).

3. Pick each protein spot of interest with an autoclaved end-cut yellow tip (~2 mm), and transfer the gel piece into a fresh 1.5-mL tube containing 1 mL of distilled water. Wash the gel piece twice by adding 100 µL of spot destaining buffer (40% (v/v) 50 mM NH_4HCO_3 in acetonitrile), shaking for 10 min and discarding the destaining buffer. Repeat this step until the Coomassie Blue G-250 dye disappears (~5 times). Add 50 µL of 100% acetonitrile, shake for 3 min, and discard the acetonitrile. Repeat this step until the gel piece turns white

Table 2
List of 43 differentially expressed proteins (normal vs. HCC) identified by nanoLC-MS/MS

Spot. #	GI #	Protein name	Ratio (HCC/Nor)	p value	MW	pI	Score	Pep. Match	Cov. (%)
Decreased									
1	gi|179619	Plasma protease (C1) inhibitor precursor	−3.23	0.01	55,182	6.09	72	5	3
2	gi|179619	Plasma protease (C1) inhibitor precursor	−2.41	0.02	55,182	6.09	174	6	9
10	gi|112910	Alpha-2-HS-glycoprotein	−7.11	0.00005	39,324	5.43	246	111	12
11	gi|112910	Alpha-2-HS-glycoprotein	−5.09	0.0017	39,324	5.43	96	3	12
12	gi|619383	Apolipoprotein D_1 apoD	−2.19	0.02	27,993	5.14	141	11	12
14	gi|130675	Serum paraoxonase/arylesterase 1	−7.27	0.0027	39,749	5.08	189	17	14
20	gi|5776545	Tax1-binding protein, TXBP151	−2.99	0.012	86,251	5.31	49	4	1
21	gi|186972736	Apolipoprotein C-III	−2.25	0.00057	8,765	4.72	176	18	45
22	gi|14277770	Apolipoprotein C-II	−2.04	0.007	8,915	4.66	272	58	69
28	gi|130675	Serum paraoxonase/arylesterase 1	−4.82	0.0018	39,749	5.08	87	16	5
30	gi|219978	Prealbumin	−2.36	0.026	15,919	5.52	104	5	9
33	gi|219978	Prealbumin	−3.00	0.014	15,919	5.52	132	10	9
34	gi|179161	Antithrombin III	−2.03	0.0067	52,618	6.32	76	5	8
35	gi|177872	Alpha-2-macroglobulin	−2.98	0.002	70,794	5.47	518	71	21
36	gi|388519	Complement factor H-related protein 1, FHR-1	−4.18	0.044	37,244	7.81	94	4	9
38	gi|182424	Alpha-fibrinogen precursor	−2.30	0.0088	69,809	8.26	145	5	9
39	gi|229479	Lipoprotein Gln I	−2.02	0.011	28,346	5.27	247	17	28
40	gi|35897	Retinol-binding protein 4	−2.04	0.00079	22,868	5.48	142	16	12
41	gi|62988821	Alstrom syndrome 1, ALMS1 protein	−2.67	0.0067	2,78,758	5.03	47	2	0

Increased									
3	gi|16418467	Leucine-rich alpha-2-glycoprotein precursor	2.49	0.00053	38,177	6.45	211	13	13
4	gi|16418467	Leucine-rich alpha-2-glycoprotein precursor	2.71	0.0046	38,177	6.45	286	33	13
5	gi|16418467	Leucine-rich alpha-2-glycoprotein precursor	2.12	0.011	38,177	6.45	356	56	15
6	gi|16418467	Leucine-rich alpha-2-glycoprotein precursor	2.14	0.0031	38,177	6.45	270	26	13
7	gi|16418467	Leucine-rich alpha-2-glycoprotein precursor	2.02	0.0027	38,177	6.45	76	2	2
8	gi|179674	Complement component C4A	2.76	0.017	1,92,859	6.65	84	22	0
9	gi|179674	Complement component C4A	3.62	0.011	1,92,859	6.65	148	32	1
13	gi|1236759	Golgin, 256 kD	3.07	0.00022	2,55,737	5.34	48	4	0
15	gi|78101271	Complement component C3C	3.38	0.022	3,9,488	4.79	379	48	22
16	gi|78101271	Complement component C3C	3.16	0.034	39488	4.79	599	102	30
17	gi|78101271	Complement component C3C	2.95	0.0049	39,488	4.79	844	332	41
18	gi|553734	T cell receptor C-alpha	2.06	0.0085	2,214	9.79	51	34	38
19	gi|4262000	14-3-3 protein/cytosolic phospholipase A2	3.81	0.0034	7,957	8.16	46	1	20
23	gi|182439	Fibrinogen gamma chain	2.21	0.0013	49,481	5.61	242	15	11
24	gi|532198	Angiotensinogen	2.40	0.01	53,155	5.78	207	12	8
25	gi|28336	Mutant beta-actin	2.59	0.013	41,812	5.22	367	30	20
26	gi|1333634	Paraoxonase-3	2.62	0.0026	38,461	4.95	71	5	7
27	gi|1230564	RNA helicase-II/Gu protein	12.48	0.0047	89,250	9.36	46	2	2
29	gi|12232634	Apolipoprotein L-I	3.42	0.0017	42,383	5.99	214	18	15
31	gi|12232634	Apolipoprotein L-I	3.44	0.0037	42,383	5.99	189	6	6
32	gi|180249	Ceruloplasmin	19.47	0.0029	97,698	5.29	143	25	2
37	gi|4557739	Mannose-binding protein C precursor	2.05	0.00081	26,143	5.39	332	16	22
42	gi|34810822	Alpha-1-antitrypsin precursor	6.16	0.00034	25,409	8.23	89	4	20
43	gi|306882	Haptoglobin precursor	18.68	0.00023	45,204	6.24	237	11	9

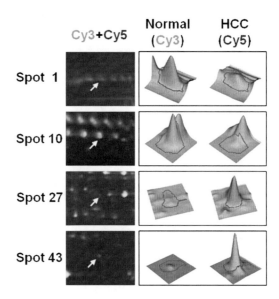

Fig. 3. Representative spot images showing the overlapped Cy3-Cy5 fluorescence image and the same data as 3-D intensity plot using DeCyder. *Green* indicates the proteins that are more abundant in "normal" plasma, while *red* indicates proteins that are more abundant in "HCC" plasma.

(~2 times). Remove the supernatant, and dry the gel piece in a speed vacuum evaporator for 10 min. Add 2.5 µL of the trypsin stock solution, and leave the gel piece on ice for 45 min. Add 17.5 µL of 50 mM NH_4HCO_3 and incubate the gel piece at 37°C for 12 h (see Note 13).

4. Transfer the supernatant to a fresh tube. Add 50 µL of acetonitrile and then shake for 3 min. Collect and combine the supernatants and repeat this step twice. Dry the combined supernatants (digest solution) in a speed vacuum evaporator for 10 min. Store the dried peptides at 4°C until performing nanoLC-MS/MS for protein identification (see Note 14).

3.4. Protein Identification by NanoLC-MS/MS

1. Identify the digest peptides by nanoLC-MS/MS with a LTQ mass spectrometer using a capillary column packed with C18 stationary phase slurry. Set the solvent gradient for the column as follows: 8% B to 35% B in 30 min, 85% B in 10 min, and 8% B in 15 min, maintaining a 300 nL/min flow rate. Acquire mass spectra using data-dependent acquisition with a full mass scan (m/z 360–1,200) followed by MS/MS scans and generate MS peak lists using the appropriate software. Set the temperature of the ion transfer tube to 120°C, the spray voltage to 1.7–2.2 kV, and for MS/MS the normalized collision energy to 32%.

2. Convert raw data into an XML file and identify peptide sequences using the software Mascot searching the NCBInr

database. The search parameter settings should be as follows: *Homo sapiens*, variable modification, oxidized at methionine residues (+16 Da), carbamidomethylated at cysteine residues (+57 Da), maximum allowed missed cleavage = 1, MS tolerance = 1.2 Da, MS/MS tolerance = 0.6 Da, and charge states = 2+ and 3+. Filter the matched peptides with a significance threshold of $p < 0.05$ and set the minimum threshold to 30 Mascot peptide score. For further details see ref. 12.

4. Notes

1. The lysis buffer must not contain DTT or BPB because DTT interferes with CyDye labeling and BPB obstructs the CyDye color checking during the labeling process.

2. The equilibrium solution must be made in the dark without the addition of distilled water, and TBP must be added freshly to the equilibrium solution prior to 2DE. The single-step treatment of TBP and acrylamide is used for efficient reduction and alkylation of cystine/cysteine residues (19).

3. Healthy donors and the HCC case control patients tested negative for HIV-1 and HIV-2 antibodies, HIV-1 antigen (HIV-1), hepatitis B surface antigen (HBsAg), hepatitis B core antigen (anti-HBc), hepatitis C virus (anti-HCV), HTLV-I/II antibody (anti-HTLV-I/II), and syphilis. The HCC patients' clinical and pathologic data were gathered at Yonsei University College of Medicine and are as follows: 70 years of age, male, and cancer grade = HCC stage II with 10% necrosis of liver tissue. Authorization for use of plasma for research purposes was obtained from the Institutional Review Board (IRB).

4. Bound proteins are eluted from the MARS column with Buffer B at a flow rate of 1 mL/min for 3.5 min. The MARS column is regenerated by equilibrating with Buffer A for 8 min at a flow rate of 1 mL/min.

5. After TCA treatment, the pellets must be resuspended in acetone into very small particles for the following reasons. First, any residual TCA remaining in the pellet may increase the amount of NaOH solution necessary to adjust to pH 8.5 for CyDye labeling, and NaOH interferes with IEF. Second, resuspension maximizes the surface of the particles and subsequent exposure to the labeling reaction. Third, this method minimizes protein loss.

6. The 50 μg of the internal pooled standard sample is prepared by combining 25 μg each of the two samples prior to Cy2 labeling.

7. The reaction time for 50 μg of protein and 400 pmol of CyDye must be kept at 30 min or overlabeling will occur, and single spot images may appear as double spots.

8. If it is not used immediately after IEF, each strip can be packed to prevent exposure to air humidity and light and then stored at −85°C until use.

9. To increase the solubility of TBP, 4% isopropanol is added into the equilibrium solution, and the sample is sonicated for 30 min at room temperature. The reaction time must be kept under 25 min. Handling of TBP should be performed in a fume hood because it is very corrosive and flammable.

10. For fluorescence scanning, the photomultiplier tube (PMT) should be adjusted equally for the three CyDye emission wavelengths with respect to total spot volume intensity for all gels with ImageQuant software. Expected PMT values are in the range of 500–600 V. A PMT value over 600 V results from undefined background signal. 2D DIGE gel plates on standby for scanning should be kept in the dark at 4°C.

11. For image analysis, an adequate estimated spot number is 2,500 for plasma or serum because of the presence of high-abundance proteins. If set over 2,500, the high-abundance spots will be split into several areas, and then each area must be manually merged. If set under 2,500, nearby spots might be assigned as one area. In this case, the spot areas cannot be split by the DeCyder program (v6.5.11). If the clinical sample is cells or tissue, an adequate estimation of spot number is usually 3,000.

12. The threshold for the fold change is usually set to more than ±1.5-fold. In the case of 2D PAGE, the ratio cutoff is over ±2.0 due to gel-to-gel variation. In our results, a threshold of ±1.5 produced very large numbers of differentially expressed protein spots.

13. The gel piece is easy to lose at this point because it becomes smaller and transparent during the destaining procedure.

14. When the digest solution is not used immediately, store at −70°C or lyophilize the solution to inhibit proteolysis.

Acknowledgments

This study was supported by a grant from the National R&D Program for Cancer Control (1120200 to YKP), National Project for the Personalized Medicine, Ministry for Health and Welfare, Republic of Korea (A111218-11-CP01 to YKP). The authors declare no conflicts of interest. Corresponding author: Young-Ki Paik (paikyk@yonsei.ac.kr).

References

1. Paik YK, Kim H, Lee EY et al (2008) Overview and introduction to clinical proteomics. Methods Mol Biol 428:1–31
2. Hochstrasser DF, Sanchez JC, Appel RD (2002) Proteomics and its trends facing nature's complexity. Proteomics 2(7):807–812
3. Cho SY, Lee EY, Kim HY et al (2008) Protein profiling of human plasma samples by two-dimensional electrophoresis. Methods Mol Biol 428:57–75
4. Cho SY, Lee EY, Lee JS et al (2005) Efficient prefractionation of low--abundance proteins in human plasma and construction of a two-dimensional map. Proteomics 5(13):3386–3396
5. Parkin DM, Bray F, Ferlay J et al (2005) Global cancer statistics. 2002. CA Cancer J Clin 55:74–108
6. Wright LM, Kreikemeie JT, Fimmel CJ (2007) A concise review of serum markers for hepatocellular cancer. Cancer Detect Prev 31:35–44
7. Daniele B, Bencivenga A, Megna AS et al (2004) Alpha-fetoprotein and ultrasonography screening for hepatocellular carcinoma. Gastroenterology 127:S108–S112
8. Sun S, Lee NP, Poon RT et al (2007) Oncoproteomics of hepatocellular carcinoma: from cancer markers' discovery to functional pathways. Liver Int 27(8):1021–1038
9. Santamaría E, Muñoz J, Fernández--Irigoyen J et al (2007) Toward the discovery of new biomarkers of hepatocellular carcinoma by proteomics. Liver Int 27(2):163–173
10. Lilley KS, Friedman DB (2004) All about DIGE: quantification technology for differential-display 2D-gel proteomics. Expert Rev Proteomics 1(4):401–409
11. Park KS, Kim H, Kim NG et al (2002) Proteomic analysis and molecular characterization of tissue ferritin light chain in hepatocellular carcinoma. Hepatology 2002 35(6):1459–1466
12. Park KS, Cho SY, Kim H et al (2002) Proteomic alterations of the variants of human aldehyde dehydrogenase isozymes correlate with hepatocellular carcinoma. Int J Cancer 97(2):261–265
13. Lee HJ, Lee EY, Kwon MS et al (2006) Biomarker discovery from the plasma proteome using multidimensional fractionation proteomics. Curr Opin Chem Biol 10(1):42–49
14. Lee HJ, Na K, Kwon MS et al (2009) Quantitative analysis of phosphopeptides in search of the disease biomarker from the hepatocellular carcinoma specimen. Proteomics 9(12):3395–3408
15. Jeong SK, Lee EY, Cho JY et al (2010) Data management and functional annotation of the Korean reference plasma proteome. Proteomics 10(6):1250–1255
16. Jeong SK, Kwon MS, Lee EY et al (2009) BiomarkerDigger: a versatile disease proteome database and analysis platform for the identification of plasma cancer biomarkers. Proteomics 9(14):3729–3740
17. Na K, Lee EY, Lee HJ et al (2009) Human plasma carboxylesterase 1, a novel serologic biomarker candidate for hepatocellular carcinoma. Proteomics 9(16):3989–3999
18. Friedman DB, Lilley KS (2008) Optimizing the difference gel electrophoresis (DIGE) technology. Methods Mol Biol 428:93–124
19. Yan JX, Kett WC, Herbert BR et al (1998) Identification and quantitation of cysteine in proteins separated by gel electrophoresis. J Chromatogr A 813:187–200

Chapter 17

2D DIGE for the Analysis of RAMOS Cells Subproteomes

Marisol Fernández and Juan Pablo Albar

Abstract

Overexpression of human polμ in a Burkitt's lymphoma-derived B cell line (RAMOS), in which somatic hypermutation (SHM) is constitutive, induced an increase in somatic mutations in the parental cell line (Nucleic Acids Res 32:5861–5873, 2004). To further study Polμ implications in SHM, a dominant-negative (DN) mutant of Polμ (Polμ-DN) was generated which showed moderated overexpression of the Polμ-DN protein. The subcellular prefractionation was used to improve the detection of low-abundance proteins contained in membrane/organelles and nuclei, which are efficiently separated from high-abundance proteins commonly found in the cytosol that might otherwise hamper analysis. Two-dimensional (2D) difference gel electrophoresis (DIGE) is a technique for comparative proteomics, which improves the reproducibility and reliability of differential protein expression analysis between samples. The standard sample included in every gel (Cy2) comprises equal amounts of each sample to be compared, and thus improves the accuracy of protein quantification between samples from different gels, allowing accurate detection of small differences in protein levels. The combination of this techniques allowed the detection in Fraction F2 (membrane/organelles) of 2,111 spots, 55 of them with significant variation (19 increased and 36 decreased in a ratio >2.0 or <−2.0), and in Fraction F3 (nuclear) of 2,416 spots, 80 of them with significant variation (51 increased and 29 decreased in a ratio >1.5 or <−1.5).

Key words: Subcellular proteome, Difference analysis, Protein quantitation, DIGE, Gel-to-gel matching

1. Introduction

The separation of complex protein mixtures is very important in proteomics to allow the detection of low-abundance proteins for quantitative and qualitative analysis (1, 2). The success in such analysis depends on standardized and reproducible sample preparation methods prior to protein separation by 2D gel electrophoresis (3, 4). Due to the complexity of eukaryotic proteomes, it is convenient to partial fractionate proteomes of a given organism in order to maximize the coverage of the sample and to increase the visualization of low-abundance proteins and at the same time make them accessible for subsequent analysis (5). Since both membrane proteins and nuclear proteins are of particular pharmacological interest, a suitable extraction procedure is desirable that selectively harvests these proteins (which are frequently present in very low quantities) into distinct subcellular fractions, separated from high-abundance proteins present in cytosol or cytoskeleton. The method of sequential extraction of cell content is based on the different solubility of certain subcellular compartments in special reagent mixtures (6). Sequential extraction results in a stepwise disintegration of the cell's structure, yielding four subcellular fractions (6). Upon treatment with the first extraction buffer, cells shrink as a result of the release of the cytosolic content but remain intact in their overall structure. After the second extraction step, due to the solubilization of membranes and membrane organelles, only isolated nuclei and the cytoskeleton remain intact. The treatment of the residual material with the third extraction buffer destroys the structure of the nucleus, thus probably solubilizing the nuclear proteins. Finally, the cytoskeleton components are liberated during the fourth and final extraction.

In the application of difference gel electrophoresis (DIGE), different samples are labeled with fluorescent dyes of similar mass and identical charge and are then separated on the same 2D gel. The standard sample included in each gel comprises equal amounts of each sample to be compared and was found to improve the accuracy of protein quantification between samples from different gels, allowing accurate detection of small differences in protein levels between samples.

Image analysis is automatically performed with minimum user intervention with DeCyder Differential Analysis Software which comprises the four modules listed below.

DIA (differential in-gel analysis): protein spot detection and quantification on a set of images, from the same gel. Features include background subtraction, in-gel normalization, and gel artifact removal.

BVA (biological variation analysis): matching of multiple images from different gels to provide statistical data on differential protein abundance levels between multiple groups.

Batch Processor: fully automated image detection and matching of multiple gels without user intervention.

XML toolbox: extraction of user-specific data facilitating automatic report generation (7, 8).

The DIA module quantifies the spot volume for each image and expresses these values as a ratio, thereby indicating changes in abundance by direct comparison of corresponding spots. This ratio parameter can be used, in a small-scale experiment, to directly evaluate changes between two labeled protein samples (e.g., control and drug-treated samples). Alternatively, the ratio can be used for protein spot quantification of a sample against a pooled internal standard to allow accurate inter-gel protein abundance comparisons.

The BVA module processes multiple gel images, performing gel-to-gel matching of spots and enabling quantitative comparisons of protein abundance across multiple gels.

The Batch Processor executes both the DeCyder DIA and BVA processes, performing fully automated spot codetection and inter-gel matching of multiple gel images. Once the Batch Processor has been set up with the necessary image files and parameters, the gels are processed sequentially without user intervention. The Batch Processor module significantly reduces the hands-on analysis time required.

2. Materials

Both cell conditions, Ramos-ev and Ramos-DN cells, were provided by Dr. A. Bernad (CNIC, Madrid).

2.1. Subcellular Fractionation

1. ProteoExtract™, subcellular proteome extraction kit (Calbiochem; see Note 1).

2.2. Chloroform/Methanol Precipitation

1. Chloroform, ultrapure grade.
2. Methanol, ultrapure grade.
3. Deionized H_2O, from a Milli-Q water purification system or equivalent.

2.3. Sample Solubilization and Quantification of Proteins

1. Lysis buffer: 30 mM Tris–HCl, pH 8.5, 7 M urea, 2 M thiourea, and 4% (w/v) CHAPS. Store in aliquots at –20°C.
2. pH Test strips 4.5–10.0.
3. RC DC protein assay kit (Bio-Rad): RC reagent I and RC reagent II (see Note 2), DC protein assay reagent A (see Note 3), protein assay reagent B (see Note 4), and protein assay reagent S (see Note 5).

2.4. CyDye DIGE Fluor Minimal Dyes Solutions for Labeling Protein Samples

1. Anhydrous dimethylformamide (DMF), 99.8% (DMF) (Sigma) (see Note 6).
2. 1-mL syringe and 25G×5/8″ (0.5×16 mm) needle (see Note 7).
3. CyDye DIGE fluor Cy3 (minimal dye) (GE Healthcare).
4. CyDye DIGE fluor Cy5 (minimal dye) (GE Healthcare).
5. CyDye DIGE fluor Cy2 (minimal dye) (GE Healthcare).
6. 10 mM lysine (Sigma). Store in aliquots at –20°C.

2.5. Isoelectric Focusing

1. Rehydration buffer: 7 M urea, 2 M thiourea, and 4% (w/v) CHAPS, stored in aliquots at −20°C. For use, add carrier ampholytes and DeStreak and/or dithiothreitol (DTT) (see Notes 8–10).
2. Immobiline DryStrips (GE Healthcare).
3. Immobiline reswelling tray (GE Healthcare).
4. Cover fluid (GE Healthcare).
5. IPGphor cup-loading strip holder (GE Healthcare).
6. Electrode pads (GE Healthcare).
7. Ettan IPGphor IEF unit (GE Healthcare).

2.6. Equilibration Solutions

1. Equilibration solution: 50 mM Tris–HCl, pH 8.8, 6 M urea, 30% (v/v) glycerol, 2% (w/v) SDS, and trace bromophenol blue. Store in aliquots at −20°C.
2. Reduction solution: equilibration solution containing 1% DTT.
3. Alkylation solution: equilibration solution containing 4% iodoacetamide (see Note 11).

2.7. SDS-Polyacrylamide Gel Solutions

1. 0.3% (w/v) agarose in running buffer 1×. Store at 4°C.
2. Water-saturated butan-2-ol (see Note 12).
3. SDS electrophoresis buffer:

   1× = (25 mM Tris–HCl, 192 mM glycine, 0.1% (w/v) SDS) for lower tank.

   2× = (25 mM Tris–HCl, 192 mM glycine, 0.2% (w/v) SDS) for top tank.

   Store at room temperature (RT) (see Note 13).
4. Monomer stock solution: 30% (w/v) acrylamide and 0.8% (w/v) methylenebisacrylamide. Electrophoresis purity reagent Bio-Rad (see Note 14). Store at 4°C in dark.
5. Resolving gel buffer: 1.5 M Tris–HCl and pH 8.8. For 1 L of solution, add approximately 33 mL of concentrated HCl (12 N), then make up to 1 L with water. Store at 4°C.
6. 10% (w/v) SDS (see Note 15).
7. Ammonium persulfate: 10% (w/v) solution in water. Prepare fresh each time.
8. N,N,N,N'-tetramethylethylenediamine (TEMED) (Sigma). Store at 4°C.
9. 12.5% acrylamide gel solution with 0.2% (w/v) SDS for four gels: 125 mL of acrylamide solution (30% acrylamide: 0.8% methylenebisacrylamide), 75 mL of 1.5 M Tris–HCl, pH 8.8, 93.5 mL of Milli-Q water, 6 mL of 10% (w/v) SDS (0.2% (w/v) final concentration), 3 mL of 10% (w/v) ammonium persulfate, and 41.5 mL of 100% TEMED.

2.8. Scanning Gels and Analyzing

1. Typhoon 9400 (GE Healthcare).
2. DeCyder V7.0 Software (GE Healthcare).

3. Methods

3.1. Subcellular Fractionation

Both cell conditions were grown in parallel, and protein extracts from the nuclear, membranes, and cytosolic fractions were prepared as described in the protocol for suspension culture cells, included in the ProteoExtract™ Subcellular Proteome Extraction Kit (Calbiochem, San Diego, USA) but with a greater number of cells for the subsequent analysis.

Briefly, cells in their logarithmic growth phase were centrifuged and washed twice with 2 mL of ice-cold wash buffer, at $300 \times g$ and 4°C for 10 min. The cell pellet was resuspended in 1 mL of ice-cold extraction buffer I and 5 μL of protease inhibitor cocktail, incubated for 10 min at 4°C under gentle agitation and centrifuged at $1,000 \times g$ and 4°C for 10 min. The supernatant was Fraction F1, and the cell pellet was resuspended in 1 mL of ice-cold extraction buffer II and 5 μL of protease inhibitor cocktail, incubated for 30 min at 4°C under gentle agitation and centrifuged at $6,000 \times g$ and 4°C for 10 min. The supernatant was Fraction F2, and the cell pellet was resuspended in 500 μL of extraction buffer III with 5 μL of protease inhibitor cocktail and 1.5 μL (>375 U) of benzonase, incubated for 10 min at 4°C under gentle agitation and centrifuged at $6,800 \times g$ and 4°C for 10 min. The supernatant was Fraction F3, and the cell pellet was resuspended in 500 μL of room temperature extraction buffer IV and 5 μL of protease inhibitor cocktail to give Fraction F4 (see Note 1).

3.2. Sample Preparation

One of the most important steps is the sample preparation which includes the elimination of all the contaminants present that could interfere with the electrophoresis, such as salts, detergents, nucleic acids, lipids, polysaccharides, phenols, etc. In this experiment in particular, the reagents of the kit used in Subheading 3.1 are incompatible with electrophoresis. Due to this fact, the following chloroform-/methanol-based protein precipitation protocol was used:

1. 1 mL of Fractions F2 and F3 are divided into 250 μL aliquots. It is recommended to perform the entire process in 1.5- or 2.0-mL polypropylene tubes.
2. Add 3 volumes of methanol and 1 volume of chloroform and vortex.
3. Add 3 volumes of Milli-Q water and vortex well.

4. Centrifuge for 10 min at 15,000×g at 4°C. Proteins should be at the interface.

5. Remove carefully the aqueous top layer and discard it.

6. Add 3 volumes of methanol and vortex well.

7. Centrifuge for 5 min at 15,000×g. Proteins should be seen as a precipitate.

8. Remove carefully as much liquid as possible, without disturbing the precipitate.

9. Allow the pellet to air-dry for 10–15 min.

10. Store the sample at –20°C until needed.

3.3. Sample Solubilization and Quantification of Proteins

These steps describe the preparation of a cell lysate compatible with CyDye DIGE fluor minimal dye labeling.

1. Resuspend the precipitated proteins in a minimal volume of the lysis buffer and vortex at room temperature for 1 h. If necessary, sonicate intermittently until the precipitate is dissolved (see Note 16).

2. Centrifuge the cell lysate for 10 min at 15,000×g in a microcentrifuge.

3. Transfer the supernatant which contains the cell lysate to a labeled tube. Discard the pellet.

4. Check the pH of the cell lysate by spotting 0.2 μL on a pH indicator strip; it should be pH 8.0–9.0. If the pH is below pH 8.0, it will need to be adjusted before labeling (see Note 17).

5. Determine the quantity of proteins by RC DC protein assay (see Note 18).

3.4. Preparation of CyDye DIGE Fluor Minimal Dyes for Protein Labeling

1. Take the CyDye DIGE fluor minimal dye and leave it to thaw for 5 min at room temperature without opening. This will prevent exposure of the dye to water condensation which may cause hydrolysis.

2. Dispense a small volume of DMF from its original container into a microfuge tube. Reconstitute the stock CyDye DIGE fluor minimal dye in DMF (see Notes 6 and 7). CyDye fluor minimal dye is supplied as a solid and is reconstituted in DMF to a concentration of 1 nmol/μL.

After reconstitution in DMF, the dye will give a deep color: Cy2-yellow, Cy3-red, and Cy5-blue. The dye stock solution (1 nmol/μL) is now prepared. The dye will need to be further diluted to the working dye solution before being used in the labeling reaction.

3.5. Calculating the Amount of CyDye DIGE Fluor Minimal Dye Required to Label a Protein Lysate

Prior to the steps in this subheading, the experimental design should be made in case it has not previously been designed (see Table 1).

It is recommended to use 400 pmol of dye for labeling 50 μg of protein.

1. Vortex and spin down the dye stock solution in a microfuge.
2. In order to achieve 400 pmol/μL of working dye solution in 5 μL, take a fresh microcentrifuge tube and add 3 μL of DMF.
3. To the 3 μL of DMF, add 2 μL of reconstituted dye stock solution. Ensure that all dye is removed from the pipette tip by pipetting up and down several times into the working dye solution.

The recommended experimental design should include a pooled internal standard from all the samples. Enough amount of internal standard must be prepared in order to be included on every gel in the experiment.

3.6. Protein Labeling with the CyDye DIGE Fluor Minimal Dyes

1. Add a volume of protein sample equivalent to 50 μg to a microfuge tube.
2. Add 1 μL of working dye solution to the microfuge tube containing the protein sample (50 μg of protein is labeled with 400 pmol of dye for the labeling reaction).

Table 1
Experimental design for labeling Fraction F2 (membrane and organelles) using biological replicates from four different extractions, randomization, and inverse labeling. Fraction F3 (nuclear) should be labeled analogously

Gel	Cy2 standard interno	Cy3	Cy5
1	50 μg (6.25 μg each/sample): WT F2 (1,2,3,4), –DN F2 (1,2,3,4)[a]	50 μg WT F2 (1)	50 μg –DN F2 (2)
2	50 μg (6.25 μg each/sample): WT F2 (1,2,3,4), –DN F2 (1,2,3,4)[a]	50 μg WT F2 (2)	50 μg –DN F2 (4)
3	50 μg (6.25 μg each/sample): WT F2 (1,2,3,4), –DN F2 (1,2,3,4)[a]	50 μg –DN F2 (1)	50 μg WT F2 (3)
4	50 μg (6.25 μg each/sample): WT F2 (1,2,3,4), –DN F2 (1,2,3,4)[a]	50 μg –DN F2 (3)	50 μg WT F2 (4)

[a]WT = RAMOS-ev; dominant-negative mutant of Polμ (–DN)

3. Mix the dye and protein sample thoroughly by vortexing.

4. Centrifuge briefly in a microcentrifuge to collect the solution at the bottom of the tube. Leave the tube on ice and in the dark for 30 min.

5. Add 1 µL of 10 mM lysine to stop the reaction. Mix and spin the tube briefly in a microcentrifuge. Leave the tube on ice and in the dark for 10 min.

Labeling is now finished. The labeled samples can be processed immediately or stored for up to 3 months at –80°C in the dark.

3.7. Isoelectric Focusing

The main difference between conventional 2D gel electrophoresis and DIGE is that the latter will enable up to three different protein samples to be run on a single 2D gel. To achieve this, you need to mix the differentially labeled protein samples before the first dimension run.

1. Rehydrate the Immobiline DryStrips for 8–22 h (gel side down), without the protein sample, in the Immobiline DryStrip reswelling tray. The total volume applied to the strip depends on its length and the manufacturer's instructions. We used 18 cm pH 3–11 NL Immobiline DryStrip, a total volume of 340 µL of rehydration buffer containing 0.5% ampholytes and 1.2% DeStreak per strip. The strip is then covered with 2 mL of IPG cover fluid.

2. After rehydration of the Immobiline DryStrip, hold the end of the strips with forceps and lift the strip out of the tray. Allow excess cover fluid to run off the Immobiline DryStrip onto a piece of tissue. Place the Immobiline DryStrip gel side up with the basic end (flat end of Immobiline DryStrip) and flush with the flat end of the IPGphor cup-loading strip holder.

3. Prepare electrode pads by cutting 5 × 10 mm pieces from the IEF electrodes strips. Place these on a clean dry surface such as a plate and soak with distilled water. Remove excess water by blotting with filter paper. Place a prepared damp electrode pad onto the acidic and basic end of the gel.

4. Clip down the electrodes firmly onto the electrode pad. Ensure that there is a good contact with the Immobiline DryStrip and the metal on the outside of the strip holder.

5. Apply at least 4 mL of DryStrip cover fluid to the IPGphor cup-loading strip holder, allowing the oil to spread so it completely covers the Immobiline DryStrip.

6. Clip the loading cup in the acidic end of the strip, so it is positioned between the two electrodes.

7. Combine the three differentially labeled samples into a single microfuge tube and mix. One of these samples should be the pooled internal standard.

8. Add a volume of rehydration buffer (7 M urea, 2 M thiourea, 4% (w/v) CHAPS, with 1% ampholytes, 20 mM DTT) and leave on ice for at least 10 min. The samples are now ready for the first dimension isoelectric focusing step. Once the rehydration buffer has been added, it is recommended that the sample is run immediately after at least 10 min on ice.

9. Load up to 100 μL of sample into the bottom of the loading cup.

10. Put the clear plastic strip cover onto the strip holder. The strip holder is now ready to be loaded onto the IPGphor IEF unit.

11. For cup loading, the following parameters were used: 20°C 50–75 μA/strip, applying an increasing voltage as follows: 300 V for 3 h, gradient increase from 300 to 1,000 V for 6 h, gradient increase from 1,000 to 8,000 V for 3 h, and a final step increase to 8,000 V for 3 h.

12. When focusing is complete, remove the strip, drain off the mineral oil on a paper towel, and place it onto a strip holder with gel side down and cover the acidic end of the strip with parafilm. You can store the strip at −20°C in the dark or continue with the second-dimension SDS-PAGE.

3.8. Second-Dimension SDS-PAGE

3.8.1. Equilibration

1. Add 4 mL of reduction solution to the strip, cover the basic end of the strip with parafilm, and incubate with agitation for 10 min.

2. Open the basic end of the strip and discard the solution, draining it off on a paper towel. Then, add 4 mL of alkylation solution, cover, and incubate in the same way as described above. After draining off the alkylation solution, briefly rinse the Immobiline DryStrip by submerging it in SDS electrophoresis running buffer.

3.8.2. SDS-PAGE

1. The second dimension is run on a 12.5% acrylamide gel with 0.2% (w/v) SDS (see Subheading 2.7, item 9, and Note 19). When pouring the gel, remember to take into account that a 1-cm gap should be left at the top of each gel to allow the IPG strip to be loaded.

2. Following IPG strip equilibration, the gel should be overlaid with hot agarose solution (see Subheading 2.7, item 1), and the strip should be pushed through the still liquid agarose.

3. The amount of SDS electrophoresis buffer required depends on the gel tank used. For the lower buffer tank of the Ettan

DALT*six* system, 5 L of 1× SDS electrophoresis buffer are required.

4. Insert the gels with the Immobiline DryStrips in place into the gel electrophoresis system.
5. Pour 2× SDS electrophoresis buffer in the top buffer tank to the fill line (1.2 L for the Ettan DALT*six* system).
6. Run the gel at 2 W/gel for 30 min and then at 5–10 W/gel at 20°C until the bromophenol blue dye front reaches the bottom of the gel.

3.9. Scanning Gels and Image Analysis

Once the gel run is finished, gels should immediately be scanned to avoid the diffusion of protein spots. If they are not immediately scanned, keep the gels moist between the glass plates in the dark and at 4°C. However, care should be taken to allow the gels to reach room temperature before scanning. Do not fix the gels in gel-fix before the gels are scanned, as this may affect the DeCyder Differential Analysis Software quantitation of the labeled proteins.

1. In the Fluorescence Setup window, select the appropriate emission filter from the emission filters list. This list displays the filters that are installed on the Typhoon Variable Mode Imager along with a description of the typical filter use. The scanner control software automatically suggests the laser to be used with the emission filter selected.
2. Prescan at 1,000 μm pixel resolution to identify a suitable PMT voltage. The prescan can be opened in ImageQuant software. Spots showing the most intense signal should be selected using one of the Object Tools such as the rectangle.
3. High-resolution (100 μm) scans must be used to collect quantitative data. This resolution is required for subsequent image analysis using the DeCyder Differential Analysis (DA) Software. The maximum pixel value should not exceed 100,000 as this indicates signal saturation has been reached and will prevent quantitative analysis from being achieved. A target maximum pixel value of 50,000–80,000 is usually suitable.
4. Once the voltage has been optimized for one gel in an experiment, these settings should be used for all similar gels within the same experiment. The maximum pixel value should be within the specified range for all gels to enable accurate quantitation of spot volumes.

The analysis with the Decyder v7.0 software allowed the detection of 2,111 spots in Fraction F2 (membrane/organelles)—55 of these with significant variation (19 increased and 36 decreased in a ratio >2.0 or <−2.0)—and 2,416 spots in Fraction F3 (nuclear)—80 of these with significant variation (51 increased and 29 decreased in a ratio >1.5 or <−1.5; see Figs. 1 and 2).

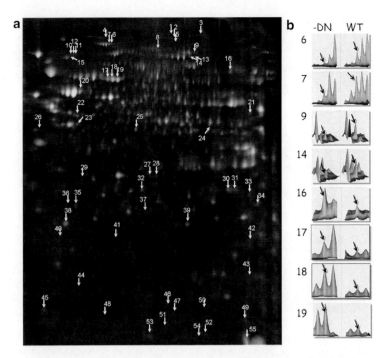

Fig. 1. Visualization of proteins from Fraction F2 (membrane/organelles) after scanning each gel for each CyDye. (**a**) The image result of merging the images of Cy3 and Cy5. *Numbers* refer to spots with increased or decreased levels in dominant-negative mutant of Polµ (−DN). (**b**) 3D representation of some of the differentially expressed proteins. WT = RAMOS-ev.

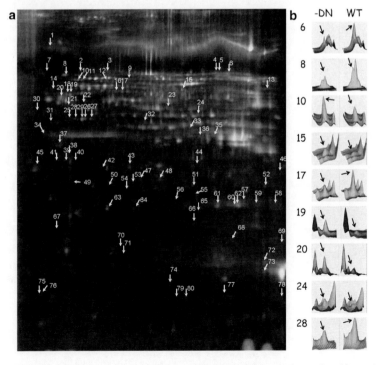

Fig. 2. Visualization of proteins from Fraction F3 (nuclear) after scanning each gel for each CyDye. (**a**) The image result of merging the images of Cy3 and Cy5. *Numbers* refer to spots with increased or decreased levels in dominant-negative mutant of Polµ (−DN). (**b**) 3D representation of some of the differentially expressed proteins. WT = RAMOS-ev.

4. Notes

1. A total of 10^7 cells were used for subcellular fractionation. Only Fractions F1, F2, and F3 were used for further analysis.
2. The RC DC (reducing agent and detergent compatible) protein assay is a colorimetric assay for protein determination in the presence of reducing agents and detergents. It is compatible with a broader range of reagents allowing direct simplified protein quantitation in complex sample solutions.
3. This is an alkaline copper tartrate solution.
4. This is a dilute Folin reagent.
5. This is a surfactant solution.
6. The DMF must be high quality and anhydrous, and every effort should be taken to ensure it is not contaminated with water. DMF, once opened, will start to degrade generating amine compounds. Amine will react with the CyDye DIGE fluor minimal dyes reducing the concentration of dye available for protein labeling.
7. The 1-mL syringe is used to take the DMF out of the flask without opening it to avoid contamination with water.
8. The rehydration buffer (7 M urea, 2 M thiourea, 4% (w/v) CHAPS) ought to be made up without the DTT and ampholytes and frozen at −20°C in 1 mL aliquots. Each aliquot will then require 0.5% of carrier ampholytes and 50 mM DTT or 1.2% (v/v) of DeStreak to be added before being used.
9. Carrier ampholytes focusing will take place at their isoelectric points and facilitate proteins focusing. DTT will reduce disulfide bridges.
10. Unspecific oxidation of protein thiol groups occurs during 2D electrophoresis of proteins, especially at pH > 7. In the resulting protein map, this is seen as horizontal streaking and extra spots. Transforming the thiol groups into a stable disulfide using DeStreak reagent prevents unspecific oxidation. The preparation of Immobiline DryStrip with DeStreak reagent will result in 2D maps with reduced streaking between spots in the pH range 7–9. It will also simplify the spot pattern as it reduces the number of spots caused by oxidation of proteins. At pH > 7, the isoelectric point of proteins may change due to the transfer of thiols or oxidation products to disulfides.
11. Iodoacetamide has to be freshly prepared and kept in darkness in order to avoid oxidation to the corresponding carboxylic acid. The cysteine residues previously reduced are alkylated by carbamidomethylation, thus preventing the reformation of disulfide bridges.

12. Mix butan-2-ol:water (70:30) thoroughly for several min. Then leave it to settle until two separated phases are formed and take the upper phase (water-saturated butan-2-ol).

13. This SDS concentration is recommended for the upper chamber of the Ettan Dalt Six for a correct migration of the proteins to avoid spots that look like tears.

14. Weigh acrylamide inside the fume hood, wearing a mask. Unpolymerized acrylamide is a neurotoxin, and care should be taken to avoid skin contact.

15. Weigh SDS inside the fume hood, wearing a mask.

16. Avoid higher temperatures in the presence of urea to prevent the carbamylation of proteins.

17. The pH of the lysate can be increased to pH 8.5 by the careful addition of dilute sodium hydroxide (1 M). Test the sample pH by spotting a small volume (0.2 µL) on a pH indicator strip. If the pH increases more, add 1 N HCl to decrease it again.

18. Protein assay RC DC Bio-Rad: pipette 25 µL of standards and samples into clean and dry microfuge tubes. Add 125 µL of RC reagent I into each tube and vortex. Incubate the tubes for 1 min at room temperature. Add 125 µL of RC reagent II into each tube and vortex. Centrifuge the tubes at $15,000 \times g$ for 3–5 min. Discard the supernatant by inverting the tubes on a clean and absorbent tissue paper. Allow the liquid to drain completely from the tubes. Add 127 µL of reagent A (5 µL of DC reagent S to each 250 µL of DC reagent A) to each microfuge tube and vortex. Incubate tubes at room temperature for 5 min, or until the precipitate is completely dissolved. Vortex before proceeding to the next step. Add 1 mL of DC reagent B to each tube and vortex immediately. Incubate at room temperature for 15 min. Then, the values for absorbance can be read at 750 nm. These will be stable for at least 1 h.

19. Low-fluorescence glass plates must be used for gel electrophoresis using DIGE. They ensure the lowest background pixel values of scanned images. Low-fluorescence glass plates must not be scratched as the scratches will appear on the image. After casting, the gel(s) should be overlaid with water-saturated butan-2-ol and left for 2 h at room temperature. The overlay can then be removed and replaced with running buffer and left until required. It is recommended that the gels are prepared at least 1 day in advance to ensure reproducible results. Gels can be stored at 4°C for up 2 weeks before use (as long as they do not dry out).

References

1. Banks, R.E., Dunn, M.J., Hochstrasser, D.F., Sanchez, J.C., Blackstock, W., Pappin, D.J., et al. (2000) Proteomics: New perspectives, new biomedical opportunities. *Lancet*, **356**, 1749–1756.
2. Westbrook, J.A., Yan, J.X., Wait, R., Welson, S.Y., Dunn, M.J. (2001) Zooming-in on the proteome: very narrow-range immobilised pH gradients reveal more protein species and isoforms. *Electrophoresis*. **22**, 2865–2871.
3. Hanash, S.M. (2000) Biomedical applications of two-dimensional electrophoresis using immobilized pH gradients: current status. *Electrophoresis*, **21**, 1202–1209.
4. Westbrook, J.A., Yan, J.X., Wait, R., Welson, S.Y., Dunn, M.J. (2001) Zooming-in on the proteome: very narrow-range immobilised pH gradients reveal more protein species and isoforms. *Electrophoresis*. **22**, 2865–2871.
5. Krapfenbauer, K., Berger, M., Friedlein, A., Lubec, G., Fountoulakis, M. (2001) Changes in the levels of low-abundance brain proteins induced by kainic acid. *Eur J Biochem*. **268**, 3532–3537.
6. Abdolzade-Bavil, A., Hayes, S., Goretzki, L., Kroger, M., Anders, J., et al. (2004) Convenient and versatile subcellular extraction procedure, that facilitates classical protein expression profiling and functional protein analysis. *Proteomics* **4**, 1397–1405.
7. Alban, A., David, S.O., Bjorkesten, L., Andersson, L., Sloge, E., Lewis, S., Currie, J. (2003) A novel experimental design for comparative two-dimensional gel analysis: two-dimensional difference gel electrophoresis incorporating a pooled internal standard. *Proteomics* **3**, 36–44.
8. Tonge, R., Shaw, J., Middleton, B., Rowilinson, R., Rayner, S., Young, J., Pognan, F., Hawkins, E., Currie, L., Davison, M. (2001) Validation and development of fluorescence two-dimensional differential gel electrophoresis proteomics technology. *Proteomics* **1**, 377–396.

Chapter 18

Application of Saturation Labeling in Lung Cancer Proteomics

Gereon Poschmann, Barbara Sitek, Bence Sipos, and Kai Stühler

Abstract

Cancer is a quite heterogeneous disease and each cancer type can be divided in different subentities. Normally this is done by pathologist using classical dye-staining protocols or considering specific biomarkers. To identify new biomarkers, allowing a more specific diagnosis clinical tissue specimen is the material of choice. But the amount of clinical material obtained by resection or biopsy is often limited. In order to perform analytical studies with such scarce sample material, a sensitive analysis method is required. Using two-dimensional electrophoresis (2DE) for the analysis of small protein amounts, protein saturation labeling using fluorescence dyes has been successfully applied. Here, we describe the application of saturation labeling in combination with microdissection for the analysis of lung tumor cells and bronchial epithelium cells. The presented study demonstrates all relevant steps of differential proteome analysis with scarce protein amount: experimental design, manual microdissection, optimization of saturation labeling, 2DE, protein identification and validation. As a result, 32 non-redundant proteins could be identified to be differentially regulated between the respective tissue types and are candidate biomarkers for describing lung cancer in more detail.

Key words: DIGE, Saturation labeling, Cancer proteomics, Microdissection, Reference proteome

1. Introduction

Cancer is a quite heterogeneous disease. In the case of lung cancer, different tumor types, e.g., squamous cell carcinoma (SCC), adenocarcinoma, small cell lung cancer, and large cell carcinoma, can be pathologically discriminated. Each tumor type differs in its behavior concerning progression, metastasis, and treatment and has its own morphological and cellular properties. Even a single tumor type offers large tissue heterogeneity: tumor and stroma areas must be distinguished, certain areas—even in a single tumor—could consist of cells of several differentiation grades (good, moderately, or poorly differentiated), and cells in central areas of the tumor could be different from cells at the tumor periphery. To identify novel biomarkers by quantitative proteome analysis, the study design is of high importance:

1. It is mandatory to select a relevant tumor subtype and control tissue. In the presented study, we analyzed SCC tissue. Because SCC typically originates from bronchial epithelium cells which are transformed in a multistep process including metaplasia and dysplasia (1), we choose bronchial epithelium as control tissue.

2. Another important point is the isolation of relevant cells (2). For this purpose, we used manual microdissection of tissue slices. We collected cells from bronchial epithelium as control tissue and separated tumor cells from surrounding stroma, blood vessels, and nonmalignant tissue. Here, cell areas of grade 2 (moderately differentiated) as well as grade 3 (poorly differentiated) were included.

The selection of relevant cells for the analysis reduces sample heterogeneity because the resulting protein sample is not "diluted" by proteins from irrelevant cells, leading to an increased experimental sensitivity.

Because microdissection is a time-consuming method, only small amount of cells can be collected in reasonable time. Therefore, DIGE saturation labeling is the method of choice analyzing small cell numbers (1–5 μg protein) (3, 4).

2. Materials

2.1. Microdissection and Sample Preparation

1. Microscope: BH2 (Olympus, Wetzlar, Germany).
2. Cryostat: Shandon Cryotome SME (Thermo Scientific, Dreieich, Germany).
3. Ultrasonic bath.
4. Hand homogenisator.
5. Injection needle: 0.65 × 25 mm.
6. Lysis buffer: 2 M thiourea, 7 M urea, 4% CHAPS (3-[(3-cholamidopropyl)dimethylammonio]-1-propanesulfonate), 30 mM Tris–HCl, pH 8.0.
7. Ethanol, 96%.
8. 25% (w/v) hematoxylin, according to Mayer (Merck).
9. 1.7% (w/v) eosin, diluted in 96% ethanol.
10. Bradford Protein Assay (Bio-Rad).

2.2. Cysteine-Specific Protein Labeling Using CyDye DIGE Fluor Saturation Dyes

1. Dye stock solution: CyDye DIGE fluor saturation dyes (GE Healthcare, Life Sciences, Freiburg, Germany). The dyes are reconstituted in high-quality anhydrous dimethylformamide (DMF), giving a concentration of 2 mM (100 nmol dye in 50 μL DMF) for analytical gels and 20 mM (400 nmol dye in 20 μL DMF) for preparative gels. The dye solutions are stable at −20°C for 3 months.
2. 2 mM Triscarboxethylphosphine (TCEP) solution for analytical gels and 20 mM TCEP solution for preparative gels: prepare freshly.
3. 1.4 M DTT: store in aliquots at −20°C.
4. Image software: ImageQuant™ TL v2005 (GE Healthcare).
5. Differential image analysis software: Decyder 2D 6.0 (GE Healthcare).
6. Fluorescence Scanner: Typhoon 9400 (GE Healthcare).

2.3. Isoelectric Focusing

1. Servalyt 2-4 (Serva, Heidelberg, Germany): store at 4°C.
2. Separation gel buffer: 3.59% (w/v) acrylamide, 0.31% (w/v) piperazindiacrylamide, 4.1% (v/v) carrier ampholines mixture (pH 2–11), 9.0 M urea, 5.1% (w/w) glycerol, and 0.06% (v/v) TEMED: store in aliquots at −80°C.
3. Cap gel buffer: 12.3% (w/v) acrylamide, 0.13% (w/v) piperazindiacrylamide, 4.1% (v/v) carrier ampholyte mixture, 9.0 M urea, 5.1% (w/w) glycerol, and 0.06% (v/v) TEMED: store in aliquots at −80°C.
4. 1.2% (w/w) Ammonium persulfate (APS): store in aliquots at −80°C.
5. Anodic buffer: 3 M urea and 0.74 M phosphoric acid: prepare freshly.
6. Cathodic buffer: 9 M urea, 5% (w/v) glycerol, and 0.75 M ethylenediamine: prepare freshly.
7. Sephadex solution: 270 mg Sephadex suspension (20 g Sephadex swollen in 500 mL of water and resuspended into 1 L of 25% (v/v) glycerol solution and filtered) with 233 mg urea, 98 mg thiourea, 25 μL of ampholine mixture, pH 2–11, and 25 μL of 1.4 M DTT: store in aliquots at −20°C.
8. Protection solution: 30% (w/v) urea, 5% (w/w) glycerol, and 2% carrier ampholytes, pH 2–4: store in aliquots at −80°C.
9. Equilibration solution: 125 mM Tris base, 40% (w/w) glycerol, 65 mM DTT, and 104 mM SDS: store in aliquots at −20°C.

2.4. Two-Dimensional Polyacrylamide Gel Electrophoresis

1. Glass plates compatible with fluorescence imaging ($25 \times 30 \times 0.4$ cm).
2. Plastic spacer ($30 \times 1 \times 0.15$ cm).

3. Apparatus for vertical SDS-PAGE (System VA Sarstedt, Nümbrecht, Germany).

4. Gel carrier grooves (homemade).

5. Gel solution: 376 mM Tris–HCl, pH 9.4, 0.03% (v/v) TEMED, 3.5 mM SDS, 15% (w/v) acrylamide, and 0.2% (w/v) bisacrylamide: store in aliquots at −20°C.

6. Running buffer: 25 mM Tris base, 192 mM glycine, and 3.5 mM SDS: store at room temperature.

7. 40% (w/w) APS: store in aliquots at −20°C.

8. Protection solution: 376 mM Tris–HCl and pH 9.4, 3.5 mM SDS: store in aliquots at −20°C.

9. Agarose solution: 1% (w/v) agarose (dissolved in running buffer) and 0.001% (w/v) bromophenol blue.

3. Methods

Manual microdissection as well as laser capture microdissection is a time-consuming process. Depending on the cell type, to harvest the collection of 1,000 cells (~1 μg extracted protein), approximately 1–2 h is needed. Therefore, high sensitive protein detection using protein saturation labeling is a method of choice for quantitative protein analysis (5). Differences between minimal labeling and saturation labeling prior to two-dimensional electrophoresis (2DE) are listed below:

- If maleimide dyes are used instead of NHS ester dyes, prior to the covalent attachment of the dye via a thioether linkage, the proteins must be reduced using TCEP. The concept is to label all cysteine residues of a protein sample; therefore, proteins without a cysteine residue cannot be detected.

- The saturation labeling reaction is performed at pH 8 and 37°C (minimal labeling: pH 8.5 at 4°C).

- As the amount of cysteine residues might vary from sample to sample, it is mandatory to determine the optimum amount of TCEP and dye for each type of sample. A low amount of TCEP and dye can result in an incomplete labeling of proteins leading to vertical streaks and nonoptimal overlay in 2D images. If the amount of TCEP and dye is too high, unspecific labeling might occur. Additional labeling of lysine groups reduces the proteins' charge and finally might lead to horizontal streaks in 2D images. Over and under-labeling can be controlled by performing a "same/same" experiment (see Subheading 3.3). In this experiment, same amounts of protein in both channels (Cy3 and Cy5) are applied, whereas the amount of dye is varied until optimal labeling conditions are found.

- In contrast to minimal labeling (Cy2, Cy3, and Cy5 dyes are available), only two dyes (Cy3 and Cy5) are available for saturation labeling. Therefore, the experimental design is different: the internal standard is labeled with Cy3, whereas the sample is labeled using Cy5, and no color swapping is performed. The use of the internal standard minimizes variations introduced by the system and corrects for quenching effects of highly labeled proteins (6) ensuring highly accurate quantification of relative protein abundance.

- For protein identification after in-gel digestion, protein amounts of a saturation labeling experiment are too low, a reference proteome on a preparative gel has to be applied with a protein spot matching rate of >90% for the analyzed sample. As a source for such a reference proteome sample, tissue resection or cell culture cells can be considered, but the selection for the optimal reference proteome has to be determined empirically. Once a sample for the generation of a reference proteome has been selected, this reference proteome is applied as the internal standard for the analytical gels. For preparative gels (e.g., 400 µg protein), a higher concentrated Cy3 dye (20 mM) is used. It should be kept in mind that the reference proteome concentration should be >2 µg/µL, thus not exceeding the maximum loading volume at the isoelectric focusing (IEF) step.

3.1. Manual Microdissection

1. Harvest relevant tissue specimens (neoplastic and non-neoplastic peritumoral samples) (see Fig. 1), freeze resections on liquid nitrogen and store at −80°C (see Note 1). For histological classification of tumor tissue, prepare frozen 5-µm sections from tissue blocks using a cryostat. Place sections briefly in 96% ethanol and stain with hematoxylin and eosin for immunohistological inspection.

2. For microdissection from subsequent tissue slides, prepare frozen 10-µm sections from tissue blocks containing the required tumor grades or non-neoplastic tissue. Place them briefly in ice cold 96% ethanol and only stain with hematoxylin (see Note 2).

3. Under microscopic observation, harvest the required number of cells (see Note 3) using a sterile injection needle directly into 50 µL of lysis buffer (4°C).

4. Sonicate (6 × 10-s pulses on ice) after each collection step and finally centrifuge the lysate for 5 min at 12,000 × g and store the supernatant at −80°C.

3.2. Determination of the Minimal Number of Microdissected Cells

1. Because microdissection is time-consuming and the number of available sample material is limited, it is necessary to find the minimal number of cells sufficient for a differential proteome analysis. Therefore, a pool of approximately 10,000 microdissected cells is required to determine the optimal number of spots by a dilution experiment.

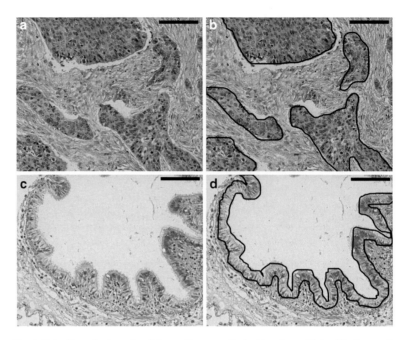

Fig. 1. Selection of relevant cell types. Tumor cells (**a, b**) or bronchial epithelium cells (**c, d**) can be identified in tissue slices after staining with hematoxylin and eosin. Tissue for collection of cells by microdissection for subsequent proteome analysis (marked in **b, d**) should be stained without the use of eosin. The *bar* represents a distance of 100 μm.

2. Prepare three aliquots containing 1,000, 4,000, and 5,000 cells in 50 μL of lysis buffer each (see Note 4).

3. Add 2 nmol TCEP, i.e., 1 μL of the 2 mM TCEP solution, to each sample and incubate for 1 h at 37°C. Label the samples by incubating them with 4 nmol saturation dye (either Cy3 or Cy5 can be used, but all samples should be labeled with the same dye) for 30 min at 37°C in the dark (see Note 5). For stopping the labeling reaction, add 5 μL of 1.4 M DTT and 5 μL of Servalyt 2-4 before 2DE.

4. Subsequent to 2DE (see Subheading 3.5) scan the gels as described elsewhere (see Note 6) and detect the number of spots using the differential in-gel analysis (DIA) module of the DeCyder™ software (see Note 7). The optimal number of cells considered for subsequent analyses is found if the spot number cannot be significantly increased by applying a higher number of cells (see Fig. 2).

3.3. Optimization of Fluorescence Dye Labeling

1. Having established the optimal number of cells for comprehensive proteome coverage, one must empirically optimize the labeling conditions. This will avoid the effects of over- and under-labeling. Therefore, a so-called "same/same" experiment titrating different dye amounts is performed. For this

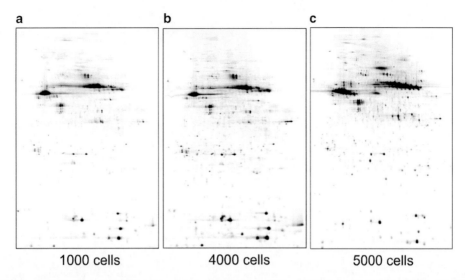

Fig. 2. Determination of the required number of cells. Protein from different numbers of cells ((**a**): 1,000 cells, (**b**): 4,000 cells, (**c**): 5,000) was labeled by saturation labeling and subsequently separated by 2DE. Using image analysis software (DeCyder 6.5, GE Healthcare), 2,638 (**a**), 3,271 (**b**), and 3,725 (**c**) protein spots were detected. Therefore, 5,000 cells were collected from each sample for subsequent experiments.

experiment, prepare six aliquots containing the optimal number of cells (see Subheading 3.2) from a pool of 30,000 cells.

2. Label three of them with 2, 4, and 8 nmol Cy3 and another three with 2, 4, and 8 nmol Cy5 (reduction of disulfide bounds is performed with 1, 2, and 4 nmol TCEP, respectively).

3. Mix samples with equal dye amounts and perform 2DE.

4. After gel scanning, analyze the overlay images in order to find the optimal ratio of protein and dye showing an accurate overlay of Cy3 and Cy5 images in ImageQuant™ TL. As shown in Fig. 3, applying too much dye results in horizontal shifts that could occur due to nonspecific labeling of the amine groups on lysine residues. This will result in an increased mass as well as a reduction of the protein's charge by 1. An insufficient amount of dye causes vertical streaks due to inadequate protein labeling.

3.4. Reference Proteome for Internal Standardization and Protein Identification

1. 2DE analysis of 5,000 microdissected cells (~4 μg) does not deliver sufficient sample material for protein identification using mass spectrometry. Therefore, a reference proteome from a closely related source has to be defined allowing protein identification by protein spot matching. Additionally, this reference proteome may be considered as an internal standard (labeled with Cy3) which reduces consumption of microdissected samples.

2. A suitable reference proteome has to be identified which not only demonstrates a high correlation to microdissected cells but also provides an adequate protein amount for subsequent

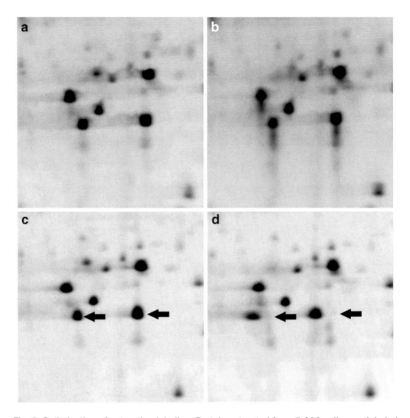

Fig. 3. Optimization of saturation labeling. Protein extracted from 5,000 cells was labeled with 4 nmol (**a, b**) and 8 nmol (**c, d**) dye, respectively: Cy3 (**a, c**) and Cy5 (**b, d**). Samples labeled with the same amount of dye were co-separated by 2DE. Whereas labeling with 4 nmol dye resulted in a good overlap of spot maps (**a, b**), the higher amount of dye resulted in unspecific additional attachment of dye to proteins. For the latter, some proteins shift in horizontal direction (*arrows* in **c, d**), probably because of unspecific dye attachment to lysine residues, which subsequently lost their charge.

identification. A reference proteome could therefore be prepared from tumor cell lines or solid tissues (e.g., sections of tumor tissue grade 2 and grade 3).

3. Prepare 25-μm-thick sections of tumor tissue and epithelial tissue (esophagus) using a cryostat. Homogenize ~100 mg of the tissue on ice in 240 μL of lysis buffer using a hand homogenizer and sonicate the sample 6 times for 10 s on ice.

4. To remove insoluble debris centrifuge the homogenate for 5 min at $12,000 \times g$.

5. Perform a Bradford assay to determine the protein concentration of the samples.

6. Label 5 μg of each lysate with 4 nmol Cy3 according to the standard labeling protocol (see Subheading 3.2, step 3).

7. Label aliquots of 5,000 microdissected cells with 4 nmol Cy5.

8. Pairwise mix the labeled tissue lysates with labeled cells and process by 2DE.

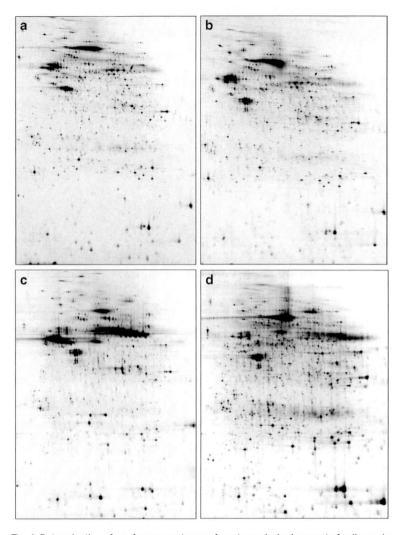

Fig. 4. Determination of a reference proteome. As not any desired amount of cells can be collected by microdissection, a reference proteome was established for subsequent protein spot identification. 2D DIGE analysis of a mixture of grade 2 and grade 3 tumor tissue harvested from 25-μm-thick sections (**d**) showed a spot matching rate of >90% to gels with separated protein from microdissected bronchial epithelium (**a**) and squamous cell carcinoma tissue (**b**). The matching rate of spot pattern produced from epithelium cells from esophagus (which can be sampled much more easily) (**c**) was about 80%, and therefore esophagus was not considered for subsequent experiments.

9. After scanning, analyze the images using the DIA module of DeCyder (see Fig. 4) and calculate the matching rate between microdissected cells and carcinoma tissue protein spots.
10. Optimize the labeling conditions for the tissue lysate showing the highest matching rate with the microdissected cells by a same/same experiment (see Subheading 3.3).

3.5. Two-Dimensional Electrophoresis

1. These instructions are modified from the 2DE technique as described by Klose and Kobalz (7). This technique is based on IEF employing carrier ampholyte tube gels. It can be easily adapted to the DIGE technique and other formats, including analytical as well as preparative gels. In spite of the fact that most of the instruments were constructed in-house, equivalent equipment is commercially available. For the second dimension, the Desaga VA300 gel system is applied.

2. Two days before running, IEF tube gels (Ø 1.5 mm, 20 cm) are prepared. Add 45 µL of the 1.2% APS solution to 2 mL of the separation gel buffer and fill the tube to the first mark (20 cm) using a syringe. Now, add 14 µL of the 1.2% APS solution to 0.7 mL of the cap gel buffer and cast the cap gel to the second mark (20.5 cm) behind the separation gel. To prevent urea crystallization, place an air cushion under the cap gel to the third mark (21 cm).

3. Before starting IEF, apply 2 mm of the sephadex solution to prevent protein precipitation onto the separation gel. Then, load the sample (dye-labeled) and overlay the sample with protection solution (approximately 5 mm) to prevent direct contact of the acidic cathodic buffer (see Note 8).

4. Fill anodic (bottom) and cathodic buffer (top) into the IEF chambers. Ensure that no air bubbles hamper IEF. Start IEF applying a stepwise voltage program (100 V for 1 h, 200 V/1 h, 400 V/17.5 h, 650 V/1 h, 1,000 V/30 min, 1,500 V/10 min, 2,000 V/5 min).

5. While the 21.5-h IEF is running, gels for the second dimension should be prepared. Clean the glass plates using a lint-free paper towel. For the first wash, use double-distilled water followed by 100% ethanol and finally 70% ethanol. To ensure correct gel dimensions, two 1.5-mm plastic spacers are placed between two plates sealed by silicon.

6. Add 288 µL of the 40% APS solution into 144 mL of separation gel buffer, cast the gel, and overlay with water-saturated isobutanol.

7. After polymerization (45 min), remove the isobutanol and wash the surface with protection solution. To protect the gel from drying, place two-dimensional polyacrylamide gel electrophoresis (2D PAGE) protection solution (see Subheading 2.4) onto the gel and store the gel at 4°C.

8. After IEF, extrude the gel by means of inserting a nylon fiber (see Note 9) into the gel groove of the IEF gel carrier and incubate with equilibration solution for 15 min to load proteins with SDS.

9. Wash the gel 3 times with 2D PAGE running buffer before applying to the second dimension.

10. For the transfer of the IEF gel onto the SDS-PAGE gel, hold the groove with the gel in contact with the edge of the glass plate and slide the gel between the glass plates using a wire suitably formed.

11. Overlay the IEF gels with agarose solution, add the running buffer to the upper and lower (15°C) chambers, and start the electrophoresis. For the entrance of the proteins into the SDS-PAGE gel, apply low current (75 mA) for 15 min. When the proteins have entered increase the current to 200 mA for approximately 5–7 h.

3.6. Analysis of Microdissected Cells from Squamous Cell Carcinoma of the Lung and Bronchial Epithelium Cells

1. This protocol uses the information obtained from the optimization procedures described above.

2. Harvest 5,000 cells of each (bronchial epithelial, small cell carcinoma grade 2 cells and small cell carcinoma grade 3 cells) by manual microdissection and incubate the cells in 50 µL of lysis buffer each.

3. For protein labeling, according to the previous optimization (see Fig. 3), reduce the proteins with 2 nmol TCEP and label the proteins of the microdissected cells with 4 nmol Cy5 dye.

4. For the preparation of the internal standard, reduce 5 µg protein (reference proteome: lysate of a pool of grade 2 and grade 3 tumor tissue sections; see Fig. 4) with 2 nmol TCEP and label the proteins with 4 nmol Cy3 dye subsequently.

5. To each reaction add 5 µL of 1.4 M DTT to stop the labeling reaction and add 5 µL of the ampholytes solution Servalyt, pH 2–4.

6. Mix the samples and analyze the mixture by 2DE as described above.

3.7. Image Analysis

1. After 2DE, leave the gels between the glass plates and acquire the images using the Typhoon 9400 scanner (Focal plane: +3 mm; see Note 6). Choose the excitation wavelengths and emission filters specific for each of the CyDyes according to the Typhoon user guide. Low resolution pre-scans should be performed to adjust the PMT for each channel to gain maximum pixel intensities in the region of interest in the range 50,000–80,000. Finally, scan the gel with a pixel size of 100 µm.

2. Before image analysis with the DeCyder software, crop the images with ImageQuant TL (see Fig. 5).

3. For intra-gel spot detection and quantification use the DIA mode of the DeCyder software. Set the estimated number of spots to 10,000. Apply an exclusion filter to remove spots with a volume smaller than 30,000.

4. Define a representative gel from your experiment as master gel and match spots between the different gels using the Biological

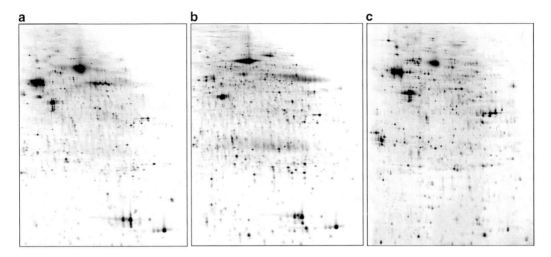

Fig. 5. Representative gels from the 2D DIGE analysis. The internal standard (reference e, labeled with Cy3) (a) comprises grade 2 and grade 3 tumor tissue. In the analysis, bronchial epithelium (b) was compared to squamous cell carcinoma (SCC) tumor cells (c) (both labeled with Cy5).

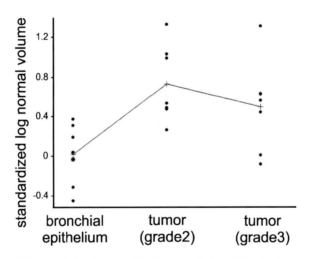

Fig. 6. Example of a differentially regulated protein found by image analysis and identified by mass spectrometry. Annexin A2 has been found to be higher abundant in tumor cells (both grade 2 and grade 3 tumors) compared to bronchial epithelium cells. The *circles* represent individual spot volumes and crosses the group means.

Variation Analysis (BVA) mode of Decyder (see Note 10). Let the software perform a statistical analysis between the control group (microdissected bronchial epithelium cells) and both disease groups (microdissected grade 2 and grade 3 tumor cells). Consider only protein spots with an expression fold-change of >3.4 and p-value (Student's t test) <0.05 as significantly regulated and review spots carefully (see Fig. 6).

3.8. Micropreparation of Significantly Regulated Protein Spots

1. After the differential analysis, protein spots are identified using mass spectrometry (MALDI MS, ESI MS) subsequent to tryptic in-gel digestion. For this analysis, 400 μg of the internal standard (reference proteome) must be labeled and applied to 2DE. A preparative label kit with 400 nmol Cy3 is available.

2. For preparative gels label 400 μg of tissue lysate with 320 nmol of Cy3 (160 nmol TCEP) (see Note 11).

3. Directly after gel scanning, isolate the protein spots of interest manually (see Note 12). Assign the positions of the spots using a print-out of the gel placed underneath the glass plate.

4. Put the isolated protein spots into sample glass containers and store them at −80°C.

5. For protein identification, different MS techniques can be applied subsequent to enzymatic in-gel digestion. Matrix-assisted laser desorption/ionization MS (MALDI MS) allows fast and sensitive identification performing peptide mass fingerprinting (PMF) (8). In cases where more sequence information is necessary (e.g., post-translational modification), liquid chromatography-coupled electrospray ionization (LC-ESI) with its high performance for peptide fragment/sequence analysis using tandem MS (MS/MS) is preferable (4, 9).

3.9. Validation of Differentially Regulated Proteins

As 2D gels separate protein isoforms, protein spots found to be differentially expressed between the groups can represent protein isoforms. To validate the identified differentially regulated proteins (also in respect of false positives), a validation with alternative methods is mandatory (see Fig. 7). Typical methods are western blotting, immunohistochemistry (providing also information about the spatial distribution in tumor tissues), and ELISA as antibody-based methods or selected reaction motoring (SRM) as MS-based method. Sample material not used in the initial analysis should be used for these experiments.

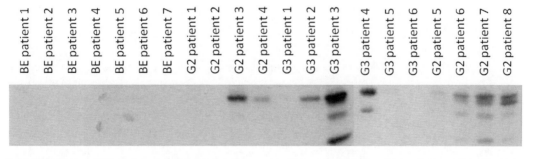

Fig. 7. Validation of a differential regulated protein using western blot analysis. Western blot analysis confirms the higher abundance of Annexin A2 in tumor cells of both grade 2 (G2) and grade 3 (G3) tumors compared to bronchial epithelium cells (BE). Some additional bands can also be detected which might represent splice variants or proteolytic products of Annexin A2.

4. Notes

1. To protect samples against tissue disruption (freezer burn) after resection the tissue is placed in tin foil at first and then stored in a cryo-tube.

2. Eosin has an adverse effect on protein recovery and should be omitted (3). An alternative stain compatible with the procedure is methylgreen (10).

3. First, count for each cell type to be harvested the required number of cells (e.g., 5,000) in series of slices and measure the area the cells occupy. These areas can be used to estimate the required cell numbers during routine microdissection.

4. Usually, depending on the IEF tube's diameter, not more than 50 µL can be applied. Therefore, we increased the volume by blowing up the glass tube so that a sample volume of 100–200 µL can be applied.

5. Prior to labeling, the pH of the sample should be checked by spotting small drop of lysate on a pH indicator strip. A pH below 7.8 or above 8.2 should be adjusted to 8 by adding lysis buffer (7 M urea, 2 M thiourea, 4% (w/v) CHAPS, and 30 mM Tris–HCl) preadjusted to pH 9.5 or 7.5, respectively.

6. Before scanning, the glass plates have to be cleaned thoroughly. First of all, use ethanol for removing of acrylamide or silicone and then clean the glass plates with water.

7. To avoid the detection of dust particles and artifacts as protein spots in the gel, an exclusion filter should be applied. We recommend excluding all spots with a volume <30,000. Alternative software solutions, e.g., Delta 2D (Decodon) or Progenesis SameSpots (Nonlinear Dynamics), can also be used for image analysis. Each software package has its pros and cons, e.g., no missing value, spot co-detection features, or alternative options for statistical analysis (11).

8. Air bubbles should be avoided by applying the solution slowly under the surface (1 mm) of the solution which has been applied before.

9. To prevent destruction of the IEF gel by the nylon fiber, (a) the thermoplastic nylon fiber should be fitted to the tube inner diameter by melting one end into the tube and (b) the gel should be extruded using the cap gel (acrylamide concentration of 12.3%) as a cushion or (c) another possibility is to polymerize a highly concentrated acrylamide solution (15%) above the extruding side after the IEF has been finished and use this gel piece as a cushion.

10. Spot detection and matching can also be performed using the "batch processor" of the software simultaneously for all gels.

11. To avoid a high sample volume for the preparative labeling, TCEP and Cy3 should have a concentration of 20 mM instead of 2 mM. The amount of TCEP and Cy3 for labeling of the preparative protein lysate has to be calculated according to labeling conditions for analytical gels.

12. Before scanning, mark the glass plate from the picking gel with fluorescence stickers. The stickers are necessary for matching the gel as the proteins are not visible and the gel image. After scanning, print the gel image in the original size. Put the print under the glass plates and align the position of the stickers on the glass plate with the image.

Acknowledgments

The authors would like to thank Anna Lendzian and Sabine Roggenbrodt for excellent technical assistance. This work was supported by a grant from the European Commission (LCVAC, COOP-CT-2004-512855).

References

1. Basbaum C and Jany B (1990) Plasticity in the airway epithelium. Am J Physiol. 259: L38–462.
2. Poschmann G, Sitek B, Sipos B et al. (2009) Cell-based proteome analysis: the first stage in the pipeline for biomarker discovery. Biochim Biophys Acta. 1794(9):1309–16.
3. Sitek B, Lüttges J, Marcus K et al (2005) Application of fluorescence difference gel electrophoresis saturation labelling for the analysis of microdissected precursor lesions of pancreatic ductal adenocarcinoma. Proteomics. 5(10):2665–79.
4. Poschmann G, Sitek B, Sipos B et al. (2009) Identification of proteomic differences between squamous cell carcinoma of the lung and bronchial epithelium. Mol Cell Proteomics. 8(5):1105–165.
5. Minden JS, Dowd SR, Meyer HE, Stühler K (2009) Difference gel electrophoresis. Electrophoresis. Jun;30 Suppl 1: 156–61.
6. Gruber HJ, Hahn CD, Kada G et al. (2000) Anomalous fluorescence enhancement of Cy3 and cy3.5 versus anomalous fluorescence loss of Cy5 and Cy7 upon covalent linking to IgG and noncovalent binding to avidin. Bioconjug Chem. 11(5):696–704.
7. Klose J and Kobalz U (1995) Two-dimensional electrophoresis of proteins: an updated protocol and implications for a functional analysis of the genome. Electrophoresis. 16(6):1034–59.
8. Stühler K, Meyer HE (2004) MALDI: more than peptide mass fingerprints. Curr Opin Mol Ther. 6(3):239–48.
9. Schaefer H, Chervet JP, Bunse C et al. (2004) A peptide preconcentration approach for nano-high-performance liquid chromatography to diminish memory effects. Proteomics. 4(9):2541–4.
10. Shekouh AR, Thompson CC, Prime W (2003) Application of laser capture microdissection combined with two-dimensional electrophoresis for the discovery of differentially regulated proteins in pancreatic ductal adenocarcinoma. Proteomics. 3(10):1988–2001
11. Albrecht D, Kniemeyer O, Brakhage AA, et al. (2010) Missing values in gel-based proteomics. Proteomics 10(6):1202–11.

Chapter 19

Proteomic Profiling of the Epithelial-Mesenchymal Transition Using 2D DIGE

Rommel A. Mathias, Hong Ji, and Richard J. Simpson

Abstract

Metastasis remains the primary cause of cancer patient death. Although the precise molecular mechanisms at play remain largely unknown, tumor progression is currently hypothesized to follow a series of sequential steps known as the metastatic cascade. An important component, thought to be involved early in this cascade, is the process known as epithelial-mesenchymal transition (EMT), whereby epithelial cells undergo morphogenetic alterations and acquire properties typical of mesenchymal cells. EMT confers a metastatic advantage to the cancer cells through the loss of cell-cell adhesion, enhanced proteolytic activity, and increased cell migration and invasiveness. This chapter describes the experimental workflow for the secretome analysis of MDCK cells undergoing oncogenic Ras, and Ras/TGF-β-mediated EMT. To enable this comparison, serum-free cell culture conditions were optimized, and a secretome purification methodology established. Secretome samples were then subjected to DIGE analysis to reveal and quantify proteins that are differentially expressed during EMT. The proteomic strategy detailed within successfully identified several EMT modulators and broadens our understanding of the extracellular facets of the EMT process.

Key words: Epithelial-mesenchymal transition, EMT, MDCK, DIGE, Ras, Secretome, Quantitative proteomics

Abbreviations

2DE	Two-dimensional gel electrophoresis
APS	Ammonium persulfate
BVA	Biological variation analysis
CM	Conditioned medium
DIA	Differential in-gel analysis
DIGE	Two-dimensional fluorescence difference gel electrophoresis
DMEM	Dulbecco's Modified Eagle Medium
DMF	Dimethylformamide
DTT	Dithiothreitol
EDA	Extended data analysis

EMT	Epithelial-mesenchymal transition
FCS	Fetal calf serum
LC	Liquid chromatography
MDCK	Madin Darby canine kidney
MS	Mass spectrometry
RT	Room temperature

1. Introduction

Metastasis is the leading cause of cancer-related mortality despite advances in early detection of the disease and new therapeutic treatments (1). Understanding the cellular mechanisms that enhance tumorigenesis has been a major challenge, and to date, the molecular details surrounding metastatic progression remain unresolved (2). The current metastatic paradigm consists of several steps, all of which are required to achieve tumor spreading (3, 4). First, cancer cells breach the basement membrane, and individual cells dissociate from the bulk of the tumor (3, 5, 6). Cells then invade neighboring tissue, intravasate into blood and lymph vessels, and transport to distant sites (3, 5, 6). Finally, cells lodge, and outgrow to form micrometastases and secondary tumors (3, 5, 6). The process underlying the initial breach of the basement membrane and release of cancer cells from their neighbors has been postulated by many to involve epithelial-mesenchymal transition (EMT) (7–9). EMT is broadly defined by diminished epithelial characteristics and increased mesenchymal attributes (10). These changes encompass many diverse alterations including loss of cell polarity, reduced cellular adhesion, and acquisition of enhanced cell migration and invasion (11–13). Importantly, EMT can promote metastasis in several ways including the loss of cell-cell adhesion to facilitate tumor cell invasion (14), induce secretion of protein degrading enzymes such as matrix metalloproteinases to remodel the extracellular matrix (15), stimulate cell signaling pathways such as Wnt and TFG-β to increase cell motility (16, 17), and induce the expression of the mesenchymal transcriptional program to modulate the EMT process (18, 19).

Over the years, classical molecular and cell biology-based approaches have helped us understand many molecular mechanisms of EMT; however, they are usually restricted by the number of targets that can be monitored in any one experiment. Interdisciplinary approaches such as transcriptomics and proteomics have the potential to further our understanding of complex regulation, as they are able to expression profile multiple targets simultaneously without bias (20, 21). For example, numerous studies have employed transcript gene expression analysis to reveal new potential markers in well-characterized EMT models (22, 23). While informative, it is now widely accepted that mRNA levels do not always correlate with translated protein amounts, due to posttranscriptional processing and miRNA repression (24, 25). In addition, transcript expression data can sometimes be misleading because it does not consider protein localization within subcellular compartments. For these reasons, proteomics emerges as a tool to substantiate expression at the protein level, shed new light on underlying EMT mechanisms, and reveal novel EMT effectors. In this regard, several studies have recently adopted a proteomic strategy to gain insights into the cellular process (for a review see ref. 26).

The majority of proteomic EMT studies to date have implemented a classical approach, coupling two-dimensional gel electrophoresis (2DE) and mass spectrometry (MS) to identify proteins in the cell lysate that are differentially expressed (26). These studies have provided useful protein identification and regulatory information, but generally on the most abundant proteins in the sample—e.g., vimentin (27, 28). This may result from the limited sample capacity and detection sensitivity of the 2DE technique (dynamic range of protein separation is typically between 10^4 and 10^5 (29)), or the absence of sample pre-fractionation for sub-proteome analysis. In addition, a major limitation of comparative 2DE analysis is the high degree of gel-to-gel variation in spot patterns, which makes it hard to distinguish any true biological variation from experimental variation (30). A significant advancement in 2DE analysis and reproducibility was made by Unlu et al., who multiplexed samples labeled with fluorescently resolvable dyes (CyDyes) within the same 2D gel (31). This approach known as two-dimensional fluorescence difference gel electrophoresis (DIGE) has several advantages over 2DE (32, 33). These include (1) the use of fluorescent labels that render DIGE much more quantitative than the standard calorimetric staining methods, both with regard to sensitivity, as less sample is required, and linearity, (2) minimizing gel-to-gel variation as multiple DIGE samples are multiplexed on the same gel, and (3) the use of an internal standard that normalizes protein abundance measurements across multiple gels and provides confidence that differences in spot intensities are purely attributed to biological and not experimental variation.

The DIGE workflow (see Fig. 1) consists of protein extracts being covalently tagged with spectrally resolvable fluorescent N-hydroxysuccinimide derivative dyes via nucleophilic substitution reactions between the fluorophore and the ε-amino group of lysine residues, forming an amide (30). Usually, a minimal labeling strategy is implemented where 1–2% of all lysines are labeled to avoid large changes in mass and avoid compromising protein solubility (34). Typically, two protein samples to be compared are labeled with either Cy3 or Cy5 dye, while a third pooled sample containing equal amounts of all samples in the experiment is labeled with Cy2 to serve as an internal standard (35). All labeled samples are

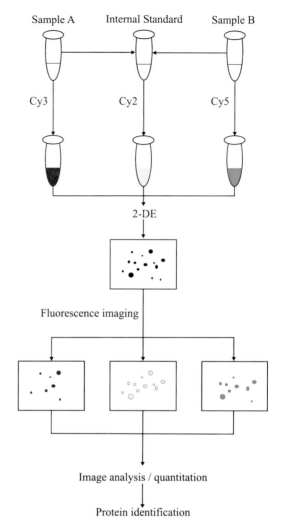

Fig. 1. Two-dimensional fluorescence difference gel electrophoresis (DIGE). Different protein samples are labeled with Cy3 and Cy5 dyes, while an internal standard consisting of equal amounts of each protein sample is labeled with Cy2. The internal standard serves to normalize for gel-to-gel variation, and thereby distinguishes biological from experimental variation. All three CyDye-labeled samples are pooled and resolved by 2DE on the same gel. Individual protein spot maps are obtained by excitation and emission of each CyDye fluorophore, and images are overlaid to reveal proteins that are differentially expressed. Color dominance represents protein abundance. For example, proteins more abundant in sample A appear *red*, those in sample B appear *blue*, and those equally expressed between samples are *purple*. Image analysis software incorporating each sample and the internal standard allows comparison across multiple gels, and enables fold-change determination of protein spots that are either up- or downregulated. Following statistical analysis, protein spots of interest are excised and subjected to LC-MS/MS for protein identification.

then mixed and subjected to 2DE on a single gel, and protein spot patterns are visualized by alternately illuminating the gel with the excitation wavelengths of each of the dyes. The resulting images are then overlaid and analyzed with software such as DeCyder (GE Healthcare) to determine proteins that are differentially expressed between samples. Following analysis, gel spots are excised and subjected to MS for protein identification.

In this chapter, we describe the experimental workflow to analyze changes in the secretome of Madin Darby canine kidney (MDCK) cells undergoing oncogenic Ras (21D1 cells), and Ras/TGF-β-mediated EMT. In order to do so, serum-free culture conditions were optimized and established, and a secretome purification methodology developed. Secretome samples from all three cell lines/conditions were obtained and subjected to DIGE analysis to reveal proteins that are differentially expressed during EMT (see Fig. 2). Following quantitative analysis, MS sequencing of gel spots of interest revealed that proteins downregulated were predominantly involved in cell-cell and cell-matrix adhesion, while upregulated proteins were proteases and factors that promote cell motility (36).

2. Materials

2.1. Cell Culture and Secretome Preparation

1. Dulbecco's Modified Eagle Medium (DMEM) (Invitrogen, Cat no: 11995065).
2. Human transforming growth factor beta (TGF-β) (BD Biosciences, Cat no: 354039).
3. T-175 Tissue culture flasks (BD Falcon, Cat no: 353112).
4. Fetal calf serum (FCS) (Invitrogen, Cat no: 10101-145).
5. Trypsin/EDTA (Invitrogen, Cat no: R-001-100).
6. 150-mm Tissue culture dishes (BD Falcon, Cat no: 353025).
7. Complete EDTA-free protease inhibitor cocktail tablets (Roche, Cat no: 04693124001).
8. 0.1-μm Supor Membrane VacuCap 60 Filter Unit (Pall, Cat no: 4631).
9. 3K NMWL Amicon Ultra Centrifugal Filter Device (Millipore, Cat no: UFC900324).
10. 2D Clean-Up Kit (GE Healthcare, Cat no: 80-6484-51).
11. 2DE sample buffer: 7 M urea, 2 M thiourea, 4% (w/v) CHAPS, 20 mM Tris–HCl, pH 8.5.
12. 2D Quant Kit (GE Healthcare, Cat no: 80-6483-56).

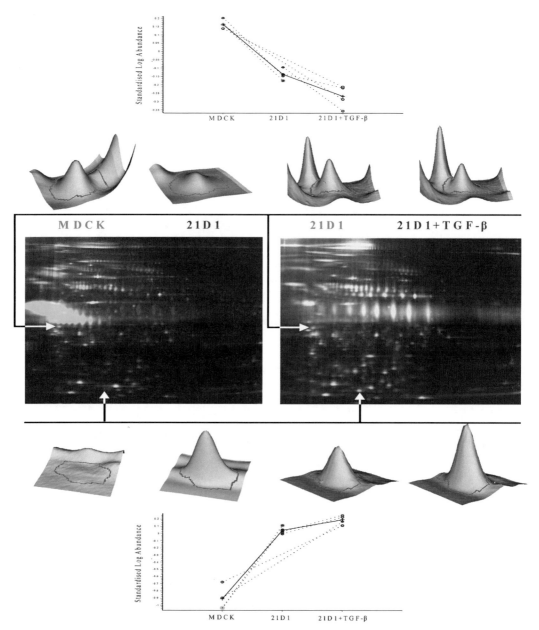

Fig. 2. DIGE analysis of MDCK cell secretome perturbations during oncogenic Ras/TGF-β EMT. Secretome was collected from MDCK, 21D1, and 21D1 cells stimulated with TGF-β for comparative proteomics. The 2D gel images are representative spot maps showing proteins that are differentially expressed as a consequence of EMT. *Green spots* on the *left gel* highlight proteins more abundant in the MDCK secretome (Cy3), while *red spots* indicate proteins that are highly expressed in the secretome of 21D1 cells (Cy5). Comparison of 21D1 cells with or without TGF-β stimulation (*right gel*) reveals that secretome protein expression remains largely unchanged. For this reason, the majority of spots appear *yellow* (i.e., co-expressed), although there are some spots that are slightly *red* (further upregulated) or *green* (further downregulated). Above and below the gels are 3D images enabling visualization of specific spot intensities, and graphical depiction of protein expression. Thus, the abundance of a particular protein can be easily assessed across all samples. For example, the *top panel* illustrates a down-regulated protein, while the *bottom panel* represents a protein that is upregulated by oncogenic Ras and TGF-β. For a detailed list of MDCK cell secretome proteins that are significantly dysregulated during oncogenic Ras-induced EMT see ref. 36.

2.2. CyDye Reconstitution and Protein Labeling

Throughout the protocol, MilliQ-quality water should be used for making all aqueous solutions, and the pH adjusted using either 5 M HCl or 1 M NaOH, where appropriate:

1. CyDye DIGE fluor, Cy2 minimal dye (GE Healthcare, Cat no: RPK0272).
2. CyDye DIGE fluor, Cy3 minimal dye (GE Healthcare, Cat no: RPK0273).
3. CyDye DIGE fluor, Cy5 minimal dye (GE Healthcare, Cat no: RPK0275).
4. Dimethylformamide (DMF).
5. Lysine.

2.3. Immobiline DryStrip Rehydration and IEF

1. IPG Buffer pH 3–10 (GE Healthcare, Cat no: 17-6000-87).
2. DeStreak rehydration solution (GE Healthcare, Cat no: 17-6003-19).
3. Immobiline DryStrip re-swelling tray (GE Healthcare, Cat no: 80-6465-32).
4. Immobiline DryStrip pH 3–10, 24 cm (GE Healthcare, Cat no: 17-6002-44).
5. Immobiline DryStrip cover fluid (GE Healthcare, Cat no: 17-1335-01).
6. Ettan IPGphor 3 isoelectric focusing unit (GE Healthcare, Cat no: 11-0033-64).
7. Paper electrode pads (GE Healthcare, Cat no: 80-6499-14).
8. IPGphor manifold ceramic tray (GE Healthcare, Cat no: 80-6498-57).
9. Electrode set (GE Healthcare, Cat no: 80-6498-76).
10. Sample cups (GE Healthcare, Cat no: 80-6498-95).
11. Equilibration buffer: 6 M urea, 30% (v/v) glycerol, 2% (w/v) SDS, 50 mM Tris–HCl, pH 8.8.
12. Dithiothreitol (DTT).
13. Iodoacetamide.

2.4. Polyacrylamide Gel Casting and SDS-PAGE

1. Low-fluorescence glass plates for Ettan DALT (GE Healthcare, Cat no: 80-6475-58).
2. DALT*twelve* Gel Caster Complete (GE Healthcare, Cat no: 80-6467-22).
3. DALT gradient maker with peristaltic pump (GE Healthcare, Cat no: 80-6067-84).
4. 8% light acrylamide solution: 120 mL of 30% (w/v) acrylamide, 113 mL of 1.5 M Tris–HCl, pH 8.8, 207 mL of water, 4.5 mL of 10% (w/v) SDS, 4.5 mL of 10% ammonium persulfate (APS), 0.96 mL of 10% TEMED.

5. 18% heavy acrylamide solution: 270 mL of 30% (w/v) acrylamide, 113 mL of 1.5 M Tris–HCl, pH 8.8, 30 mL of water, 31 mL of glycerol, 4.5 mL of 10% (w/v) SDS, 2.3 mL of 10% APS, 0.14 mL of 10% TEMED.

6. Displacing solution: 0.375 M Tris–HCl, pH 8.8, 50% (v/v) glycerol, trace amount of bromophenol blue (Bio-Rad, Cat no: 161-0404).

7. Gel storage solution: 0.38 M Tris–HCl, pH 8.8.

8. TG-SDS running buffer (Amresco, Cat no: 0783-42).

9. Ettan DALT*twelve* separation unit (GE Healthcare, Cat no: 80-6466-27).

10. DALT cassette rack (GE Healthcare, Cat no: 80-6467-98).

11. DALT blank cassette insert (GE Healthcare, Cat no: 80-6467-03).

2.5. Gel Scanning, Image Acquisition, and Protein Quantification

1. Typhoon 9410 (GE Healthcare, Cat no: 63-0055-80).
2. Kimtech Science Kimwipes (Kimberly-Clark Professional, Cat no: 34120).
3. DeCyder 2D 7.2 Software package (GE Healthcare, Cat no: 28-9757-78).

3. Methods

3.1. Cell Culture and Secretome Preparation

Four individual secretome samples (designated a–d) are prepared from each of the MDCK, 21D1, and 21D1+TGF-β cell lines for DIGE analysis:

1. Culture MDCK, 21D1, and 21D1+TFG-β cells in DMEM using a T-175 culture flask containing 10% (v/v) FCS, at 37°C in 10% CO_2 as described in ref. 17.
2. Passage upon reaching 80% confluence with Trypsin/EDTA.
3. Seed cells into ten 150-mm tissue culture dishes and culture until 70% confluent.
4. Gently wash each dish with 10 mL of DMEM medium (see Note 1).
5. Repeat step 4 twice.
6. Add 15 mL of DMEM to each dish, and culture for 24 h (see Note 2).
7. Collect the conditioned medium (CM) from each dish and pool (see Note 3).
8. Centrifuge the CM at $480 \times g$ at 4°C for 5 min to pellet intact cells, and transfer the supernatant to a new tube.

9. Centrifuge the CM at 2,000×*g* at 4°C for 10 min to remove cell debris, and transfer the supernatant to a new tube.
10. Add protease inhibitor tablets to the CM according to the manufacturer's instructions.
11. Filter the CM through a 0.1-μm membrane filter and collect the filtrate in a new tube.
12. Concentrate the filtrate to a final volume of 1 mL using centrifugal filtering devices (see Note 4).
13. Precipitate proteins using the 2D Clean-Up Kit according to the manufacturer's instructions (see Note 5).
14. Re-solubilize secretome proteins in 2DE sample buffer.
15. Determine protein concentration using the 2D Quant Kit, following the manufacturer's instructions (see Note 6).
16. Snap freeze the purified secretome sample and store at −80°C, or proceed with protein labeling.

3.2. CyDye Reconstitution and Protein Labeling

1. Prepare the stock dye solutions (1 nmol/μL) by reconstituting 25 nmol of Cy2, Cy3, and Cy5 dyes each with 25 μL of DMF (see Note 7).
2. Vortex vigorously to dissolve the dye, which should produce deep colors of Cy2-yellow, Cy3-red, and Cy5-blue.
3. Prepare the working dye solution (200 pmol/μL) by diluting 1.5 μL of each stock dye solution with 6 μL of DMF.
4. Store the remaining stock Cy dye solutions in a light-excluding container at −20°C (see Note 8).
5. Ensure that each secretome sample is between pH 8.0 and 9.0 prior to protein labeling, and adjust as necessary.

   Each secretome sample is labeled twice with Cy3 and twice with Cy5 to eliminate dye specific bias, and separated across six 2D gels. The internal standard labeled with Cy2 is generated by combining all secretome samples, and run on every gel to coordinate inter-gel spot matching and reduce gel-to-gel variation (see Note 9).

6. Label 25 μg of each secretome sample (MDCK (a), MDCK (c), 21D1 (b), 21D1 (d), 21D1+TFG-β (a), 21D1+TFG-β (c)) with 1 μL of Cy3 working dye solution.
7. Label 25 μg of each secretome sample (MDCK (b), MDCK (d), 21D1 (a), 21D1 (c), 21D1+TFG-β (b), 21D1+TFG-β (d)) with 1 μL of Cy5 working dye solution.
8. Combine 12.5 μg from each of the 12 secretome samples, and label with 6 μL of Cy2 working dye solution (see Note 10).
9. Store the remaining working Cy dye solutions in a light-excluding container at −20°C (see Note 11).

10. Vortex secretome sample and dye to mix thoroughly, and leave on ice in the dark for 30 min.

11. Add 1 μL of 10 mM lysine to each Cy3/Cy5 tube and 6 μL to the internal standard tube to stop the reaction. Mix and leave on ice in the dark for 10 min.

12. Snap freeze the labeled secretome samples and store at −80°C, or proceed with IEF.

3.3. Immobiline DryStrip Rehydration and IEF

The Immobiline DryStrip gels must be rehydrated prior to IEF, usually the night before as a minimum of 10 h is required (see Note 12):

1. Add 15 μL of IPG buffer to 3 mL of DeStreak rehydration solution and vortex to mix.

2. Aliquot 450 μL of rehydration solution into six slots of the re-swelling tray (see Note 13).

3. Remove the protective plastic cover from the DryStrip gels and gently place the gels gel-side facing down into the re-swelling tray. Ensure there are no bubbles between the gel and the tray.

4. Cover each of the strips in each slot with 2 mL of DryStrip cover fluid to prevent evaporation.

5. Replace the re-swelling tray lid and allow the strip gels to rehydrate overnight at room temperature (RT).
   Prepare the IPGphor 3 system for sample loading and first dimension IEF.

6. Cut electrode pads (5 × 15 mm) and wet with distilled water.

7. Remove the DryStrip gels from the re-swelling tray using tweezers and place them onto the ceramic manifold, positioning the basic end of the strip toward the negative end of the IPGphor unit.

8. Cover both acidic and basic ends of the strip gel with damp electrode pads.

9. Fasten electrodes firmly onto the manifold and ensure good contact with electrode pads.

10. Clip on sample loading cups at the acidic gel end.

11. Cover the strip gels and manifold with cover fluid, and check for a good seal (see Note 14).
    Prepare the secretome samples for loading and IEF. Each gel should contain 25 μg Cy3-labeled sample, 25 μg Cy5-labeled sample, and 25 μg Cy2-labeled internal standard.

12. Combine each of the Cy3- and Cy5-labeled secretome sample pairs as outlined in Table 1, and add one aliquot (i.e., 1/6) of the total volume of the Cy2 internal standard.

13. Add an equal volume of 2× sample buffer to each of the six tubes.

14. Load samples into the loading cups.
15. Cover samples with 3–4 drops of cover fluid.
16. Close the lid and commence IEF. The following parameters have been successfully used:
    (a) Hold at 300 V for 3 h.
    (b) Ramp to 1,000 V over 6 h.
    (c) Ramp to 8,000 V over 3 h.
    (d) Hold at 8,000 V for 8 h.
17. Remove strip gels from the ceramic manifold, and place them in individual tubes.
    Prepare equilibration buffer (see Note 15).
18. Dissolve 0.6 g DTT in 60 mL of equilibration buffer, and add 10 mL to each tube.
19. Incubate for 20 min at RT with gentle agitation.
20. Dissolve 1.5 g iodoacetamide in 60 mL equilibration buffer, and add 10 mL to each tube.
21. Incubate for 20 min at RT with gentle agitation.
22. Proceed immediately with SDS-PAGE. Do not freeze equilibrated strip gels.

3.4. Polyacrylamide Gel Casting and SDS-PAGE

Polyacrylamide gels must be cast at least 1 day prior to running SDS-PAGE to allow polymerization to occur. Detailed step-by-step instructions for gel casting can be found in the Ettan DALT II system user manual and can be obtained from the GE Healthcare website (http://www.gelifesciences.com/aptrix/upp00919.nsf/content/479E615AE5FE9261C1257628001CE63D?OpenDocument&Path=Catalog&Hometitle=Catalog&entry=12&newrel&LinkParent=C1256FC4003AED40-A905385FF0BEA0FAC125701900490803_RelatedLinksNew-C21BEC677D8448

Table 1
DIGE experimental design for the analysis of secretome changes during oncogenic-Ras/TGF-β-mediated EMT

Gel number	Cy3 (25 µg)	Cy5 (25 µg)	Cy2 (25 µg)
1	MDCK (a)	21D1 (a)	Internal standard
2	21D1 + TGF-β (a)	MDCK (b)	Internal standard
3	21D1 (b)	21D1 + TGF-β (b)	Internal standard
4	MDCK (c)	21D1 (c)	Internal standard
5	21D1 + TGF-β (c)	MDCK (d)	Internal standard
6	21D1 (d)	21D1 + TGF-β (d)	Internal standard

BC1256EAE002E3030&newrel&hidesearchbox=yes&modul eid=164458). Below are brief instructions to cast 8–18% SDS-PAGE gels:

1. Clean the gel casting cassettes with water and methanol and assemble cassettes into the gel caster.
2. Prepare light and heavy acrylamide solutions and mix thoroughly (see Note 16).
3. Pour the light solution into the right side of the gradient maker (the chamber which is narrower at the bottom).
4. Allow the light solution to fill the tubing and "Y-connector" between the light and heavy chambers.
5. Close all three pinch clamps on the light and heavy chamber exit tubes and feed tube.
6. Pour the heavy solution into the left side of the gradient maker (the chamber is wider at the bottom) until the liquid level reaches a level 2 cm below the top level of the light solution in the light chamber.
7. Connect the gel caster to the gradient maker.
8. Add 75 mL of displacing solution into the balance chamber. Ensure the feed tube is tightly connected to the grommet seal to prevent leakage of displacing solution into the caster.
9. Open the pinch clamps on the feed tube and light chamber exit tube.
10. Start the peristaltic pump and open the heavy chamber exit tube clamp as soon as the light solution level falls from 2 to 1 cm above the level of the heavy solution.
11. Stop the pump when the gradient maker is empty (see Note 17).
12. Pull the feed tube out of the balance chamber grommet and allow the displacing solution to flow into the V-well at the bottom of the gel caster.
13. Spray SDS running buffer immediately onto the top of each gel in the caster. The level of buffer should be 0.5–1 cm above the top of the gel.
14. Allow the gel to polymerize at RT for at least 2 h but preferably for 24 h.
15. Unload the gels from the caster and wash any excess acrylamide on the outside of each cassette off with water.
16. Store the gels in an airtight container at 4°C with a small amount of gel storage solution to keep the gels from drying out, or proceed with strip gel loading.

    Place the polyacrylamide gels in the cassette rack and prepare for strip gel loading and SDS-PAGE.
17. Remove strip gels from the equilibration buffer and rinse in SDS running buffer.

18. Carefully slide each strip gel into a separate glass cassette and slide the strip gel down to the top of the polyacrylamide gel using a spacer/spatula, so the gels sit flush together. Avoid trapping bubbles between the gels.

Prepare the DALT*twelve* separation unit for the second-dimension SDS-PAGE.

19. Fill the lower buffer tank with SDS running buffer.
20. Switch the control unit and pump on and set the temperature to 23°C.
21. Insert the loaded cassettes into the tank and insert blank cassettes into any free slots (see Note 18).
22. Fill the upper buffer tank with SDS running buffer.
23. Close the lid and commence electrophoresis. The following parameters have been successfully used:
    (a) 5 W per gel for 30 min.
    (b) 90 V at 23°C until the dye front reaches the bottom of the cassette.
24. Once electrophoresis is complete, scan immediately (see Note 19).

3.5. Gel Scanning and Image Acquisition

1. Initialize scanner and prepare for imaging.
2. Dry cassettes with Kimwipes and ensure the complete removal of fingerprints and dust.
3. Place gel into the scanner and set scanning parameters:
    (a) Set the acquisition mode to fluorescence.
    (b) Set excitation/emission wavelengths for Cy2 (488/520 nm), Cy3 (532/580 nm), and Cy5 (633/670 nm).
    (c) Set the PMT voltage.
    (d) Select the orientation of the gel.
    (e) Select to press sample.
    (f) Select focal plane to +3 mm.
4. Define the area to be scanned (i.e., gel boundaries).
5. Perform a pre-scan with pixel size set to 1,000 μm resolution (see Note 20).
6. Perform final scan at 100 μm resolution (see Fig. 2).
7. Scan all gels.

3.6. Image Analysis and Protein Quantification

The images can now be processed by software such as DeCyder 2D to assist detection, quantitation, matching, and analysis of DIGE gels. A full description of the software is beyond the scope of the chapter; however, an extensive and a very useful DeCyder manual can be obtained from the GE Healthcare website (http://www.gelifesciences.com/aptrix/upp00919.nsf/Content/65B08B3DA

677AA7AC1257628001D512B/$file/28941447AA.pdf). Briefly, the software merges the gel images into one image for spot detection, and the images are then re-separated and the spot boundaries applied to each image, followed by calculation of spot area, volume, and peak height. Using the Batch Processor tool is an easy way to automate the linkage of differential in-gel analysis (DIA) and biological variation analysis (BVA) modules together. The gel containing the highest number of spot features is assigned the master gel, and manual spot matching can then be performed to correctly match the remaining five Cy2 gel images with the Cy2 master. In DIA, spot boundaries and volumes are co-detected for Cy3, Cy5, and Cy2 channels on each gel, and protein spot abundance is expressed as a standard/sample ratio. In BVA, protein abundance is compared across multiple samples using the internal standard to normalize between gels (see Fig. 2), and statistical analysis is performed to provide average ratio and one-way ANOVA values between samples. The DeCyder software also contains a module called extended data analysis (EDA), which provides more detailed examination and interpretation of results including principal component analysis (PCA), supervised and unsupervised clustering, and discriminant analysis.

3.7. Protein Identification and Validation

Spots of interest determined by DeCyder-based protein quantification can then be selected for LC-MS/MS protein identification. Preparative gels containing 500 μg secretome sample are separated by 2DE, and subjected to the same IEF and SDS-PAGE conditions as the DIGE gels (see Notes 12 and 21). Following electrophoresis, the preparative gels are fixed in 40% aqueous methanol and 7% aqueous acetic acid for 30 min, washed with deionized water 3 times for 10 min, and proteins are visualized by incubating the gels in a Coomassie Blue stain such as Imperial Protein Stain (Pierce) for 1 h. Gel spots are excised, and subjected to automated in-gel reduction, alkylation, and tryptic digestion as described previously (37). Extracted peptides are fractionated by reverse phase liquid chromatography (LC) and analyzed by MS to reveal protein identity (15).

4. Notes

1. Ensure that the medium used in washing steps is pre-warmed to 37°C. The addition of medium should take place along the dish wall to minimize the disruption of adherent cells. Washing is done by gently swirling the medium around the dish several times. This stringent washing procedure is employed to remove the FCS present during normal culturing conditions.

2. The secretome collection time (i.e., duration of serum-free culture) must be optimized to minimize artifacts collected as a consequence of cell lysis. To do so, cell viability and proliferation should be measured and monitored using the Trypan blue dye exclusion and 3-(4,5-dimethylthiazol-2-yl)-2,5-diphenyltetrazolium bromide (MTT) assays (38, 39). Viability and proliferation must not diminish greater than 5% during secretome collection.

3. The CM should be kept on ice throughout the duration of secretome collection and concentration.

4. The CM concentration time required is usually cell type and density dependent.

5. GE Healthcare recommends performing this step to quantitatively precipitate proteins while leaving interfering substances, such as detergents, salts, lipids, phenolics, and nucleic acids, in solution. The 2D Clean-Up Kit increases labeling efficiency and protein abundance for DIGE analysis, and improves the quality of 2DE, by reducing streaking, background staining, and other artifacts due to interfering contaminants. We performed this step, as the CM was extremely viscous following concentration, and the DIGE images were streaky and spots unresolved if the 2D Clean-Up Kit was not used.

6. The purified secretome concentration is usually 3 µg/µL, with a total yield of 300 µg from 5×10^7 cells.

7. High-quality anhydrous DMF must be used for dye reconstitution. Avoid contamination with water.

8. GE Healthcare advises that the stock dye solution should be stable at −20°C for 2 months or until the expiry date on the container, whichever is sooner.

9. The Ettan DIGE System user manual recommends labeling and running 50 µg of each sample, as well as the internal standard. We have found that 25 µg of protein sample is sufficient for DIGE analysis, and this condition reduces the cost of the experiment as less dye is required.

10. It is particularly useful to prepare and label one additional aliquot of the internal standard sample. For example, prepare enough for seven gels, so that it can be spiked into and run on the preparative gel. This will assist correct spot mapping and gel excision for protein identification (see Subheading 3.7).

11. GE Healthcare advises that the working dye solutions are only stable for 2 weeks at −20°C.

12. We recommend rehydration in the absence of protein sample for the analytical gels, and rehydration in the presence of protein sample for the preparative gels due the large sample volumes.

13. Ensure the tray is flat using the spirit level on the tray. Apply the solution evenly across the entire slot without bubbles.
14. Fill the cup with 2× sample buffer and observe that the level of buffer does not diminish, which would signify a leak. If there is a leak, remove the cup and reposition again.
15. Gels can be stored after IEF in 10 mL of equilibration buffer at −20°C.
16. Acrylamide, TEMED, APS, and SDS are extremely hazardous. Please consult the manufacturer's material safety data sheets for correct handling of these chemicals. The 10% APS solution should be made freshly prior to use.
17. Pay special attention to stop air entering the feed tube. Once all the acrylamide solution is out of the maker, clean it by filling 2 L of water into the chambers, and pump the water through the feed tube to waste.
18. Lubricate the cassettes with SDS running buffer to facilitate loading into the tank.
19. After second dimensional SDS-PAGE is complete, the gels should be scanned as soon as possible, to minimize the diffusion of protein spots.
20. Avoid saturation, and adjust the PMT voltage accordingly. The PMT used is usually between 500 and 600 V.
21. It is advisable to run an aliquot of the internal standard on the preparative gels to assist with correct selection of spots for protein identification.

Acknowledgments

This work was supported, in part, by the National Health & Medical Research Council of Australia (program grant #487922 (R.J.S)), and funds from the Operational Infrastructure Support Program provided by the Victorian Government of Australia.

References

1. Weigelt B, Peterse JL, and van't Veer LJ (2005) Breast cancer metastasis: markers and models. Nat Rev Cancer **5**:591–602
2. Chen EI, and Yates JR, 3rd (2007) Cancer proteomics by quantitative shotgun proteomics. Mol Oncol **1**:144–159
3. Chambers AF, Groom AC, and MacDonald IC (2002) Dissemination and growth of cancer cells in metastatic sites. Nat Rev Cancer **2**:563–572
4. Woodhouse EC, Chuaqui RF, and Liotta LA (1997) General mechanisms of metastasis. Cancer **80**:1529–1537
5. Eccles SA, and Welch DR (2007) Metastasis: recent discoveries and novel treatment strategies. Lancet **369**:1742–1757
6. Gupta GP, and Massague J (2006) Cancer metastasis: building a framework. Cell **127**:679–695

7. Christofori G (2006) New signals from the invasive front. Nature 441:444–450
8. Geiger TR, and Peeper DS (2009) Metastasis mechanisms. Biochim Biophys Acta 1796:293–308
9. Thiery JP (2002) Epithelial-mesenchymal transitions in tumour progression. Nature Reviews: Cancer 2:442–454
10. Bonnomet A, Brysse A, Tachsidis A, Waltham M et al (2010) Epithelial-to-mesenchymal transitions and circulating tumor cells. J Mammary Gland Biol Neoplasia 15:261–273
11. Berx G, Raspe E, Christofori G, Thiery JP et al (2007) Pre-EMTing metastasis? Recapitulation of morphogenetic processes in cancer. Clin Exp Metastasis 24:587–597
12. Hay ED (1995) An overview of epithelio-mesenchymal transformation. Acta Anat (Basel) 154:8–20
13. Thiery JP, and Chopin D (1999) Epithelial cell plasticity in development and tumor progression. Cancer Metastasis Rev 18:31–42
14. Perl AK, Wilgenbus P, Dahl U, Semb H et al (1998) A causal role for E-cadherin in the transition from adenoma to carcinoma. Nature 392:190–193
15. Mathias RA, Chen YS, Wang B, Ji H et al (2010) Extracellular remodelling during oncogenic Ras-induced epithelial-mesenchymal transition facilitates MDCK cell migration. J Proteome Res 9:1007–1019
16. Sabbah M, Emami S, Redeuilh G, Julien S et al (2008) Molecular signature and therapeutic perspective of the epithelial-to-mesenchymal transitions in epithelial cancers. Drug Resist Updates 11:123–151
17. Chen YS, Mathias RA, Mathivanan S, Kapp EA et al (2010) Proteomic profiling of MDCK plasma membranes reveals Wnt-5a involvement during oncogenic H-Ras/TGF-{beta}-mediated epithelial-mesenchymal transition. Mol Cell Proteomics
18. Spaderna S, Schmalhofer O, Wahlbuhl M, Dimmler A et al (2008) The transcriptional repressor ZEB1 promotes metastasis and loss of cell polarity in cancer. Cancer Res 68:537–544
19. Vandewalle C, Comijn J, De Craene B, Vermassen P et al (2005) SIP1/ZEB2 induces EMT by repressing genes of different epithelial cell-cell junctions. Nucleic Acids Res 33:6566–6578
20. Zhang X, Wei D, Yap Y, Li L et al (2007) Mass spectrometry-based "omics" technologies in cancer diagnostics. Mass Spectrom Rev 26:403–431
21. Walther TC, and Mann M (2010) Mass spectrometry-based proteomics in cell biology. J Cell Biol 190:491–500
22. Jechlinger M, Grunert S, Tamir IH, Janda E et al (2003) Expression profiling of epithelial plasticity in tumor progression. Oncogene 22:7155–7169
23. Yang J, Mani SA, Donaher JL, Ramaswamy S et al (2004) Twist, a master regulator of morphogenesis, plays an essential role in tumor metastasis. Cell 117:927–939
24. Bartel DP (2009) MicroRNAs: target recognition and regulatory functions. Cell 136:215–233
25. Gregory PA, Bracken CP, Bert AG, and Goodall GJ (2008) MicroRNAs as regulators of epithelial-mesenchymal transition. Cell Cycle 7:3112–3118
26. Mathias RA, and Simpson RJ (2009) Towards understanding epithelial-mesenchymal transition: a proteomics perspective. Biochim Biophys Acta 1794:1325–1331
27. Wei J, Xu G, Wu M, Zhang Y et al (2008) Overexpression of vimentin contributes to prostate cancer invasion and metastasis via src regulation. Anticancer Res 28:327–334
28. Willipinski-Stapelfeldt B, Riethdorf S, Assmann V, Woelfle U et al (2005) Changes in cytoskeletal protein composition indicative of an epithelial-mesenchymal transition in human micrometastatic and primary breast carcinoma cells. Clin Cancer Res 11:8006–8014
29. Wu L, and Han DK (2006) Overcoming the dynamic range problem in mass spectrometry-based shotgun proteomics. Expert Rev Proteomics 3:611–619
30. Alban A, David SO, Bjorkesten L, Andersson C et al (2003) A novel experimental design for comparative two-dimensional gel analysis: two-dimensional difference gel electrophoresis incorporating a pooled internal standard. Proteomics 3:36–44
31. Unlu M, Morgan ME, and Minden JS (1997) Difference gel electrophoresis: a single gel method for detecting changes in protein extracts. Electrophoresis 18:2071–2077
32. Timms JF, and Cramer R (2008) Difference gel electrophoresis. Proteomics 8:4886–4897
33. Friedman DB, and Lilley KS (2008) Optimizing the difference gel electrophoresis (DIGE) technology. Methods Mol Biol 428:93–124
34. Minden J (2007) Comparative proteomics and difference gel electrophoresis. BioTechniques 43:739, 741, 743 passim
35. Van den Bergh G, and Arckens L (2004) Fluorescent two-dimensional difference gel

36. Mathias RA, Wang B, Ji H, Kapp EA et al (2009) Secretome-Based Proteomic Profiling of Ras-Transformed MDCK Cells Reveals Extracellular Modulators of Epithelial-Mesenchymal Transition. J Proteome Res **8**:2827–2837

37. Simpson RJ, Connolly LM, Eddes JS, Pereira JJ et al (2000) Proteomic analysis of the human colon carcinoma cell line (LIM 1215): development of a membrane protein database. Electrophoresis **21**:1707–1732

38. Mosmann T (1983) Rapid colorimetric assay for cellular growth and survival: application to proliferation and cytotoxicity assays. J Immunol Methods **65**:55–63

39. Phillips HJ, and Terryberry JE (1957) Counting actively metabolizing tissue cultured cells. Exp Cell Res **13**:341–347

(continued from previous: electrophoresis unveils the potential of gel-based proteomics. Curr Opin Biotechnol **15**:38–43)

Chapter 20

Method for Protein Subfractionation of Cardiovascular Tissues Before DIGE Analysis

Athanasios Didangelos, Xiaoke Yin, and Manuel Mayr

Abstract

Difference gel electrophoresis (DIGE) (Electrophoresis 18, 2071–2077, 1997, 1) is widely used in cardiovascular research. However, the dynamic range limitations stemming from contaminating plasma proteins and highly abundant extracellular matrix components can make cardiovascular tissues difficult to analyze. Here we describe a novel methodology for biochemical subfractionation of cardiovascular tissues before DIGE analysis.

Key words: Cardiovascular, Heart, Proteomics, Vascular, Vessel

1. Introduction

Cardiovascular diseases are a major cause for morbidity and mortality in the Western world. Cardiovascular studies using proteomics have mainly focused on the analysis of whole tissue lysates or subcellular organelle (2–8). Consequently, little is known about changes of proteins in the extracellular space of cardiovascular tissues. The extracellular space is composed of the extracellular matrix (ECM), including, collagens, proteoglycans and glycoproteins and proteins associated with the ECM such as growth factors, cytokines, and proteinases. The characterization of these extracellular proteins is difficult for two reasons: First, proteins that are bound to the ECM are scarce and their identification is hampered by the presence of cellular proteins, and second, ECM proteins are usually cross-linked in tight aggregates and are therefore difficult to extract and solubilize. Moreover, they are subject to extensive posttranslational modifications, in particular glycosylation, which alter their molecular mass, charge, and electrophoretic properties. These characteristics not only render the proteins difficult to identify by mass spectrometry

(MS) but they are also responsible for many of the technical problems associated with two-dimensional gel electrophoresis (2-DE) such as incomplete isoelectric focusing and poor resolution during electrophoresis.

To analyze cellular as well as extracellular proteins, we have recently applied a three-step methodology to cardiovascular tissues, which allows a biochemical subfractionation of proteins based on their different extractability properties (9): (1) Solubilization of extracellular space proteins and plasma contaminants, which are "loosely" bound to the ECM using nondenaturing, ionic conditions; (2) Decellularization and retrieval of cellular proteins for analysis by 2-DE; (3) Extraction and solubilization of the remaining heavily cross-linked proteins which constitute the mature ECM. The ECM and its associated proteins represent relatively simple subproteomes and can be analyzed by gel-LC-MS/MS or shotgun proteomics. The refined cellular proteome is largely devoid of plasma proteins and extracellular matrix whereas the other fractions contain less cellular proteins facilitating the identification of scant ECM components. This three-step protein subfractionation procedure can be readily applied to clinical samples, including small tissue biopsies, and allows the interrogation of changes in extracellular as well as cellular proteins.

2. Materials

Prepare all solutions/buffers using double-distilled, sterile-filtered deionized water (ddH$_2$O) (>18.2 MΩ resistance) and analytical grade reagents (AnalR). Filter all solutions/buffers through 0.2-μm filters. All washing and extraction buffers must be supplemented with 25 mM ethylenediaminetetraacetic acid (EDTA) and include broad-spectrum proteinase inhibitors: 2 mM of phenylmethylsulfonyl fluoride (PMSF) (serine proteinase inhibitor), aprotinin (serine proteinase, including trypsin and related enzymes, inhibitor), pepstatin (aspartic proteinase inhibitor), and leupeptin (cysteine, serine, and threonine proteinase inhibitor). Alternatively, commercially available proteinase inhibitor cocktails containing a range of proteinase inhibitors can be used.

2.1. General Solutions

1. 250 mM EDTA, pH 8.0, stock solution. Adjust pH using 12 M NaOH.
2. 100 mM Tris–HCl stock solution, pH 7.5. Adjust pH with 12 M HCl.
3. 1× phosphate buffered saline (PBS): 150 mM NaCl, 1.7 mM KH$_2$PO$_4$, and 5 mM Na$_2$HPO$_4$. Adjust carefully to pH 7.4 using 6 M NaOH.

2.2. Extraction Buffers

1. *Extraction buffer 1*: 0.5 M NaCl and 10 mM Tris–HCl, pH 7.5 (dilute the 100 mM Tris–HCl stock solution).

2. *Extraction buffer 2*: 1% (35 mM) sodium dodecyl sulfate (SDS). In order to facilitate dissolution of SDS, warm up the buffer under hot tap water. SDS readily crystallizes at <20°C. Warm up to redissolve before use.

3. *Extraction buffer 3*: 4 M guanidine HCl and 50 mM sodium acetate. pH should be adjusted to 5.8 using 12 M NaOH.

4. *Interstitial collagen extraction buffer*: 1 M solution of pure acetic acid. Add 10 μg/mL ultrapure pepsin from porcine gastric mucosa (Sigma-Aldrich, Poole, UK).

5. *4× deglycosylation buffer*: 600 mM NaCl, 200 mM Na_2HPO_4, in ddH_2O, pH 6.8. Prior to deglycosylation, 0.05 U of the following enzymes are added to the 1× *deglycosylation buffer*:

    Chondroitinase ABC from *Proteus vulgaris*, which catalyzes the removal of polysaccharides containing 1→4-β-D-hexosaminyl and 1→3-β-D-glucuronosyl or 1→3-α-L-iduronosyl linkages to disaccharides containing 4-deoxy-β-D-gluc-4-enuronosyl groups. It acts on chondroitin 4-sulfate, chondroitin 6-sulfate, and dermatan sulfate glycosaminoglycan side chains (Sigma-Aldrich).

    Keratanase from *Bacteroides fragilis*, which cleaves internal 1→4-β-galactose linkages in unbranched, repeating poly-N-acetyl-lactosamine and keratan sulfate (Sigma-Aldrich).

2.3. DIGE Analysis

1. *Protein precipitation and sample clean-up:* 2DE clean-up kit (Bio-Rad, Hercules, CA).

2. *DIGE lysis buffer:* 8 M urea, 30 mM Tris–HCl, pH 8.5, 4% (w/v) CHAPS, and protease inhibitors. To prepare 100 mL of DIGE lysis buffer, dissolve 48 g urea in 90 mL of ddH_2O. After urea is dissolved, add 3 mL of 1 M Tris and 4 g CHAPS. Carefully adjust pH to 8.5 with diluted HCl while the buffer is still cold. Add two predissolved protease inhibitor cocktail tablets (Roche, Burgess Hill, UK) and top up to 100 mL with ddH_2O. Mix well, then freeze 1-mL aliquots at −80°C.

3. *DIGE 2× buffer:* 8 M urea, 2% (v/v) immobilized pH gradient (IPG) buffer (GE Healthcare), 2% (w/v) DTT, 4% (w/v) CHAPS. To prepare 100 mL of DIGE 2× buffer, dissolve 48 g urea in 90 mL of ddH_2O. Once urea is dissolved, add 2 mL of IPG buffer (matching the pH of the IPG strips), 2 g DTT, 4 g CHAPS and top up to 100 mL with ddH_2O. Mix well, then freeze 0.5-mL aliquots at −80°C.

4. *Rehydration buffer:* 8 M urea, 0.5% (w/v) CHAPS, 0.2% (w/v) DTT, 0.5% (v/v) IPG buffer, trace of bromophenol blue. To prepare 100 mL of rehydration buffer, dissolve 50 g urea in water and top up to 100 mL. Once urea is dissolved, add 1 g

of Amberlite, stir for 10 min, and filter through a 0.2-μm filter. Add 0.5 g CHAPS, 0.2 g DTT, 0.5 mL of IPG buffer (matching the pH of the IPG strips), trace of bromophenol blue and top up to 100 mL with ddH$_2$O. Mix well, then freeze 1-mL aliquots at −80°C.

5. *Equilibration buffer:* 6 M urea, 2% (w/v) SDS, 30% (v/v) glycerol, 50 mM Tris–HCl, bromophenol blue. To prepare 500 mL of equilibration buffer, mix 150 mL of glycerol and 150 mL of ddH$_2$O and dissolve 180.17 g urea. Separately dissolve 10 g SDS in 50 mL of ddH$_2$O and filter through a 0.2-μm filter. Add 16.75 mL of 1.5 M Tris–HCl buffer, pH 8.8, and a trace of bromophenol blue to the glycerol/water/urea solution. Finally, add the SDS solution and top up to 500 mL with ddH$_2$O. Store in 40-mL aliquots at −20°C.

6. Equilibration buffer with 1% (w/v) SDS.

7. Equilibration buffer with 4.8% (w/v) iodoacetamide.

8. *IPG strips:* Immobiline DryStrip gels (IPG, GE Healthcare).

9. *Gel buffer:* 1.5 M Tris–HCl, pH 8.8. To prepare 1 L of gel buffer, dissolve 181.7 g Tris-base in 800 mL of water. Add 20 mL concentrated HCl and stir for 3–4 h. Carefully adjust pH to 8.8 with diluted HCl. Fill up to 1 L with ddH$_2$O and filter through a 0.2-μm filter. Store at 4°C and keep away from light.

10. Acrylogel 2.6 (40%) solution (37.4% acrylamide, 2.6% *N,N′*-methylenebisacrylamide (VWR, Lutterworth, UK)). Store at 4°C in the dark.

11. *Gel polymerization:* To cast large-format polyacrylamide gels (12% T (total acrylamide concentration), 2.6% C (degree of cross-linking)), mix 187.5 mL of Acrylogel 2.6 (40%) solution with 150 mL of 1.5 M Tris–HCl, pH 8.8, and 253.5 mL of ddH$_2$O. Finally, add 6 mL of 10% SDS, 252 μL of tetramethylethylenediamine (TEMED) and 2.4 mL of 10% ammonium persulfate (APS). Stir for about 10 s, then pour the gel solution into the gel caster. Leave enough space at the top for placing the IPG strip.

12. *Overlay solution:* Two butanol and ddH$_2$O are mixed at a 3:2 ratio. Wait until there are two phases. Use the upper phase as overlay solution.

13. *Gel storage solution:* 50 mL of 1.5 M Tris–HCl, pH 8.8, and 2 mL of 10% SDS are mixed and topped up to a final volume of 200 mL with ddH$_2$O.

14. *SDS running buffer:* Dilute 10× Tris-glycine running buffer (Invitrogen, Carlsbad, CA; LC2675) with ddH$_2$O to make 1× and 2× running buffer.

15. *Agarose sealing solution:* 1% (w/v) low-melting point agarose in SDS running buffer with trace of bromophenol blue.

3. Methods

Samples stored in liquid nitrogen can be used, but snap-freezing causes cell rupture and may also damage the conformation of the ECM. Thus, it is preferable to process samples immediately without freezing. The protein subfractionation methodology is summarized in Fig. 1.

3.1. Sample Preparation

Typically, 50 mg of tissue is sufficient to obtain an adequate amount of proteins in each extraction step for proteomics analysis. Prior to extraction, tissue samples must be weighed and diced into smaller

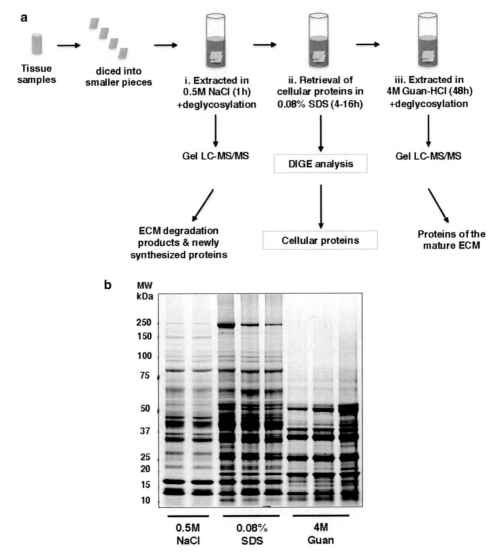

Fig. 1. Biochemical subfractionation of the cardiac proteome: (**a**) schematic representation of the biochemical subfractionation for improved proteomics analysis of cardiovascular tissues (adapted from: Didangelos et al. (9)), (**b**) a representative Coomassie blue-stained 1-D gel comparing the NaCl, SDS, and guanidine extracts from different murine cardiac samples. Note that the SDS extracts are the most complex and are subjected to further analysis by 2-DE (see Fig. 2).

pieces. Proteins will not be extracted efficiently from large specimen. Frozen samples should be defrosted but always kept on ice until dissection and weighing are complete. Cardiovascular samples are invariably contaminated with plasma proteins. To decrease this contamination, place samples in 10 mL of 1× PBS, 25 mM EDTA plus proteinase inhibitors. Mix gently and place on a rotating platform for a few minutes. Change the solution and repeat the procedure five times.

3.2. Protein Extraction

1. *Extraction buffer 1 (0.5 M NaCl): Nondenaturing, ionic.* The salt ions induce displacement of ionic interactions between proteins, thus enabling the extraction of weakly bound proteins or protein fragments present in the extracellular space. These include newly synthesized ECM proteins, which are not yet cross-linked with the mature ECM, proteolytic fragments as well as plasma proteins such as lipoproteins, enzymes, and coagulation factors (10).
2. Place samples in 2-mL tubes with a screw cap to avoid leakage.
3. Add extraction buffer 1. The buffer volume is adjusted to the weight of the tissue at a ratio of 10:1, i.e., 100 mg of tissue are incubated in 1 mL of buffer.
4. Gently vortex samples for 1 h at room temperature (RT). The optimal incubation time has to be determined experimentally for each type of tissue (see Note 1).
5. After incubation, remove the buffer and centrifuge at $16,000 \times g$. Freeze supernatant for later use or proceed with deglycosylation step (see Subheading 3.4).
6. Rinse the tissue samples once with ddH$_2$O supplemented with EDTA and proteinase inhibitors before proceeding with the second extraction step.
7. *Extraction buffer 2 (0.08% SDS): Decellularization.* SDS has been previously shown to be very effective in solubilizing cytoplasmic and nuclear membranes, thereby selectively retrieving cellular proteins from tissues (11). Unlike other decellularizing agents, such as nonionic detergents (Triton X-100) (12), SDS preserves the ECM and its associated proteins. However, SDS can cause protein denaturation and disrupt native ECM proteins (13) if used above its critical micelle concentration, which is 0.24% weight per volume or 8 mM, in water.
8. Dilute 1% SDS to 0.08% using ddH$_2$O and add EDTA and proteinase inhibitors. There is no need to adjust the pH of the SDS buffer.
9. Add the 0.08% SDS buffer to tissue samples at a volume-to-tissue weight ratio of 10:1.

10. Incubate samples for 4–16 h at RT (see Note 2) with gentle vortexing to avoid mechanical disruption of the ECM in the presence of SDS.

11. The extraction buffer is centrifuged at $16,000 \times g$ and stored at $-20°C$ for later use. The processing of the decellularized tissue samples continues with the third extraction step.

12. *Extraction buffer 3 (4 M guanidine-HCl)*: Strongly denaturing. 4 M guanidine is very effective in solubilizing most of the heavily cross-linked, aggregated ECM components, including large proteoglycans (versican, aggrecan, etc.), small proteoglycans (decorin, biglycan, etc.), cell attachment matrix glycoproteins (such as type VI collagen, fibronectins, and laminins), and basement membrane components (perlecan, type IV collagen, etc.). Guanidine induces disaggregation of ECM components by destabilizing their ionic, disulfide-dependent protein conformation (14). 4 M guanidine only partially extracts interstitial collagen. The buffer-to-tissue weight ratio is kept constant at 10:1.

13. Incubate samples in extraction buffer 3 for 48 h at RT and vortex vigorously to facilitate the mechanical disruption of the ECM and the extraction and solubilization of ECM proteins.

14. Remove the supernatant after 48 h, centrifuge at $16,000 \times g$ and either store the extracts at $-80°C$ for later use (4 M guanidine does not freeze at $-20°C$) or proceed with the removal of 4 M guanidine (recommended, see Subheading 3.3).

15. *Interstitial collagen extraction (1 M acetic acid and pepsin)*: Wash tissue samples with ddH$_2$O three times to remove any remaining guanidine remnants.

16. Add 1 M interstitial collagen extraction buffer and incubate samples at RT for 48–72 h with vigorous vortexing (see Note 3).

17. After incubation, vacuum dry the extraction buffer to remove the acetic acid. The resulting collagen-rich pellet can be redissolved in gel loading buffer.

3.3. Removal of 4 M Guanidine

1. For removal of 4 M guanidine from the tissue extracts (see Note 4), mix the extraction buffer with 100% ethanol at $-20°C$ for 16 h at an ethanol-to-tissue extraction buffer volume ratio of 5:1.

2. Precipitate proteins by centrifugation ($16,000 \times g$ for 45 min).

3. Wash pellets with 90% ethanol and air dry.

3.4. Deglycosylation Step (see Note 5)

1. For the 0.5 M NaCl extracts, add 4× deglycosylation buffer to a final concentration of 1×.

2. For the 4 M guanidine extracts, dissolve the precipitated protein pellets in 1× deglycosylation buffer (see Note 4).

294   A. Didangelos et al.

3. Supplement the deglycosylation buffer with chondroitinase ABC and keratanase as described above.

4. Incubate samples for 16 h at 37°C.

3.5. DIGE Analysis of SDS Extracts (see Note 6 and Fig. 2)

1. Prior to the DIGE analysis, SDS has to be removed by protein precipitation (see Note 5). We use a commercial 2DE clean-up kit (Bio-Rad, Hercules, CA). Resuspend protein pellets in DIGE lysis buffer.

2. The labeling reaction with CyDye DIGE fluor minimal dyes is performed according to the manufacturer's instruction (GE Healthcare). However, a lower dye-to-protein ratio (200 pmol dye/50 μg protein) is sufficient because the SDS extracts are already depleted of high-abundant plasma and ECM proteins.

3. The labeled samples are mixed with the same volume of DIGE 2× buffer. Top up to a final volume of 450 μL with rehydration buffer. Vortex and centrifuge at $16,000 \times g$ for 1 min.

4. Rehydrate Immobiline DryStrip gels (IPG, GE Healthcare) in a reswelling tray according to the manufacturer's instruction.

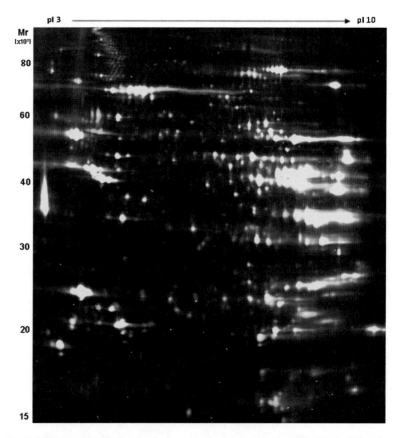

Fig. 2. DIGE image of SDS extracts from murine hearts (pH 3–10NL, 18-cm IPG strip, 12% gel).

Table 1
IEF program for 18-cm IPG strip (3–10 NL)

Step	Voltage (V)		Duration	Vh
Optional	30	Step and hold	12h	360
1	150	Step and hold	2 h	300
2	300	Step and hold	2 h	600
3	600	Step and hold	2 h	1,200
4	1,500	Step and hold	8 h	12,000
5	8,000	Gradient	30 min	2,375
6	8,000	Step and hold	2 h	16,000
Optional	500	Step and hold	2 h	1,000

5. Strips are focused at a maximum of 0.05 mA per IPG strip at 20°C overnight in the dark. An optional desalting step at 30 V can be added. Table 1 provides an example for an IEF program suitable for the analysis of cardiovascular tissues using an 18-cm IPG strip. Once IEF is complete, a constant voltage of 500 V is applied to prevent protein diffusion.

6. Focused strips are sealed in plastic bags and stored in X-ray film cassettes at −80°C or processed immediately.

7. To cast large-format polyacrylamide gels (12% T (total acrylamide concentration), 2.6% C (degree of cross-linking)), mix 187.5 mL Acrylogel 2.6 (40%) solution with 150 mL of 1.5 M Tris–HCl, pH 8.8, and 253.5 mL of ddH$_2$O.

8. Degas for at least 1 h with continuous stirring.

9. Add 6 mL of 10% SDS, 252 μL of TEMED, and 2.4 mL of 10% APS. Stir for about 10 s, then pour the gel solution into the gel caster. Leave enough space at the top for placing the IPG strip.

10. Overlay the top surface of the gel solution with overlay solution (upper phase; see Subheading 2.3).

11. Cover the gel caster with cling film.

12. After 1.5 h, discard the overlay solution, wash briefly with ddH$_2$O, and overlay with gel storage solution. Leave to polymerize overnight at RT. If the gels are used at a later time, seal them with gel storage solution in plastic bags and keep at 4°C.

13. Equilibrate the IPG strips in equilibration buffer with 1% (w/v) DTT for 15 min. Shake gently.

14. Discard DTT solution and replace with equilibration buffer plus 4.8% (w/v) iodoacetamide. Incubate for another 15 min.
15. Briefly wash the strips with ddH$_2$O, then place them on top of the gels and add 3 mL of agarose sealing solution. Ensure that there are no bubbles between the IPG strips and the gel surface. Let the agarose set.
16. Run the second dimension. We run the gels on an Ettan DALT*six* Electrophoresis Systems (GE Healthcare) at 2 W/gel for 15 min, 2.5 W/gel for 30 min before the power is increased to 30 W/gel. The run is continued until the bromophenol blue dye front migrates off the lower end of the gels.

4. Notes

1. If the tissue samples remain for a prolonged period of time in the hypertonic solution, cell membranes may collapse as a result of the continuous loss of water from the cells, thereby causing loss of cellular proteins in the buffer. In our experience, vascular samples, especially arteries, which are rich in ECM, withstand 4 h of incubation with 0.5 M NaCl. In contrast, the incubation period should be restricted to 1 h for cardiac tissues. Also, buffer-to-tissue weight ratio is crucial; if the ratio is too low, then the buffer will be quickly saturated with extracted proteins, and the effectiveness of the extraction will be reduced. The extracted protein concentration is difficult to predict, but in practice, a 10:1 ratio provides good extraction efficiency without excessive dilution of the sample.
2. The optimal incubation time is strongly dependent on the cellularity of the tissue. In our experience, a 4-h incubation is sufficient for vascular tissues, but 16 h are required for retrieval of cellular proteins in cardiac tissues. Yet, some myofilament proteins will still remain, and we have used other protocols to study this particular cardiac subproteome (8).
3. Notes on collagen solubilization: Interstitial collagen is very difficult to solubilize. Acidity is essential together with the action of pepsin, which cleaves collagen at the telopeptide region, thereby increasing its extraction efficiency (15).
4. Removal of the chaotropic agent is essential for deglycosylation. Its denaturing properties would inhibit enzymatic activity. Also, the ionic strength of the buffer would negatively affect separation by gel electrophoresis. Guanidine is readily soluble in ethanol, which will also precipitate the extracted proteins. Most of the heavily glycosylated ECM proteins are in the 4 M guanidine extracts.

5. It is not mandatory to deglycosylate the cellular (SDS) extracts. If deglycosylation is desired, it has to be done after protein precipitation to remove SDS (see Subheading 3.5).

6. Because of the differential solubilization of extracellular proteins, the 0.5 M NaCl and the 4 M guanidine extracts represent relatively simple subproteomes and can be effectively analyzed by gel LC-MS/MS. Alternatively, they could be analyzed by DIGE, following the cleaning steps described above (see Subheading 3.5, step 1).

Acknowledgements

M.M. is a senior fellow of the British Heart Foundation.

References

1. Unlu, M., Morgan, M. E., and Minden, J. S. (1997) Difference gel electrophoresis: a single gel method for detecting changes in protein extracts, *Electrophoresis 18*, 2071–2077.
2. Arrell, D. K., Neverova, I., and Van Eyk, J. E. (2001) Cardiovascular proteomics: evolution and potential, *Circ Res 88*, 763–773.
3. Didangelos, A., Simper, D., Monaco, C., and Mayr, M. (2009) Proteomics of acute coronary syndromes, *Curr Atheroscler Rep 11*, 188–195.
4. Mayr, M., Liem, D., Zhang, J., Li, X., Avliyakulov, N. K., Yang, J. I., Young, G., Vondriska, T. M., Ladroue, C., Madhu, B., Griffiths, J. R., Gomes, A., Xu, Q., and Ping, P. (2009) Proteomic and metabolomic analysis of cardioprotection: Interplay between protein kinase C epsilon and delta in regulating glucose metabolism of murine hearts, *J Mol Cell Cardiol 46*, 268–277.
5. Mayr, M., Madhu, B., and Xu, Q. (2007) Proteomics and metabolomics combined in cardiovascular research, *Trends Cardiovasc Med 17*, 43–48.
6. Mayr, M., Zhang, J., Greene, A. S., Gutterman, D., Perloff, J., and Ping, P. (2006) Proteomics-based development of biomarkers in cardiovascular disease: mechanistic, clinical, and therapeutic insights, *Mol Cell Proteomics 5*, 1853–1864.
7. McGregor, E., and Dunn, M. J. (2006) Proteomics of the heart: unraveling disease, *Circ Res 98*, 309–321.
8. Yin, X., Cuello, F., Mayr, U., Hao, Z., Hornshaw, M., Ehler, E., Avkiran, M., and Mayr, M. (2010) Proteomics analysis of the cardiac myofilament subproteome reveals dynamic alterations in phosphatase subunit distribution, *Mol Cell Proteomics 9*, 497–509.
9. Didangelos, A., Yin, X., Mandal, K., Baumert, M., Jahangiri, M., and Mayr, M. (2010) Proteomics characterization of extracellular space components in the human aorta, *Mol Cell Proteomics 9*, 2048–2062.
10. Mason, R. M., and Mayes, R. W. (1973) Extraction of cartilage protein-polysaccharides with inorganic salt solutions, *Biochem J 131*, 535–540.
11. Dahl, S. L., Koh, J., Prabhakar, V., and Niklason, L. E. (2003) Decellularized native and engineered arterial scaffolds for transplantation, *Cell Transplant 12*, 659–666.
12. Korossis, S. A., Wilcox, H. E., Watterson, K. G., Kearney, J. N., Ingham, E., and Fisher, J. (2005) In-vitro assessment of the functional performance of the decellularized intact porcine aortic root, *J Heart Valve Dis 14*, 408–421; discussion 422.
13. Gilbert, T. W., Sellaro, T. L., and Badylak, S. F. (2006) Decellularization of tissues and organs, *Biomaterials 27*, 3675–3683.
14. Sajdera, S. W., and Hascall, V. C. (1969) Proteinpolysaccharide complex from bovine nasal cartilage. A comparison of low and high shear extraction procedures, *J Biol Chem 244*, 77–87.
15. Bannister, D. W., and Burns, A. B. (1972) Pepsin treatment of avian skin collagen. Effects on solubility, subunit composition and aggregation properties, *Biochem J 129*, 677–681.

Chapter 21

Application of DIGE and Mass Spectrometry in the Study of Type 2 Diabetes Mellitus Mouse Models

Celia Smith, Davinia Mills, and Rainer Cramer

Abstract

Knowledge of the differences between the amounts and types of protein that are expressed in diseased compared to healthy subjects may give an understanding of the biological pathways that cause disease. This is the reasoning behind the presented protocol, which uses difference gel electrophoresis (DIGE) to discover up- or down-regulated proteins between mice of different genotypes, or of those fed on different diets, that may thus be prone to develop diabetes-like phenotypes. Subsequent analysis of these proteins by tandem mass spectrometry typically facilitates their identification with a high degree of confidence.

Key words: Difference gel electrophoresis, Mass spectrometry, Proteomics, Quantitation, Type 2 diabetes mellitus

1. Introduction

Type 2 diabetes mellitus (T2DM) has reached epidemic proportions globally, and the number of sufferers is expected to more than double in the period 2000–2030 to 366 million worldwide(1, 2). The protocol presented here combines difference gel electrophoresis (DIGE) separation and quantitation with mass spectrometric identification of the proteins that are differentially expressed in lean and obese mice when fed standard and high-fat diets. It is hoped that the identification of these proteins will help to unravel the complex biological pathways involved in glycaemic control and obesity, which are closely related to T2DM.

Five biological replicates of each mouse genotype/diet were chosen to obtain some statistical validity. Frozen liver samples from these animals were solubilised and labelled with one of two fluorescent dyes (red, Cy3, and blue, Cy5). Equal amounts of each of the underivatised samples were pooled and labelled with a third fluorescent dye (yellow, Cy2). The samples were then mixed and subjected to two-dimensional gel electrophoresis (2DE) so that per gel there were two samples of different fluorescent labels, plus an aliquot of the labelled pooled internal standard. Subsequent fluorescence scanning of the gels at each of the dyes' excitation/emission wavelengths yields gel images that can be superimposed, and in this image spots of red and blue indicate the more predominant proteins (see Fig. 1).

Use of the pooled internal standard across all of the gels allows the gel imaging software to normalise the response for each gel. Thus, comparisons can be made to identify the proteins that are reproducibly either up- or down-regulated between samples. Protein spots of interest from one of the DIGE gels or a preparative gel, which is run simultaneously without fluorescent dye but with an increased amount of protein, can then be picked, subjected to tryptic digestion and analysed by LC-MS/MS. Protein identification is then made by matching actual peptide and fragment masses with theoretical equivalents from a sequence database.

Fig. 1. Fluorescent DIGE image highlighting the differential proteins expressed for the Swiss Jim Lambert genotype fed on either a standard (Cy3-labelled, *pink*) or a high-fat (Cy5-labelled, *blue*) diet (gel 4 of Table 1).

2. Materials

2.1. Equipment

The 2DE equipment employed was an Ettan DALT*twelve* System from GE Healthcare (Little Chalfont, Buckinghamshire, UK) and most of the associated hardware, consumables and manuals/protocols were from the same supplier.

2.1.1. Protein Extraction from Mouse Liver Tissue

1. Pestle and mortar.
2. Ice bath.
3. Spectrophotometer.

2.1.2. Protein Labelling

1. Ice bath.
2. Glass syringe, 500 µL.

2.1.3. Rehydration of IPG Strips

1. IPG strips (3-10NL, 24 cm; GE Healthcare).
2. Rehydration tray and lid (GE Healthcare).

2.1.4. Isoelectric Focusing

1. Ettan IPGphor II with electrodes (GE Healthcare).
2. IEF ceramic tray and wicks (GE Healthcare).
3. Two pairs of tweezers.
4. Spirit level.

2.1.5. Second-Dimension Gel Electrophoresis

1. Rehydration tray and lid (GE Healthcare).
2. Two pairs of tweezers.
3. Thin but blunt long-bladed spatula.
4. Microwave oven.
5. Ettan DALT*twelve* separation unit (GE Healthcare).
6. Rehydration tray and lid (GE Healthcare).

2.1.6. Gel Scanning and Image Analysis

1. Typhoon 9400 plus blue laser module and control software (GE Healthcare).
2. Gel alignment guide (GE Healthcare).
3. ImageQuant v5 software (GE Healthcare).
4. Progenesis SameSpots (Non-Linear Dynamics, Newcastle upon Tyne, UK).

2.1.7. Visualisation and Picking of Protein Spots of Interest

1. Staining trays.
2. Light box.
3. Manual spot picker (e.g. One Touch Spot Picker, available from the Gel Company) and compatible tips with 1.5 mm-orifice.
4. Fine nose tweezers.

2.1.8. In-Gel Digestion and NanoLC-MS/MS Analysis

1. Sample concentrator (SpeedVac; Thermo Fisher Scientific, Hemel Hempstead, UK).
2. Dionex Ultimate 3000 with NCS-3500RS pump. NanoLC system comprising vacuum degasser, nanocapillary pump, loading pump, temperature-controlled column compartment, autosampler and Chromeleon Express control software (Dionex, Amsterdam, Holland).
3. LTQ-Orbitrap XL mass spectrometer with nanoelectrospray (nESI) source and Xcalibur control software (Thermo Fisher Scientific).
4. Acclaim PepMap 100 LC column, 75 μm ID × 15 cm, 3 μ C_{18} packing, and a precolumn cartridge (300 μm ID × 0.5 cm) packed with Acclaim PepMap 100 (Dionex).
5. Proxeon ES508 stainless steel emitters (Thermo Fisher Scientific).
6. Proteome Discoverer 1.0 software (Thermo Fisher Scientific).
7. In-house Mascot search engine (Matrix Science, London, UK).

2.2. Consumables

All solvents should be of HPLC-grade or equivalent unless stated otherwise.

2.2.1. Protein Extraction from Mouse Liver Tissue

1. Liquid nitrogen.
2. Lysis buffer at pH 8.3 (8 M urea, 2 M thiourea, 4% CHAPS (w/w), 10 mM Tris, 0.5% NP40).
3. Protein Assay (Bio-Rad, Hemel Hempstead, UK).
4. Bovine serum albumin (BSA) standards in the range of 0–2 mg/mL in water.
5. Disposable semi-microcuvettes.

2.2.2. Protein Labelling

1. *N,N*-dimethylformamide, anhydrous (DMF) (see Note 1).
2. CyDye DIGE Fluor minimal labelling kit, 2 or 5 nmol (GE Healthcare).
3. 10 mM solution of L-lysine monohydrochloride.
4. Lysis buffer at pH 8.3 (8 M urea, 2 M thiourea, 4% CHAPS (w/w), 10 mM Tris, 0.5% NP40).

2.2.3. Rehydration of IPG Strips

1. 1.3 M dithiothreitol (DTT; Sigma-Aldrich, Gillingham, UK).
2. IPG buffer with the same pH interval as the IPG strips (3–10NL; GE Healthcare).
3. DeStreak rehydration solution (GE Healthcare).
4. DryStrip cover fluid (GE Healthcare).
5. IPGphor strip holder cleaning solution (GE Healthcare).

2.2.4. Isoelectric Focusing

1. Lint-free wipes.
2. DryStrip cover fluid (GE Healthcare).
3. IPGphor strip holder cleaning solution (GE Healthcare).

2.2.5. Second-Dimension Electrophoresis	1. SDS-PAGE gels. 2. DTT. 3. Iodoacetamide (IA; GE Healthcare). 4. Equilibration buffer (aqueous solution containing 50 mM Tris, 6 M urea, 30% glycerol, 2% sodium dodecyl sulphate). 5. Agarose (ReadyPrep Overlay Agarose; Bio-Rad). 6. Tris-glycine-SDS run buffer, 10 times concentrate (National Diagnostics Protogel, obtainable from Fisher Scientific, Loughborough, UK).
2.2.6. Gel Scanning and Image Analysis	1. Water for plate cleaning.
2.2.7. Visualisation and Picking of Proteins of Interest	1. Fixing solution (40% ethanol, 10% acetic acid). 2. Destaining solution (1% acetic acid). 3. Ammonium sulphate. 4. Brilliant Blue G250. 5. Phosphoric acid. 6. Methanol. 7. Ethanol.

3. Methods

To exclude the possibility of sample contamination with non-mouse proteins or other contaminants, it is advisable to carry out as many of the laboratory procedures as possible in a designated clean room or area. If this is not possible, great care must be taken at every step to avoid contamination, especially from human keratin which tends to be ubiquitous.

3.1. Protein Extraction from Mouse Liver Tissue

This protocol assumes that mouse liver tissues are frozen at −80°C. It is critically important to keep the liver tissues frozen until the lysis buffer has been added to the ground samples, to prevent proteases degrading the proteins.

The fact that liver tissue samples contain high levels of protein gives the advantage that it is not necessary to undertake additional sample clean-up steps. By taking only a small quantity of the liver lysate for DIGE, levels of possible interfering compounds will be low:

1. Add the mouse liver tissue to the pre-cooled mortar containing liquid nitrogen. Grind the frozen tissue to a fine powder, topping up the liquid nitrogen as it evaporates so that the tissue does not thaw.

2. Add a small amount (ca. 15–20 mg) of the powdered tissue to an Eppendorf tube and immediately add 1 mL of lysis buffer. Vortex and then sonicate the Eppendorf tube until the powdered tissue is dissolved.

3. Centrifuge the tube at 6°C and $12,000 \times g$ for 10 min to pellet the cell debris. Pipette off the supernatant, aliquot and store (as required) at –80°C.

4. Estimate the concentration of protein in the supernatant using a modification of the Bio-Rad Protein Assay, which is based on the Bradford method (3), by diluting the colour reagent 5 times with water, then adding 20 μL of each sample or standard to 1.6 mL of the diluted colour reagent in a semi-microcuvette (see Note 2).

3.2. Protein Labelling

As two samples can be run per gel, the minimum number of gels will be half of the total number of samples to be analysed. Thus, if there are 20 samples to be run, it will be necessary to run ten gels. It is useful to prepare a randomised gel plan in advance, showing which samples will be labelled with the Cy3 and the Cy5 dyes, respectively, and how these will be combined on the gels. Table 1 shows such a plan for four different combinations of genotypes and diets and five biological replicates for each combination. Each gel is typically loaded with a total of 300 μg of protein: 100 μg from each of the two samples and a further 100 μg from the pooled internal standard, which is labelled with the Cy2 dye:

Table 1
Example gel plan showing the labelling scheme for two genotypes (Gen1 and Gen2), two diets (Diet1 and Diet2) and five biological replicates of each possible combination of genotype and diet

Gel no.	Cy2	Cy5	Cy3
1	Pool	Gen1/Diet1-1	Gen1/Diet2-1
2	Pool	Gen1/Diet2-2	Gen2/Diet2-1
3	Pool	Gen2/Diet1-1	Gen1/Diet1-2
4	Pool	Gen2/Diet2-2	Gen2/Diet1-2
5	Pool	Gen1/Diet1-3	Gen2/Diet1-3
6	Pool	Gen2/Diet2-3	Gen1/Diet2-3
7	Pool	Gen2/Diet2-4	Gen1/Diet1-4
8	Pool	Gen1/Diet2-4	Gen2/Diet1-4
9	Pool	Gen2/Diet1-5	Gen2/Diet2-5
10	Pool	Gen1/Diet1-5	Gen1/Diet2-5

1. Normalise the sample protein levels to 3 mg/mL by the addition of lysis buffer as necessary. This step greatly simplifies the protein labelling procedure. Check that the pH of each sample is between 8 and 9 (see Note 3).

2. To prepare the internal standard for labelling, pool an equal volume (amount) of 20 μL (60 μg) of each sample in a suitable tube to give a sufficient volume (amount) of internal standard for all gels. Ensure the tube contents are well-mixed (see Note 4).

3. To prepare the individual samples for labelling, transfer for each individual sample twice the volume used in the previous step (40 μL) into an Eppendorf tube for labelling with either Cy3 or Cy5 dye (see Notes 4 and 5).

4. Keep the samples on ice until you are ready to label them.

5. Prepare the CyDyes by addition of an appropriate volume of anhydrous DMF to each tube of the CyDye kit (see Note 6). The dye concentration of the resultant solution should be 400 pmol/μL.

6. Usually 100 μg of protein is labelled with 800 pmol of dye. In the example presented, the amount of protein actually taken was 120 μg, and so 2.4 μL of the appropriate CyDye solution is added to 40 μL of the sample.

7. As the total volume of the pooled internal standard is 20×20 μL = 400 μL (1,200 μg), this is then labelled by the addition of 24 μL of Cy2 dye.

8. The tubes containing the labelled samples and internal standards are vortexed briefly, spun and kept on ice in the dark for 30 min.

9. To stop the labelling reaction at least an equal volume of 10 mM lysine solution compared to the CyDye solution volume of steps 6 and 7 should be added to each tube. However, for convenience a higher volume, i.e. 5.6 and 56 μL respectively, should be chosen to increase the total volume by 20% to 48 and 480 μL, respectively. All sample solutions now provide a sufficient number of 40 μL aliquots, each containing 100 μg of sample, for ten gels to be run with two individual samples and one internal standard.

10. The samples are mixed, spun and kept on ice in the dark for a further 10 min and can then be either frozen and stored at this stage, or further prepared for the rehydration of the IPG strips (see Subheading 3.3).

3.3. Rehydration of IPG Strips

In this procedure, the labelled samples are firstly mixed in accordance with the gel plan (see Table 1) and then reduced with DTT. Modified DeStreak solution is added to make a total volume that is compatible with the IPG strip size and method of rehydration.

In this example, where the 24 cm-IPG strips are rehydrated in the presence of the protein sample, a maximum volume of 450 μL is used:

1. Pipette 40 μL of the Cy5-labelled sample into a 0.5 mL Eppendorf tube. Add 40 μL of the Cy3-labelled sample and 40 μL of the Cy2-labelled internal standard. Repeat for all samples as detailed in the gel plan.
2. Add 22 μL of the 1.3 M-DTT solution to each of the sample tubes and mix.
3. Modify the DeStreak solution by the addition of the IPG buffer—added here at 2% (see Note 7)—to give a sufficient volume for all samples, i.e. for the ten gels prepared in this example more than $(450-22-(3 \times 40))\,\mu L \times 10 = 3.08$ mL is needed.
4. Add 308 μL of modified DeStreak solution to each tube, mix the tube contents and stand on ice in the dark for 30 min.
5. Level the clean and dry rehydration tray in a convenient position on the bench and pipette each sample into an empty lane and record which sample is in which lane.
6. Using tweezers, carefully peel off the backing strip for each IPG strip, and place gel-side down on top of the sample. Ensure there are no air bubbles trapped between the sample and the gel.
7. If the strips are the correct way up, it should be possible to read the serial numbers of the IPG strips. Record these on the Gel Plan (see Table 1).
8. Cover each lane with approximately 2 mL of DryStrip cover fluid, then slide the lid into place.
9. Protect from light and allow the strips to rehydrate overnight (see Note 8).

3.4. Isoelectric Focusing

This is the first dimension of the separation, where voltage is applied across the strips to separate the proteins according to their pI. The voltage is increased in either a stepwise or gradient fashion to firstly remove ionic material to the ends of the strips and then to gradually move the proteins to their pI. The IEF programme detailed in Table 2 was found to be suitable for use with mouse livers. However, for any new study, the programme must be optimised empirically as there are many factors that can influence the separation, such as protein loading and concentration of the IPG buffer.

IEF is conveniently carried out overnight. However, focusing for too long can cause horizontal streaking in the gels. In the programme detailed in Table 2, the sole purpose of the last stage (S6 in Table 2) is to keep the proteins focused at low voltage until it is convenient to remove them from the IPGphor. The programme

Table 2
IPGphor programme used in the isoelectric focusing of mouse liver lysates

Stage	Step or gradient	Voltage (V)	Duration or kV-hours
S1	Step	150	2 h
S2	Step	500	2 h
S3	Gradient	1,000	3 h
S4	Gradient	8,000	5 h
S5	Step	8,000	96,000 kVh
S6	Step	500	2,000 kVh

can be stopped at any stage during S6, and the strips removed promptly before they are processed further (see Note 8):

1. Level the IPGphor II and ensure that the ceramic tray is completely clean and dry (see Note 9).
2. Place the ceramic tray onto the IPGphor II and carefully add approximately 108 mL of DryStrip cover fluid, ensuring that the fluid is evenly distributed across the lanes.
3. Using two pairs of tweezers, carefully remove the IPG strips from the rehydration tray and drain off any surplus cover fluid onto a lint-free wipe. Ensure that the strip is placed gel-side up—the serial nos. should not be readable.
4. Place each strip, gel-side up, into a lane of the ceramic tray so that the end of the strip marked with "+" is furthest away from you.
5. Dampen each wick with 150 µL of deionised water and place them at both ends of each IPG strip, so that they slightly overlap the ends of the gel (see Note 10).
6. Place each electrode assembly on top of the wicks at either end of the IPG strips. The assembly should make electrical contact with the IPG strips through the wicks, and with the IPGphor.
7. Add more DryStrip cover fluid to the tray so that the strips are completely covered, close the lid and protect the strips from light.
8. Set the IPGphor to run at 20°C and 75 µA/strip. The programme detailed in Table 2 has been successfully used for mouse livers.
9. Start the focusing programme (see Table 2) and enter the number of IPG strips when prompted.

10. When focusing is complete (see Note 11), stop the focusing programme, remove the strips and drain the DryStrip cover fluid onto lint-free paper without delay.

11. If not proceeding to the second-dimension immediately, it is possible to freeze the strips at −80°C (see Note 8).

3.5. Second-Dimension Electrophoresis

Before the proteins can be separated according to their molecular weight, it is first necessary to break down the three-dimensional structure of the proteins and to saturate the strip with sodium dodecyl sulphate (SDS). This is done by equilibrating the strips in a cocktail, which includes a pH buffer, urea, SDS and a dye for monitoring the solvent front (see Subheading 2.2.5). Any disulphide bonds are reduced with DTT and then alkylated with IA. These equilibration steps are conveniently carried out in the rehydration tray:

1. If the strips have been frozen, allow them to thaw to room temperature.

2. If using the rehydration tray for the equilibration steps, the total amount of equilibration buffer needed will be 2 * [no. of IPG strips] * 3 mL as there are two equilibration stages. This is then split into two equal aliquots.

3. Solid DTT is added to the first aliquot at 0.5% (w/v), and solid IA is added to the second aliquot at 4.5% (w/v). Both aliquots are placed on the roller shaker to dissolve the added contents; the IA-containing aliquot must be protected from light.

4. Place the IPG strips gel-side up in the rehydration tray and cover each strip with at least 2 mL of the equilibration buffer containing DTT. Ensure all of the strips are completely covered with the buffer.

5. Cover the tray with the lid and place on the orbital shaker for 15 min.

6. Remove the strips from the tray and immerse each one briefly in a measuring cylinder containing diluted run buffer (see Subheading 2.2.5), then drain onto lint-free paper.

7. Clean out the rehydration tray and ensure it is completely clean and dry.

8. Place the IPG strips gel-side up in the rehydration tray and cover each strip with at least 2 mL of the equilibration buffer containing IA. Ensure all of the strips are completely covered with the buffer.

9. Repeat steps 5 and 6.

10. Add 750 mL of concentrated run buffer to the Ettan DALT*twelve* separation unit and add deionised water to the "7.5-L" mark. Close the lid and turn on the pump to ensure thorough mixing of the contents.

11. With the second-dimension gel assembly in an upright orientation add 1 mL of overlay agarose to the top of the gel using a pasteur pipette (see Note 12).

12. Lay the second-dimension gel assembly horizontally on the bench with the opening facing you and the smaller of the two plates on top.

13. Using tweezers, place the IPG strip gel-side up on the inner face of the larger of the two glass plates, with the "+" end facing towards the left as you look at it.

14. Quickly stand the gel cassette upright with the smaller glass plate in front (the "+" end of the IPG strip should now be pointing to the right).

15. Carefully use the spatula to push the IPG strip down into the agarose between the two glass plates (see Note 13).

16. Add more agarose if needed to completely seal the IPG strip into place. Allow the agarose to cool and set.

17. Repeat this procedure for all of the other second-dimension gels and IPG strips.

18. Introduce the gel cassettes into the separation unit (see Note 14).

19. Dilute a further 200 mL of the concentrated run buffer to 2 L with deionised water. Add this to the top of the unit, ensuring that the final liquid level is between the MIN and MAX fluid levels.

20. The Ettan DALT*twelve* can conveniently be programmed to run overnight. It is advisable to focus the proteins at a low constant power level (e.g. 0.5 W per gel) for the first 2 h, and then increase the wattage per gel to effect the separation. Optimum conditions will be found empirically, but in the work described in this chapter, a value of 2 W per gel was used (see Note 15).

21. The dye front appears as a thin blue line travelling slowly down the gel. When the dye front is close to the bottom of the gel the Ettan DALT*twelve* should be stopped manually, and the gels scanned without delay.

3.6. Gel Scanning and Image Analysis

Gels must be scanned as soon as possible after the electrophoretic separation. Any delay could lead to diffusion of the proteins away from their tightly focused positions, which would broaden the size of the spot and increase the chance of contamination with other proteins.

Dust fluoresces and scatters light, which can cause artefacts on images and interfere with the subsequent quantitation. To minimise the chance of this happening, gels should only be handled wearing gloves, and these should be rinsed regularly with deionised water. The scanner should be completely clean and dust-free.

Conditions for the fluorescent scanning of the gels are first optimised quickly at low resolution, then all of the gels are scanned under ideal conditions at high resolution. Before scanning, gels are kept in the separation unit maintained at the sub-ambient temperature used for electrophoresis:

1. Turn on the Typhoon scanner 30 min before you intend to use it.
2. Clean the platen and position the gel alignment guide in place.
3. Two gels can be scanned at the same time. Gels are positioned with the top opening towards the left and the smaller of the two glass plates in contact with the platen. See the manufacturer's manual for details and typical scanner settings.
4. To optimise the conditions, scan one of the gels at low resolution, typically using a pixel size of 500 µm. Review this scan using ImageQuant software, and quantify the intensity as denoted by "Max. Val(pos)" (see Fig. 2). "Max. Val(pos)" must be less than 100,000 to avoid saturation. Optimum value for this parameter is in the range 50,000–80,000. However, the PMT settings for the individual channels must be optimised so that broadly similar intensities are obtained for each channel.
5. Once the settings have been optimised, scan each gel in turn at high resolution, typically using a pixel size of 100 µm.
6. Store scanned gels damp at 4°C in plastic bags until ready to proceed with the visualisation and spot picking, which should commence as soon as possible after image analysis in order to reduce further protein diffusion and contamination.

There are several 2D image analysis software packages that can be used to analyse the images from DIGE experiments and identify proteins that are either up- or down-regulated. We have found Progenesis SameSpots to be fairly straightforward to use, and as it has plenty of on-screen help, only the major steps will be covered in this protocol.

7. Open the programme and load the gel images for DIGE analysis. The software permits cropping, flipping and rotating of images, and carries out several checks on the quality of the images before proceeding further. Images are organised according to the gels they originated from.
8. Select a reference image. All of the other images will then be aligned to this image, so it is advisable to pick as reference an image that is representative and free from streaking and distortion. Mask any area that is to be excluded from the analysis, for instance the very bottom of the gel that includes the dye front.
9. Align the images to the reference image. Each internal standard image (Cy2) is matched to the Cy2 image of the reference gel;

21 Application of DIGE and Mass Spectrometry in the Study... 311

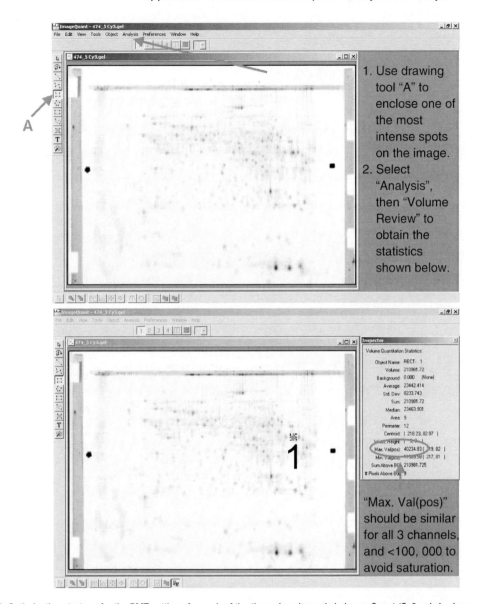

Fig. 2. Optimization strategy for the PMT settings for each of the three dye channels in ImageQuant (Cy3 only is shown here).

the Cy3 and Cy5 images are then aligned to their respective Cy2 images. This is by far the most important step in the workflow. In practice, an automatic alignment can be carried out as a first step; this can then be reviewed using the transition and checkerboard panes and any necessary modification can be made to the vectors.

10. Filter out any of the spots that you do not want to be included in the analysis such as reference markers that are placed at the edge of the gels.

11. Group the individual images together into their respective classes (e.g. diet 1 and diet 2, genotype 1 and genotype 2, control and treated).

12. Review each of the detected spots. There are certain criteria to be met in deciding which gel spots truly reflect a change in expression levels between any two conditions (diet, genotype, etc.). Tag all of the spots where the ANOVA p-value is <0.05 and fold change is greater than an appropriate level, for example, 2.

13. From the "Progenesis Stats" icon, additionally tag those spots where statistical power is greater than 80% (the generally accepted threshold level). It is thus possible to scrutinise and reduce the list of detected proteins and by selecting only those peaks that have, for example, $p < 0.05$, fold change >2, and power >0.8.

14. Visually examine these significant spots on the gel image to ensure that they can realistically be picked—and are not due to artefacts such as streaking.

3.7. Visualisation and Picking of Proteins of Interest

Once the protein spots of interest have been identified they can be excised robotically or stained, visualised and excised manually. It is possible to excise from either one of the gels used in the DIGE experiment, or from a preparative gel that has been prepared and run simultaneously but without the use of fluorescent dyes, and with a much higher protein loading.

In this example, protein spots of interest were manually excised from the analytical gels. The staining method used has high sensitivity similar to that of silver staining—yet it is compatible with mass spectrometry (4). It is advisable to pick the spots from a gel where the protein spots of interest have a high intensity—for this reason it may be necessary to stain more than one gel:

1. Clean and dry the exterior of the gel cassette and lay it onto lint-free paper with the smaller of the two glass plates in contact with the paper.

2. Prise off the top glass plate carefully to avoid damaging the gel. Remove the IPG strip and agarose, and place the gel assembly gel-side up in a clean staining tray.

3. Cover the gel with fixing solution (see Subheading 2.2.7) and place foil over the staining tray. Fix the gel overnight on a shaker.

4. Pour off the fixative and rinse the gel 3 times with deionised water.

5. To prepare 1 L of colloidal Coomassie stain, dissolve 100 g of ammonium sulphate in 500 mL of deionised water on a stirrer, and add 1.2 g of G250 Coomassie blue and 118 mL of phosphoric acid. When it has all dissolved, make the solution up to

800 mL with deionised water. Just before use add 200 mL of methanol dropwise with stirring.

6. Cover the fixed and rinsed gel with stain and place on a shaker for 16–40 h.
7. Pour off the stain and rinse the gel 3 times with deionised water.
8. Add 250 mL of destaining solution and place on the shaker for at least 2 h.
9. Once the gels have destained, scan them again on the Typhoon so that the Coomassie image can be matched with the DIGE images in SameSpots (see Note 16 and manufacturer's manual for details).
10. Import the Coomassie image(s) into the SameSpots software by selecting "Add images" from the Image QC screen. The software automatically checks the quality of the image before allowing further manipulation.
11. Select the "spot picking" screen and from there opt for "picking manually".
12. The software then presents an image of the gel with the proteins of interest clearly identified. This image can be printed and used as a visual aid in the process of manual spot picking.
13. Choose the gel for spot excision and shake it in distilled water for 2 h to remove any residual destaining solution.
14. Pre-wash the spot-picker tips with ethanol to remove any potentially leachable material that could interfere with the LC-MS/MS analysis. Allow the tips to dry.
15. Dry the back of the glass to which the gel is bonded to improve the view of the gel spots. Place it gel-side up onto the light box (see Note 17).
16. Drop pure water onto the spot that is about to be picked.
17. Carefully lower the tip of the spot picker onto the gel so that the tip encircles the spot of interest. Press down slightly to pierce the gel, and then gently rotate the tip by one quarter turn to break the surface tension between the gel and the plate.
18. Carefully transfer the gel spot to an appropriately coded Eppendorf tube, and freeze at −80°C until needed.
19. Repeat steps 16–18 for all of the spots of interest. Blanks and spots of major known proteins can also be excised as controls for the subsequent workflow.
20. Rescan the gel after picking the spots, as a record of where the spots were excised from.

3.8. In-Gel Digestion and NanoLC-MS/MS Analysis

Methods for in-gel digestion are well documented in the literature; the method used here is based on the method published by Bindschedler et al. (5). Briefly, gel spots are repeatedly washed in ammonium bicarbonate solution with increasing amounts of acetonitrile, to remove the Coomassie stain. After drying, the spots are reduced with DTT, and cysteines are then alkylated with IA. After further washing and lyophilisation, the gel spots are digested overnight with trypsin. Peptides are then extracted into a mixture of acetonitrile and trifluoroacetic acid and are dried again and stored at −80°C until ready for nanoLC-MS/MS analysis.

Typical experimental conditions for the nanoLC-MS/MS analysis are detailed ubiquitously in the literature. See Subheading 2.1, item 10 for details of the instrumentation used in this example and Table 3 for typical results obtained from the comparison of different mouse genotypes. Note that in one case in Table 3 more than one gel was chosen for the analysis of the same protein spot.

4. Notes

1. It is important to use anhydrous DMF for dissolving the fluorescent dyes. The DMF bottle should be sealed with a septum (to minimise ingress of air). If it was opened more than 3 months ago, it should not be used.

2. Dilution of the mouse liver lysates in water will be needed to bring the sample absorbance within the range of the calibration standards. Generally a 10 times dilution will often suffice. This step will also help to avoid any interference in the colorimetric technique from high levels of urea present in the lysate.

3. Optimum pH values for labelling the sample are between 8 and 9. No pH adjustment was found to be necessary when using the lysis buffer described in Subheading 2.2.1 for the presented example. However, the pH of each set of samples should be checked before proceeding.

4. The amount of protein of each sample that will be pooled is equal to no. of gels*100 μg protein/no. of samples, which in the presented example equals 50 μg per sample. If the protein concentration has been normalised at 3 mg/mL, this is then equivalent to a nominal value of 16.67 μL of each individual sample to be labelled with Cy2. To avoid problems due to volumetric losses, in the presented example, this nominal volume of 16.67 μL is increased to 20 μL (60 μg). Volumetric losses can easily occur due to factors such as sample frothing, sample adhesion to pipette tips and tubes, evaporation and the cumulative effect of very slight pipettor volumetric inaccuracies. However, it is essential that all of the gels contain the same level of internal standard. Analogously, the volume (amount) of individual samples taken for Cy3 or Cy5 dye labelling is increased to 40 μL (120 μg).

Table 3
Partial Mascot results list of significant protein identifications for the differentially expressed proteins in T2DM mouse models

Spot	Prot_desc	Prot_score	Prot_mass	Prot_matches	Prot_cover	Prot_pi	Protein name (from UniProtKB)	Fold change[a]	p-value[a]
359	Q8VC12	241	75.227	24	7.4	7.27	Urocanate hydratase	2.4 NZO down	3.87E-06
420	P32020	605	59.715	23	30.5	7.16	Non-specific lipid-transfer protein	2.9 NZO down	2.16E-08
	Q8CHT0	479	62.228	11	17.1	8.58			
	P24270	467	60.013	13	22.6	7.72	Delta-1-pyrroline-5-carboxylate dehydrogenase		
	Q99L88	135	58.444	4	9.1	8.51			
	Q03265	114	59.830	2	4	9.22	Catalase		
	P24549	89	55.060	2	5.6	7.92	Beta-1-syntrophin ATP synthase subunit alpha, mitochondrial Retinal dehydrogenase 1		
481	P11679	1,648	54.531	194	54.3	5.7	Keratin, type II cytoskeletal 8	2 NZO up	4.51E-13
	Q63836	925	53.147	118	36.7	5.78	Selenium-binding protein 2		
	Q91XD4	347	59.529	11	15	5.79	Formimidoyltransferase-cyclodeaminase		
	Q61035	102	57.893	3	5	5.68	Histidyl-tRNA synthetase, cytoplasmic		
481W	Q63836	344	53.147	13	18.6	5.78	Selenium-binding protein 2	2 NZO up	4.51E-13
	P30416	188	51.939	3	9.2	5.54	Peptidyl-prolyl cis-trans isomerase		
	P11679	148	54.531	4	8.6	5.7	Keratin, type II cytoskeletal 8		
	Q91XD4	75	59.529	3	4.6	5.79	Formimidoyltransferase-cyclodeaminase		

(continued)

Table 3 (continued)

Spot	Prot_desc	Prot_score	Prot_mass	Prot_matches	Prot_cover	Prot_pi	Protein name (from UniProtKB)	Fold change[a]	p-value[a]
481E	P11679	1,118	54.531	30	47.3	5.7	Keratin, type II cytoskeletal 8	2 NZO up	4.51E-13
	Q63836	884	53.147	32	45.1	5.78	Selenium-binding protein 2		
	Q91XD4	179	59.529	3	6.5	5.79	Formimidoyltransferase-cyclodeaminase		
	Q61035	114	57.893	2	4.9	5.68	Histidyl-tRNA synthetase, cytoplasmic		
	P07724	124	70.700	2	5.75	4	serum albumin		
484	Q63836	1,667	53.147	113	71.4	5.78	Selenium-binding protein 2	5.2 NZO up	1.11E-15
	P17563	1,290	53.051	102	55.7	5.87	Selenium-binding protein 1		
	Q91XD4	657	59.529	14	29.2	5.79	Formimidoyltransferase-cyclodeaminase		
	P11679	236	54.531	6	11.2	5.7	Keratin, type II cytoskeletal 8		
651	Q9Z0S1	541	33.517	47	35.7	5.54	3′(2′),5′-bisphosphate nucleotidase 1	2.9 NZO down	2.24E-08
	P21278	95	42.283	3	8.4	5.7	Guanine nucleotide-binding protein subunit alpha-11		
654	Q9Z0S1	489	33.517	45	28.2	5.54	3′(2′),5′-bisphosphate nucleotidase 1	5.5 NZO up	3.16E-10
864	P16015	542	29.633	33	51.2	6.89	Carbonic anhydrase 3	2 NZO down	1.82E-12
865	P16015	241	29.633	10	20	6.89	Carbonic anhydrase 3	2 NZO down	3.30E-10

All identifications have at least two unique peptide matches in Mascot. Human and associated mouse keratins are excluded

[a] These values were obtained from Progenesis SameSpots; proteins are either significantly up- or down-regulated for the genotype New Zealand Obese (NZO) in comparison to the genotype Swiss Jim Lambert (SJL)

5. It is helpful to colour-code the tops of the tubes with marker pens which are the same colour as the dye which will be added later. (Cy3 is red, Cy5 is blue and Cy2 is yellow). This helps to minimise the chance of costly mix-ups at the gel preparation stage.

6. For the 2 nmol-CyDye kit, add 5 μL of anhydrous DMF, and for the 5 nmol-CyDye kit, add 12.5 μL of anhydrous DMF. In order to label ten gels as described, it will be necessary to have two of the 5 nmol- and one of the 2 nmol-CyDye kits.

7. In some cases, it is more appropriate to add the IPG buffer at 0.5% as higher concentrations can limit the maximum voltage attainable during focusing and increase the time needed for this step. It is added here at 2% to maximise the solubility of the proteins.

8. If it is not convenient to process the strips directly after the rehydration or the IEF stage, they can be frozen at −80°C. However, they cannot be stored once the focussed strips have been equilibrated at the stage described in Subheading 3.5.

9. The ceramic tray must be cleaned with a detergent at neutral pH (e.g. the proprietary cleaning solution sold for this purpose—listed in Subheading 2.2.4). Other detergents could strip off the surface of the tray, which has been treated to minimise protein adsorption.

10. During rehydration, the gels swell and take on the blue colour of the modified DeStreak solution. It is thus easy to see the ends of the (previously colourless) gels.

11. The blue colouration from the bromophenol blue (present in the DeStreak solution) quickly clears during focusing but this does not indicate that focusing is complete.

12. If ReadyPrep Overlay Agarose is used it can conveniently be warmed up in the microwave oven. If many gels are being prepared, it may be necessary to reheat the agarose to prevent it becoming too viscous too soon.

13. It is important not to damage either gel or strip. The spatula should only touch the backing of the IPG strip, not the gel itself. The IPG strip is in the correct place when it is touching the top of the second-dimension gel and there are no air bubbles, only agarose between the two gels.

14. The Ettan DALT*twelve* will hold up to 12 gel cassettes. If there are fewer than 12 gels to be run, the vacant places must be filled with blank cassette inserts. Wet the outside of each gel cassette with the diluted run buffer in the separation unit, thus facilitating its correct placement in the Ettan DALT*twelve*.

15. When programming the Ettan DALT*twelve*, ensure that the pump is turned on, and that a suitable temperature (for instance, 10°C) is used. A temperature of less than 10°C should not be used.

16. It is possible to scan the colloidal Coomassie images using a normal flat-bed scanner. However, the process of matching the Coomassie image to the DIGE image is simplified if the Typhoon scanner is used throughout.

17. The light box used for gel-picking should be placed in an ergonomically advantageous position, and the environment should be as clean and dust-free as possible. Any contamination that is introduced after electrophoretic separation is likely to interfere with the nanoLC-MS/MS analysis of the proteins.

Acknowledgements

This work was supported by the EU 6th Framework Programme "System-wide analysis and modelling of protein modification". The authors would additionally like to thank colleagues at the University of Reading (The BioCentre, Department of Chemistry and ICMR) and former colleagues at University College London (Proteomics Unit) for their invaluable advice.

References

1. Wild, S., Roglic, G., Green, A., Sicree, R., and King, H. (2004) Global prevalence of diabetes: estimates for the year 2000 and projections for 2030, *Diabetes Care* **27**, 1047–1053.
2. WHO. (2009) Fact Sheet No. **312**: Diabetes.
3. Bradford, M. M. (1976) A rapid and sensitive method for the quantitation of microgram quantities of protein utilizing the principle of protein-dye binding, *Anal Biochem* **72**, 248–254.
4. Candiano, G., Bruschi, M., Musante, L., Santucci, L., Ghiggeri, G. M., Carnemolla, B., Orecchia, P., Zardi, L., and Righetti, P. G. (2004) Blue silver: a very sensitive colloidal Coomassie G-250 staining for proteome analysis, *Electrophoresis* **25**, 1327–1333.
5. Bindschedler, L. V., Palmblad, M., and Cramer, R. (2008) Hydroponic isotope labelling of entire plants (HILEP) for quantitative plant proteomics; an oxidative stress case study, *Phytochemistry* **69**, 1962–1972.

Chapter 22

Evaluating the Efficacy of Subcellular Fractionation of Blast Cells Using Live Cell Labeling and 2D DIGE

Yin Ying Ho, Megan Penno, Michelle Perugini, Ian Lewis, and Peter Hoffmann

Abstract

Labeling of exposed cell surface proteins of live cells using CyDye DIGE fluor minimal dyes is an efficient strategy for cell surface proteome profiling and quantifying differentially expressed proteins in diseases. Here we describe a strategy to evaluate a two-step detergent-based protein fractionation method using live cell labeling followed by visualization of the fluorescently labeled cell surface proteins and fractionated proteins within a single 2D gel.

Key words: Live cell labeling, 2D difference gel electrophoresis, Detergent-based fractionation, Cell surface proteins, Membrane proteins

1. Introduction

Cell surface proteins play an important role in running cellular processes and may represent diagnostic and therapeutic targets for many diseases. Due to the low abundance and highly hydrophobic nature of membrane proteins, detection and identification is challenging. Accordingly, highly efficient methods for the enrichment of membrane proteins are critical. The major requirement for an efficient enrichment strategy is maximal recovery of membrane proteins with minimal cytosolic contamination of the membrane fraction. This can be evaluated by live cell labeling (LCL) using impermeable fluorescent dyes that selectively label cell surface exposed proteins. Ideally, the process labels only the cell surface proteins, which should be enriched in the membrane fraction. LCL has been tested by Mayrhofer et al. (1) to assess the compatibility

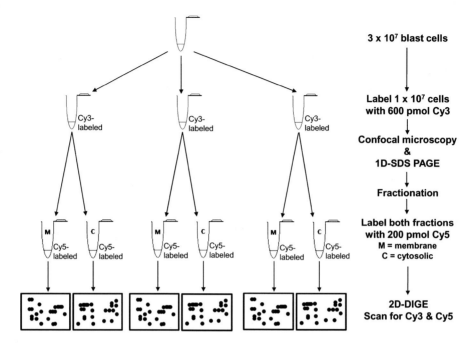

Fig. 1. Overview of the labeling workflow for the LCL/2D DIGE approach.

of CyDye DIGE fluor minimal dye labeling of human cell lines in vitro as well as complex biological system in vivo. The effectiveness of labeling cell-surface-exposed proteins can be assessed by confocal microscopy. The consistency, however, of LCL needs to be assessed due to the fact that the optimal pH for CyDye labeling is between pH 8 and 9, and incubating human cells in nonphysiological conditions over the labeling period may interfere with cell viability. Decreased cell viability during LCL will affect the evaluation process as compromised membranes become accessible to CyDyes resulting in possible labeling of cytosolic proteins.

Here, we demonstrate by means of LCL in association with 2D difference gel electrophoresis (2D DIGE), a method for assessing the quality of a membrane fractionation process that resulted in a majority of the live cell labeled proteins becoming enriched in the membrane fraction. The observation was consistent over three technical replicates. An overview of the experimental procedure is presented in Fig. 1. In brief, a depleted population of human cells was split into three groups, labeled with Cy3 in vitro and then fractionated using a commercially available kit into membrane and cytosolic fractions. Following Cy5 labeling of each fraction, labeled proteins were separated by 2D electrophoresis (2DE). The resulting gels were imaged for Cy3 (LCL) and Cy5.

A recommendation for the inclusion of an internal standard in the experimental design is discussed (see Note 1). The inclusion of an internal standard could minimize experimental variation by providing a means to standardize the data between analytical gels while improving spot matching and quantitation.

2. Materials

All solutions should be prepared using high-purity water (resistivity $\geq$ 18.2 MΩ, TOC $\leq$ 1 ppb, at 25°C). Reagents are analytical grade unless specified otherwise. Prepare all cell-culture-related reagents in laminar flow hood for the purpose of sterility. Prior to use, ensure all buffers are thawed and mixed well by vortexing. In the provided example, immature blast cells are initially isolated from bone marrow by removal of CD^{3+} T-cell and CD^{19+} B-cell populations, as blast cells are a clinically relevant cell type for investigating a patient's response to chemotherapy. This step is not specifically required for the combined LCL/2D DIGE approach.

2.1. Magnetic Cell Sorting

1. Culture media: Iscove's Modified Dulbecco's Medium (IMDM) (Invitrogen, Carlsbad, CA, USA) supplemented with 20% (v/v) fetal calf serum (FCS), penicillin/streptomycin antibiotics, L-glutamine, and DNase I. Bring media to 37°C before use by incubating in a 37°C water bath. Use DNase I at 2,500 U/50 mL.

2. Trypan blue solution (Sigma-Aldrich, St. Louis, MO, USA).

3. Phosphate-buffered saline (PBS): To prepare 1 L of 20× solution, weigh 160 g of sodium chloride, 4 g of potassium chloride, 4 g of potassium dihydrogen phosphate, and 23 g of disodium hydrogen phosphate and transfer to a 2-L conical flask containing ~800 mL of ultrapure water. Mix for 20 min. Make up the volume to 1 L with water and adjust pH to pH 7.4. Autoclave the solution prior to use.

4. MACS buffer: Add 2.5 mL of MACS BSA (Miltenyi Biotec GmbH, Bergisch Galdbach, Germany) to 1,450 mL of autoMACS rinsing solution (Miltenyi Biotec GmbH). Keep buffer at 4°C.

5. MicroBeads conjugated to monoclonal anti-human CD3 antibodies (Miltenyi Biotec GmbH). CD3 MicroBeads are supplied as a suspension containing stabilizer and 0.05% sodium azide. Store protected from light at 4°C.

6. LD column (Miltenyi Biotec GmbH).

7. Pre-separation filters (Miltenyi Biotec GmbH).

8. Column adaptor (Miltenyi Biotec GmbH), required to insert column into the MidiMACS separator.

9. MicroBeads conjugated to monoclonal anti-human CD19 antibodies (Miltenyi Biotec GmbH). CD19 MicroBeads are supplied in buffer containing stabilizer and 0.05% sodium azide. Store protected from light at 4°C.

10. 50-mL Falcon polypropylene conical centrifuge tube (BD, Franklin Lakes, NJ, USA) (see Note 2).

2.2. Live Cell Labeling

1. PBS buffer (see Subheading 2.1, item 3).
2. Binding buffer: 7 mM dithiothreitol (DTT) and 1 M urea in Hank's Balanced Salt Solution (HBSS) (Invitrogen), pH 8.5. Keep solution at 4°C.
3. CyDye DIGE fluor Cy3 minimal dyes (GE Healthcare, Uppsala, Sweden): 600 pmol CyDye working solution per labeling in anhydrous 99.8% N,N-Dimethylformamide (DMF). Store protected from light at 4°C for current use, otherwise store aliquots at −80°C (see Note 3).
4. 10 mM L-lysine. Store at −20°C.
5. NuPAGE 4–12% Bis-Tris gel (Invitrogen).
6. 2× LDS buffer: Dilute NuPAGE 4× LDS sample buffer (Invitrogen) 1:1 with water.
7. MES SDS running buffer: 50 mM MES pH 7.2, 50 mM Tris-base, 0.1% SDS, and 1 mM EDTA in water (pH 7.3; do not adjust with acid or base).

2.3. Protein Preparation (Fractionation) and Labeling

1. ProteoExtract native membrane protein extraction kit (Calbiochem, Merck KGaA, Darmstadt, Germany). Store kit at 4°C.
2. Extraction buffer I (provided in extraction kit (see Subheading 2.3, item 1)): Thaw and mix well before use.
3. Extraction buffer II (provided in extraction kit (see Subheading 2.3, item 1)): Thaw and mix well before use.
4. Protease inhibitor cocktail (provided in extraction kit (see Subheading 2.3, item 1)): Thaw and mix well before use. Keep at room temperature until needed.
5. ProteoExtract protein precipitation kit (Calbiochem).
6. Precipitation agent: To one bottle of Precipitant 1, add 1.7 mL of Precipitant 2, 1.7 mL of Precipitant 3, and 1.7 mL of Precipitant 4. Mix well. Store precipitation agent at −20°C for up to 2 months.
7. Wash solution: Add 150 mL of ethanol to the provided wash solution (Calbiochem ProteoExtract protein precipitation kit). Mix well. Store at −20°C up to 1 year.
8. CyDye DIGE fluor Cy5 minimal dye (GE Healthcare): 200 pmol CyDye working solution per labeling in DMF. Store protected from light at 4°C for current use, otherwise store aliquots at −80°C (see Note 3).
9. 10 mM L-lysine (see Subheading 2.2, item 4).
10. 1.5-mL Microcentrifuge tubes.
11. 2-mL Microcentrifuge tubes.
12. 15-mL Falcon polypropylene conical centrifuge tube (BD) (see Note 2).

13. Lysis buffer: 7 M urea, 2 M thiourea, 4% (w/v) 3-[(3-cholamidopropyl)dimethylammonio]-1-propanesulfonate (CHAPS), and 30 mM Tris in water at pH 9.5. Store aliquots at −20°C and thaw on ice when required. Once this lysis buffer is mixed with the protein sample, the pH will decrease to at about pH 8.5.

14. EZQ protein quantitation kit (Invitrogen).

15. EZQ protein quantitation reagent (Component A). Store protected from light.

16. EZQ rinse buffer: 10% methanol and 7% acetic acid in water.

17. Plastic staining tray.

18. pH indicator paper, pH 8.2–10.

19. Ovalbumin protein quantification standard: Prepare a 2 mg/mL stock solution by solubilizing 2 mg provided with EZQ protein quantitation kit in 1 mL of lysis buffer (see Subheading 2.3, item 10). Store 10-µL aliquots at −80°C and thaw on ice when required.

20. Typhoon Trio fluorescent scanner (GE Healthcare).

2.4. Two-Dimensional Electrophoresis

1. Rehydration buffer: 7 M urea, 2 M thiourea, 30 mM Tris, 2% (w/v) CHAPS, 0.5% (v/v) IPG buffer pH 3–11 NL (GE Healthcare), 1.2% (v/v) DeStreak reagent (GE Healthcare), and 0.05% (v/v) saturated Bromophenol blue solution.

2. 24-cm IPG strips, pH 4–7 (GE Healthcare). Store at −20°C.

3. PlusOne dry strip cover fluid (GE Healthcare). Store at room temperature.

4. Strip holder cleaning reagent (GE Healthcare). Store at room temperature.

5. Reswelling tray (GE Healthcare).

6. Paperwicks (GE Healthcare).

7. Loading Cups (GE Healthcare).

8. Ceramics IPG strip tray (GE Healthcare).

9. IPGPhor II isoelectric focusing apparatus (GE Healthcare).

10. 2DGel DALT 12.5% NF (Serva, Heidelberg, Germany). Store at 4°C.

11. DTT solution: To make 1 M solution, weigh 0.1542 g of DTT (Sigma-Aldrich) and dissolve in 1 mL of water. Mix well. Store aliquots at −80°C and thaw at room temperature when required.

12. DTT equilibration solution: Dissolve 0.05 g of DTT and 1.8 g urea in 5 mL of IPG equilibration buffer (Serva). This recipe is designed for one strip. Adjust the amount and volume according to the number of strips.

13. IAA equilibration solution: Dissolve 0.125 g iodoacetamide (IAA) and 1.8 g of urea in 5 mL of IPG equilibration buffer (Serva). This recipe is designed for one strip. Adjust the amount and volume according to the number of strips.
14. Ettan DALT*twelve* electrophoretic separation apparatus (GE Healthcare).
15. Anode buffer: Dissolve the content of the anode buffer bag (Serva) in 7.5 L of water.
16. Cathode buffer: Dissolve the powder in the cathode buffer bag (Serva) in 5 L of water.
17. Low-melt agarose: Weigh 1 g of low-melt agarose (Quantum Scientific, Queensland, Australia) and dissolve in 100 mL of cathode buffer (see Subheading 2.4, item 14) on a heated magnetic stirrer (70°C).
18. ImageQuant TL image analysis software (GE Healthcare).

3. Methods

Once the CyDye is introduced to the sample, protect from light by covering in aluminum foil.

3.1. Preparation of Cells for LCL

1. Thaw vials of cells (stored in liquid nitrogen) in a 37°C water bath. Add 20 mL of culture media slowly in a dropwise fashion then pellet cells by centrifugation at $400 \times g$ for 4 min at room temperature. Remove supernatant, resuspend cells in 25 mL of culture media, and transfer cells into a T75 culture flask. Recover cells by incubating cells overnight in a humidified incubator with 4% CO_2 at 37°C.
2. Transfer suspension cells into a new 50-mL Falcon tube and centrifuge at $400 \times g$ for 4 min at room temperature. Remove supernatant and resuspend cells in 20 mL of culture media. Assess viability using trypan blue exclusion and hemocytometer cell counting (2).
3. Wash cells by resuspending the cell pellet in 5 mL of MACS buffer. Centrifuge cells at $400 \times g$ for 4 min at room temperature. Remove supernatant and resuspend cells in 80 μL of MACS buffer per 1×10^7 cells. Label CD3+ cells by incubating 20 μL of CD3-MicroBeads per 1×10^7 cells in the dark for 15 min at 4°C.
4. Place LD column in the MidiMACS separator and rinse column with 2 mL of MACS buffer. Attach a pre-separation filter to the top of the LD column and wet the filter with 0.5 mL of MACS buffer and allow buffer to flow through into a waste tube.

5. Wash cells by adding 2 mL of MACS buffer per 1×10^7 cells. Spin cells at $400 \times g$ for 4 min at room temperature. Remove supernatant and resuspend up to 1.25×10^8 cells in 500 µL of MACS buffer.

6. Add cell suspension to the pre-separation filter and allow cells to flow through to the LD column to a new collection tube. Rinse the filter and column with 1 mL of MACS buffer. Repeat rinsing once. Dispose of the column and filter.

7. Make the volume of cells up to 5 mL and perform viability cell count using trypan blue. Spin cells at $400 \times g$ for 4 min at room temperature. Remove supernatant and resuspend cells in 80 µL of MACS buffer per 1×10^7 cells. Add 20 µL of CD19-MicroBeads per 1×10^7 cells and incubate 15 min at 4°C protected from light.

8. Prepare LD column and pre-separation filter (see Subheading 3.1, step 4).

9. Wash cells by adding 2 mL of MACS buffer per 1×10^7 cells. Spin cells at $400 \times g$ for 4 min at room temperature. Remove supernatant and resuspend up to 1.25×10^8 cells in 500 µL of MACS buffer.

10. Add cell suspension to the pre-separation filter and allow cells to flow through to the LD column to a new collection tube. Rinse the filter and column with 1 mL of MACS buffer. Repeat column rinsing once. Dispose of the column and filter.

11. Make up the volume of cells to 5 mL and perform viability cell count using trypan blue.

3.2. CyDye DIGE Fluor Minimal Labeling of Blast Cells

1. Wash 1×10^7 cells in 10 mL of PBS. Spin cells at $400 \times g$ for 4 min at room temperature. Repeat wash step twice. After the last wash, resuspend cells in 100 µL of binding buffer and transfer cell suspension to a 1.5-mL microcentrifuge tube. Add 600 pmol of Cy3 to the cells (see Note 3). Mix well by thorough pipetting and centrifuge briefly in a microcentrifuge. Leave on ice for 30 min in the dark. Quench the labeling reaction by adding 1 µL of 10 mM lysine. Mix by pipetting and spin briefly in a bench-top centrifuge. Leave for 10 min on ice in the dark.

2. Wash cells in 1 mL of PBS. Spin cells at $400 \times g$ for 2 min at room temperature. Repeat twice. Resuspend cells in 100 µL of PBS.

3. To evaluate the consistency and effectiveness of LCL using 1D SDS PAGE, transfer 5 µL of cell suspension into a tube and add 13 µL of 2× LDS sample buffer to lyse the cells. Mechanically break up the DNA/RNA with a needle and syringe. Sonicate sample in ice-cold water for 15 min. Spin tube at $10,000 \times g$ for 10 min. Transfer supernatant into a new microcentrifuge tube. Add 2 µL of 0.5 M DTT into the cell lysate to reduce disulfide

bonds. Separate cell lysate using a NuPAGE 4–12% Bis-Tris gel in MES SDS running buffer at 180 V and 100 mA/gel until the dye front reaches the bottom of the gel (~45 min). Image gel using the Typhoon Trio (GE Healthcare) with laser excitation at 532 nm and emission filter at 580 nm. Detectable bands indicate the labeling process has been successful.

3.3. Proteomic Extraction and Labeling of Membrane and Cytoplasmic Fractions

1. Transfer cells into a 2-mL microcentrifuge tube. Spin cells at $400 \times g$ for 2 min at room temperature. Remove supernatant carefully and completely without disturbing the pellet.

2. Add 10 μL of protease inhibitor cocktail to the wall of the tube and immediately add 2 mL of ice-cold extraction buffer I. Resuspend the cell pellet carefully and thoroughly by pipetting. Incubate tube for 10 min at 4°C on a rotary shaker.

3. Spin tube at $16,000 \times g$ at 4°C for 15 min to pellet insoluble material. Transfer supernatant to a new 15-mL Falcon tube labeled cytosolic proteins. Keep fraction on ice for same day use. Protein can be stored at −20°C at this stage.

4. Add 5 μL of protease inhibitor cocktail to the wall of the tube and immediately add 1 mL of ice-cold extraction buffer II to the cell pellet. Mix thoroughly by pipetting. Incubate tube at 4°C for 30 min on a rotary shaker.

5. Spin tube at $16,000 \times g$ at 4°C for 15 min. Transfer supernatant to a new 15-mL Falcon tube labeled membrane proteins. Keep fraction on ice for same day usage. Otherwise, store at −20°C (see Note 2).

6. To concentrate the proteins used in 2D DIGE, the ProteoExtract protein precipitation kit is used. Mix by briefly vortexing the sample with four volumes of ice-cold precipitation agent. Incubate tube overnight at −20°C. Spin down the precipitated proteins at room temperature for 10 min at $10,000 \times g$.

7. Carefully remove supernatant completely without disturbing the pellet. Wash the protein pellet by resuspending in 1 mL of fresh ice-cold wash solution and vortex briefly. Spin tube for 2 min at room temperature at $10,000 \times g$.

8. Carefully remove supernatant completely without disturbing the pellet. Repeat the wash step in 0.5 mL of ice-cold wash solution. Aspirate the supernatant without disturbing the pellet. Air-dry the pellet for 1 h at room temperature.

9. Resuspend protein lysate in a small volume of lysis buffer to start with (i.e., 10 μL). Add the volume of lysis buffer in a small interval until no visible precipitated protein is observed. Sonicating in ice-cold water bath greatly improves resolubilization of protein lysate.

10. To determine the protein concentration prior to 2D DIGE, an EZQ protein quantitation assay is performed. Prepare the ovalbumin standards by dissolving the lyophilized protein in the buffer used for solubilizing the sample (i.e., lysis buffer) at 2 mg/mL, then preparing serial twofold dilutions of this stock down to 0.125 mg/mL.

11. Place a sheet of assay paper over the microplate and insert the stainless steel backing plate into the microplate. Ensure that the paper is securely in place.

12. Apply 1 µL of each protein standard and experimental sample to individual wells of the microplate assembly. Include 1 µL of buffer alone as a background control. Load each standard and sample in triplicate. Allow samples to dry completely on the paper. Remove the assay paper from the cassette and immerse into a plastic staining tray containing 30 mL of methanol. Incubate at room temperature for 5 min with gentle agitation. After washing, air-dry the paper completely.

13. Place the assay paper into the EZQ protein quantitation reagent (Component A) and agitate gently on an orbital shaker for 30 min. Cover the tray with aluminum foil to protect from light. After staining, rinse the assay paper for 2 min in EZQ rinse buffer. Repeat twice.

14. Image the assay paper while still wet using a Typhoon Trio with 488-nm laser (excitation) and a 610-nm band pass emission filter and PMT of 300 V. Quantify the fluorescence, measured as volumes of each spot corresponding to an experimental sample, standard and background using ImageQuant TL (GE Healthcare). Subtract the average volume of the triplicate background spots from each of the standards and experimental sample spots. Determine the average of the triplicate corrected volumes for all standards and samples. Plot a standard curve of the standards against the corresponding protein concentration (0.125–2 mg/mL). Determine a line of best fit based on a quadratic equation, then calculate the sample concentrations by solving the quadratic equation for x (i.e., the unknown concentration) whereby y = average volume of the triplicate spots.

15. Label 100 µg of membrane and cytosolic proteins separately with 200 pmol of Cy5. Mix well by pipetting and pulse spin. Incubate tube on ice for 30 min in the dark. Add 1 µL of lysine to quench the labeling reaction. Mix by pipetting and pulse spin. Leave the tube on ice for 10 min in the dark.

3.4. Two-Dimensional Electrophoresis

1. Prior to isoelectric focusing, rehydrate IPG strips by pipetting evenly 450 µL of rehydration solution into the groove of the reswelling tray without introducing bubbles. Peel off the plastic backing that protects the IPG gel. Without touching the gel,

lay the strip down such that the gel is facing the solution. Ensure the buffer is evenly distributed under the strip. Add 1 mL of cover fluid on top of the strip to prevent dehydration of strip and evaporation of the buffer. Incubate strips overnight at room temperature.

2. After CyDye labeling of the samples, add 0.75 μL of IPG buffer pH 3–11 NL, 1.5 mL of 1 M DTT stock solution, and 0.75 μL of BPB solution, then make up the volume of each sample to 150 μL with lysis buffer.

3. Position the ceramic IPG strip tray on the Ettan IPGphor II (GE Healthcare). Align a rehydrated IPG strip along the groove inside the manifold with the gel side facing up. Soak two paper wicks per IPG strip in 150 μL of deionized water. Place one wet paper wick between the IPG strip and the anode and the other between the cathode and the IPG strip. Position the loading cup right below the anode wick and click in using the IPG cup pusher to form a good seal between the cup and the strip. Cover the IPG strips and around the cups by pouring 100 mL of cover fluid into the manifold. Ensure the cups are not leaking. Add the 150 μL of sample into the cup. Cover the sample with 50 μL of cover fluid to prevent sample from drying out.

4. Conduct isoelectric focusing according to the protocol outlined in Table 1.

5. Prepare the DTT and IAA equilibration solutions. Place the IPG strip in a long, rod-shaped, plastic tube. Incubate the IPG strip in the DTT equilibration solution for 15 min on an orbital shaker in a flat position. During this incubation time, prepare the anode and cathode buffers.

6. Pour off the DTT equilibration solution and add the IAA equilibration solution and incubate for 15 min at room temperature with gentle agitation. Cover solution with aluminum

Table 1
Isoelectric focusing program on the IPGPhor II

Step	Type	Voltage (V)	Duration (h)	Accum (Vh)
1	Step	300	2	600
2	Step	500	2	1,600
3	Step	1,000	2	3,600
4	Gradient	8,000	5	28,600
5	Step	8,000	5	68,600
6	Step	500	10	–

foil as IAA is light sensitive. Pour off the IAA equilibration solution and add 5 mL of cathode buffer to rinse off equilibration solution.

7. Clean the glass of the flat cassette thoroughly with ethanol and water and lint-free wipes. Apply a small volume of deionized water onto the glass plate along the closure. Remove the plastic coversheet of a 2DGel DALT 12.5% NF precast gel (see Note 4) and with the gel facing down, lay the gel down toward the glass plate from the closure toward the hinge. Use a roller to remove any air bubbles and excess liquid between the gel and the glass plate. With the cassette laying flat on the bench, close the cassette and press the edge of closure tightly to ensure the gel is placed firmly in the cassette.

8. Melt the low-melt agarose in a microwave oven for 1 min on HIGH. Pipette 5 mL of agarose into the cassette using needle and syringe. Place the IPG strip on the cassette and push it down to the gel with the IPG strip pusher. Remove any trapped air bubbles in between the gel and IPG strip by pressing the strip toward the gel, but do not damage the gel edge. Allow agarose to set at room temperature.

9. While agarose is gelling, set up the Ettan DALT*twelve* system. Pour 7.5 L of anode buffer into the separation unit, turn the pump on and ensure the system is in circulation. Spray 0.1% SDS solution on the silicone rubber seals to facilitate loading of the gel cassettes. Place a blank cassette after a gel cassette into the sealing manifold of the unit. Ensure the seals do not bend downwards by pulling up and down the cassette.

10. Pour a portion of the cathode buffer into the upper buffer chamber to ensure there is a good seal between the cathodic and the anodic reservoirs. Pour the remaining cathode buffer in the chamber. Close the lid and run the apparatus according to the program outlined in Table 2.

Table 2
Second dimension electrophoresis program on the Ettan DALT*twelve*

Phase	Current (mA)/gel	Power (W)/gel	Voltage (V)	Duration (h)
1	5	0.5	50	2
2	10	0.5	100	1
3	30	2.5	250	14

3.5. Image Acquisition and Analysis

1. Stop the SDS PAGE when the dye front reaches the bottom of the gel (approximately 16 h). Open the lid and begin disassembling the tank by removing a blank cassette from the rear or the apparatus. Place the cassettes on a gel rack and rinse the cassettes thoroughly with deionized water. Lay the gel cassette on the bench and open the cassette carefully. The gel should still attach to the glass plate. Hold the edges of the plastic backing and start pulling the gel away from the glass plate. Put the gel onto the plastic coversheet that comes with the gel. This helps transporting the gel and reducing exposure to contaminants (such as keratin).

2. Scan the gels using the Typhoon Trio scanner. Clean the glass scanner surface with ethanol and water. Squeeze a small volume of water on the glass surface. Remove the gel coversheet, hold the edges of the gel plastic backing, and slowly lay the gel downward toward the scanner glass surface without introducing bubbles between the glass surface and the gel. Place a clean low-fluorescent glass plate on top of the gel to hold the gel flat. Scan using the settings outlined in Table 3.

3. The images can be qualitatively visualized using ImageQuant TL. The brightness and contrast settings for each gel require adjustment to obtain an optimal picture. By examining the relative Cy3 (LCL) and Cy5 (post-fractionation labeling) patterns for the membrane and cytosolic fractions, the efficacy of subcellular fractionation can be assessed. As shown in Fig. 2, the majority of the Cy3-labeled proteins were enriched in the membrane fraction with very limited Cy3-labeling of proteins isolated in the cytosolic fraction. Furthermore, the similarities observed between the Cy5- and Cy3-labeled components of the membrane fraction indicated minimal cytosolic contamination. Potential improvements to this experimental design that may allow for the quantification of differences between live-cell-labeled membrane and cytosolic proteins are outlined below (see Note 1 and Fig. 3).

Table 3
Scanner setting for the Typhoon Trio

CyDye	Excitation (nm)	Emission (nm)	Resolution (μm)	PMT (V)
Cy3	532 (green)	580	100	700
Cy5	633 (red)	670	100	450

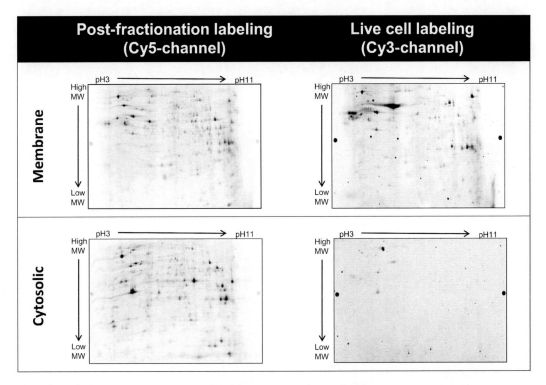

Fig. 2. Representative 2D DIGE gels of the membrane and cytosolic fractions showing live cell labeled proteins (Cy3) and proteins labeled after fractionation (Cy5). A majority of the live-cell-labeled proteins are concentrated in the membrane fraction.

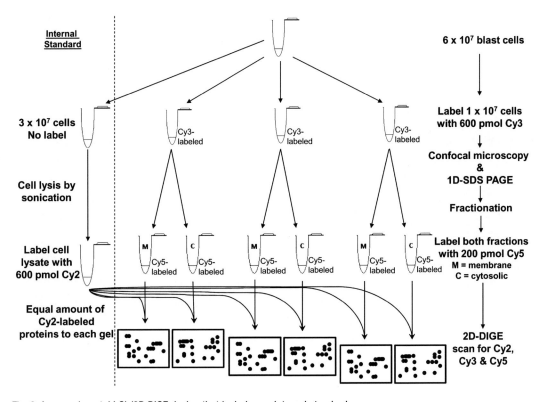

Fig. 3. An experimental LCL/2D DIGE design that includes an internal standard.

4. Notes

1. Figure 3 shows an experimental design with the inclusion of an internal standard. This internal standard is composed of all proteins from both the membrane and cytosolic fractions from the same number of cells. Such an experiment can be analyzed using DIGE-specific software, such as DeCyder (GE Healthcare), enabling meaningful information such as the standardized abundance values to be calculated and a series of statistical tests to be performed.

2. Only use BD Falcon tubes as the material from which it is constructed (polypropylene) is compatible with precipitation agent. Polystyrene tubes are not compatible with this reagent.

3. Resuspend the lyophilized 25 nmol DIGE fluor minimal Cy2, Cy3, and Cy5 CyDyes in 125 µL of anhydrous DMF to produce 200 pmol/µL stock solutions. Aliquot the stocks into clearly labeled screw-top microfuge tubes (1 µL), which should be flushed with argon before capping. The aliquots can be stored at −80°C for >12 months without appreciable loss of sensitivity. When required, simply add the sample directly to the tube of dye.

4. Keep the coversheet that is used to protect the precast gels prior to electrophoresis. It is useful when it comes to storing the gels in the freezer.

References

1. Mayrhofer C, Krieger S, Allmaier G, Kerjaschki D (2006) DIGE compatible labeling of surface proteins on vital cells *in vitro* and *in vivo*. Proteomics 6:579–585.

2. Strober W (2001) Monitoring cell growth. In: Coligan JE, Bierer BE, Margulies DH, Sherach EM, Strober W (eds). Current Protocols in Immunology, 5th edn. Wiley, New York.

Part IV

Applications in Animal, Plant, and Microbial Proteomics

Chapter 23

DIGE Analysis of Plant Tissue Proteomes Using a Phenolic Protein Extraction Method

Christina Rode, Traud Winkelmann, Hans-Peter Braun, and Frank Colditz

Abstract

Two-dimensional difference gel electrophoresis is an invaluable technique for the analysis of plant proteomes. However, preparation of protein fractions from plant tissues is challenging due to the special features of plant cells: a robust cell wall, large vacuoles which often contain high concentrations of organic acids and a broad range of secondary metabolites like phenolic compounds and pigments. Therefore, protein preparation for difference gel electrophoresis (DIGE) analyses has to be adapted. Here, we describe both a phenolic protein extraction method for plant tissues and an adapted protocol for DIGE labeling of the generated fractions.

Key words: Difference gel electrophoresis, Phenol extraction, Plant cell disruption, Plant proteomics, Protein extraction

1. Introduction

Two-dimensional difference gel electrophoresis (2D DIGE) technology allows separation of two differentially labeled protein samples on one gel, thus eliminating gel-to-gel variability. A major advantage of this technique is the possibility to identify even very small or marginal differences in protein migration patterns of two different fractions (occurring at the pI or Mr level) which are hardly detectable by 2D gel electrophoresis on the basis of separate gels (1). Furthermore, the fluorophores used in differential protein labeling not only are very sensitive but also allow precise protein quantification. In addition, this labeling allows a very sensitive detection of proteins on gels.

In plant research, 2D DIGE was applied successfully for analyses of protein samples from various species and different tissues. For instance, the technique was used to investigate different plant developmental processes and organogenesis (2–4), plant organelle proteomes (5, 6), plants under abiotic stress conditions (7, 8), analyses of plants in associations to symbiotic and/or pathogenic microbes (9–13), and posttranslational protein modifications in plants (14).

Due to the presence of a cell wall and very special biochemical properties, disruption of plant cells and isolation of protein fractions are challenging but of major importance to obtain high-resolution gels. To find a very efficient protein purification technique, four different protein extraction protocols were tested in our laboratory: (1) a rapid protocol without precipitation as described by Gallardo et al. (15), (2) a protocol utilizing TCA precipitation in combination with acetone (16), (3) a protocol described by Corcke and Roberts (17) which relies on boiling of protein fractions in Laemmli buffer (18) and subsequent protein precipitated with acetone, and (4) an extraction method using phenol which is combined with an ammonium acetate in methanol precipitation as described by Hurkman and Tanaka (19), modified by Colditz et al. (20). Optimal protein purification and resolution on IEF-SDS gels were achieved using the modified Hurkman and Tanaka (19) protocol even for low amounts of tissue.

2. Materials

All buffers, solutions, and reagents are given in the order of usage according to the methods in Subheading 3. Prepare all solutions freshly using analytical grade chemicals in combination with pure deionized water.

2.1. Protein Extraction with Phenol

1. Extraction buffer: 700 mM sucrose, 500 mM Tris, 50 mM EDTA, 100 mM KCl, pH 8.0 (HCl). Directly before usage, 2% β-mercaptoethanol is added. Extraction buffer is stored at 4°C or frozen at −20°C in aliquots.
2. Water-saturated phenol (pH 6.6/7.9; Amresco, Solon, USA), stored at 4°C.
3. Precipitation solution: 0.1 M ammonium acetate in methanol, stored at 4°C.

2.2. DIGE Sample Preparation

1. Dithiothreitol (DTT).
2. Lysis buffer: 8 M urea, 4% [w/v] CHAPS, 40 mM Tris base. Lysis buffer is stored at −20°C (see Note 1).
3. CyDye™ fluor minimal labeling reagents: Cy2™, Cy3™, and Cy5™ (GE Healthcare, Munich, Germany). The fluorophores

(400 μM per labeling reaction) are diluted in dimethylformamide (DMF) according to the manufacturer's instructions. Diluted CyDyes are stored at −20°C and should be used within 3 months.

4. Lysine stock solution: 10 mM, stored at 4°C.

5. Rehydration buffer: 8 M urea, 2% [w/v] CHAPS, a spatula tip of bromophenol blue, DTT 20–100 mM (added directly before usage), 0.5% [v/v] IPG buffer (added directly before usage; corresponding to the IPG strip used for IEF (GE Healthcare, Munich, Germany)). Rehydration buffer is stored at −20°C.

3. Methods

3.1. Protein Extraction with Phenol

This protocol is based on the protein extraction method according to Hurkman and Tanaka (19) modified by Colditz et al. (20). Freshly ground samples are transferred to 2-mL Eppendorf vessels and directly frozen in liquid nitrogen. For subsequent 2D DIGE analysis, proteins from two different plant samples are extracted in parallel.

1. Take 200 mg pulverized plant material (see Note 2) from the two protein fractions to be compared and add 750 μL extraction buffer. Incubate the samples for 10 min on ice.

2. Add 750 μL of water-saturated phenol (4°C) to each sample and vortex. Incubate the samples for 30 min on a table mixer (Eppendorf Thermomixer Compact; Eppendorf, Hamburg, Germany) at 1,000 rpm at room temperature (see Note 3–5).

3. Centrifuge samples at $11,000 \times g$ for 10 min, 4°C.

4. After centrifugation, transfer the phenol phase (generally the upper phase) of each sample into a new 2-mL Eppendorf vessel and dilute it in an equal volume of ice-cold extraction buffer. Vortex samples (see Note 6).

5. Centrifuge the samples at $11,000 \times g$ for 10 min, 4°C.

6. Collect again the phenol phase (the upper phase) of each sample and transfer it completely into a new preweighted 2-mL Eppendorf vessel (see Note 7). Add ice-cold precipitation solution up to a final volume of 2 mL to each sample. Invert the vessels several times back and forth and let the proteins precipitate for at least 4 h at −20°C.

7. Centrifuge the samples at $17,000 \times g$ for 3 min, 4°C. Discard the supernatants and dilute the protein pellets in 1 mL of ice-cold precipitation solution.

8. Repeat step 7 two times (see Note 8), then dilute the protein pellet of each sample in 1 mL of ice-cold 80% [v/v] acetone. Centrifuge the samples at 17,000×g for 3 min, 4°C.

9. Discard the supernatants and dry the protein pellets under an extractor hood at room temperature (see Note 9).

10. Finally, determine the weight of the vessels including the pellets and subtract the value of the empty vessels for calculation of the pellet weights (see Note 10).

3.2. Sample Preparation for DIGE

The following protocol is exemplified on the comparative analysis of two related protein fractions using the CyDye™ Fluor labeling reagents (GE Healthcare, Munich, Germany). Both fractions should contain 100 µg protein and should be labeled with different fluorophors. Afterwards, the samples are combined and loaded onto a single protein gel. Additionally, a 1:1 mixture of the two protein samples is labeled with a third CyDye fluorophor and is used as an internal standard to allow a comparison of relative protein spot intensity between both individual protein samples analyzed. The following protocol uses 2D IEF/SDS PAGE for protein separation:

1. Resuspend protein samples, 100 µg or more each, separately in a minimum volume of 30 µL of CyDye labeling compatible lysis buffer (pH 8.5) in absence of DTT. Shake the samples for 30 min at room temperature (see Note 11).

2. Centrifuge the samples at 17,000×g for 5 min, 4°C. Collect the supernatant.

3. Labeling reaction: Transfer a volume corresponding to a protein amount of 100 µg (usually in the range of 10–20 µL) of each sample separately to a new 1.5-mL Eppendorf vessels. A minimal volume of 10 µL for each protein sample is required for labeling reaction. For the internal reference sample (third sample), prepare a mixture of the two samples by combining the samples at a ratio of 1:1. Add 1 µL of diluted CyDye solution to each sample. Label the reference sample with the remaining third CyDye (e.g., Cy2). Centrifuge them briefly and incubate for 30 min on ice in the dark (see Note 12).

4. Stop reaction: Add 10 mM lysine stock solution (one-tenth of the volume with respect to the individual samples) to each labeling reaction to obtain a final concentration of ~1 mM lysine per sample (e.g., add 1.1 µL of 10 mM lysine solution to 10 µL sample labeled with 1 µL CyDye). Incubate the samples for 10 min on ice in the dark.

5. Add to each sample the equal volume of lysis buffer containing the double amount of DTT (i.e., 7.5 mg DTT per mL lysis buffer, see Note 1; e.g., 12.1 µL of lysis buffer with ~91 µg DTT to 12.1 µL of CyDye-labeled sample).

6. Combine all three CyDye-labeled protein samples in one reagent tube.

7. Add the remaining volume of rehydration buffer containing DTT and IPG buffer to the final volume required for IEF separation.

8. Perform 2D DIGE in the dark. For the second dimension SDS PAGE, the Laemmli or the Schägger protocol can be applied (18, 21).

9. After finishing the electrophoretic separation, the gel is immediately scanned using a fluorescence scanner (Typhoon Fluorescence Scanner; GE Healthcare, Munich, Germany). Keep the gel at 4°C and in the dark before starting the scanning procedure. Using the CyDye fluorophores, gels have to be scanned at 50–100-μm resolution at the appropriate excitation wavelengths (488 nm for Cy2™, 532 nm for Cy3™, and 633 nm for Cy5™). Digital gel images can be visualized using the ImageQuant analysis software (GE Healthcare, Munich, Germany) (Fig. 1). Quantification of relative differences of individual protein abundances can be carried out using specific software (e.g., Delta 2-D (Decodon, Greifswald, Germany) or DeCyder™ (GE Healthcare, Munich, Germany)).

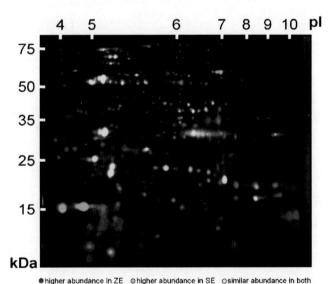

Fig. 1. Two-dimensional DIGE analysis of total protein extracts from zygotic embryos (ZE) and somatic embryos (SE) from *Cyclamen persicum*. Proteins of each tissue were extracted as described in this chapter. Protein fractions of ZE and SE were loaded onto one gel, 100 μg of each fraction. The ZE protein fraction was labeled with Cy3™ (*red*), the SE protein fraction with Cy5™ (*green*). Spots with similar abundance in both tissues are colored yellow. An internal standard was not included.

4. Notes

1. In step 1 of the protocol in Subheading 3.2, a lysis buffer without DTT addition is used. At step 5, a lysis buffer containing the twofold amount of DTT is used. Since the concentration of DTT in the lysis buffer is variable (20–100 mM), addition of 7.5 mg DTT per mL is adequate at this step.

2. Protein amount can extremely vary in different plant tissues. Thus, the amount of tissue used for sample preparation has to be adjusted: In our hands, proteins were extracted from 500 mg *Medicago* root tissue (20), 200 mg *Vigna* leaf tissue (22), and 80 mg *Cyclamen* seeds (4). A thoroughly prepared sample obtained by grinding at low temperature (<4°C) improves the quality of protein extraction. The use of a bead mill (e.g., Retsch MM 400, Retsch, Haan, Germany) for the pulverization of plant material is recommended. However, grinding time and volume of grinding tubes have to be optimized for specific plant material. Low amounts of soft plant material, like leaves, roots, and embryos, require short grinding times (1–5 min) and can be performed directly in Eppendorf vessels using steal balls. Large amounts of hard tissue may be grinded for longer time periods (up to 30 min). Long storage periods of frozen plant samples even at −80°C conditions before protein preparation minimize the yield of extracted total protein.

3. The 2-mL Eppendorf vessels should be filled up to three-quarter with fine ground plant tissue powder. When adding the extraction buffer and subsequently the phenol, the volume of the sample decreases.

4. All work with β-mercaptoethanol should take place under the extractor hood!

5. The quality of the phenol is of high importance. Phenol of insufficient quality can cause dramatic losses in protein yield. When using a two-phase phenol, it is important to take up the liquid phase only from the lower phenol phase. Do not shake the phenol in order to mix both phases. Work with phenols should only take place under the extractor hood!

6. After centrifugation (step 3), normally, an aqueous phase is obtained at the bottom of the Eppendorf vessels and a phenolic phase at the top. Plant cell walls and membrane compounds precipitate at the interface. In rare cases, aqueous and phenol phases are inverted after centrifugation. In this situation, the upper phase should be carefully removed with a pipette before collecting the phenol phase. Generally, transfer the phenol phase in several small volume steps (e.g., in 2–6 steps each collecting 100 µL).

7. The second uptake of the phenol phase (step 6) should be achieved without contamination of the liquid phase. Again, collection by several small volume steps is recommended.

8. To obtain a very pure pellet, the washing step (step 7) can be repeated several times.

9. Do not dry the protein pellet for too long under the extractor hood, because otherwise its resuspension in the lysis buffer might become difficult (1–5 min are usually sufficient). The final yield of pellet should be approximately between 5 and 20 mg protein; the protein yield can be determined using a common protein quantification method.

10. The weight of the dried pellet is close to the total amount of protein extracted.

11. For an equal protein concentration in the labeling reaction, the volume for resuspension may vary depending on the protein amounts of the samples.

12. CyDye labeling most effectively takes place if the pH value of the lysis buffer is in the range of 8.5. The pH of the sample in lysis solution can be tested via pH test strips. In case the pH value is significantly below 8.5, a lysis buffer of higher pH values (e.g., pH 9.0) should be used for resuspension of proteins. Always use CyDyes from one reaction kit diluted with DMF of one batch to assure comparative labeling conditions. Consume the CyDyes after diluting them in DMF within 3 months.

Acknowledgments

The authors would like to thank Jennifer Klodmann, Institute for Plant Genetics, Leibniz Universität Hannover, for critically reading the manuscript.

References

1. Raggiaschi R., Lorenzetto C., Diodato E., Caricasole A., Gotta S. and Terstappen G.C. (2006) Detection of phosphorylation patterns in rat cortical neurons by combining phosphatase treatment and DIGE technology. *Proteomics* 6, 748–756.

2. Lyngved R., Renaut J., Hausman J.F., Iversen T.H., and Hvoslef-Eide A.K. (2008) Embryo-specific Proteins in *Cyclamen persicum* analyzed with 2-D DIGE. *J Plant Growth Regul* 27, 353–369.

3. Gomez A., Lopez J.A., Pintos B., Camafeita E., and Bueno M.A. (2010) Proteomic analysis from haploid and diploid embryos of *Quercus suber* L. identifies qualitative and quantitative differential expression patterns. *Proteomics* 9, 4355–4367.

4. Rode C., Braun H.P., Gallien S., Heintz D., Van Dorsselaer A., and Winkelmann T. (2010) Establishment of proteome reference maps for somatic and zygotic embryos of *Cyclamen persicum* Mill. *Acta Hort* 855, 239–242.

5. Lilley K.S., and Dupree P. (2006) Methods of quantitative proteomics and their application to plant organelle characterization. *J Exp Bot* 57, 1493–1499.

6. Heinemeyer J., Scheibe B., Schmitz U.K., and Braun H.P. (2009) Blue native DIGE as a tool for comparative analyses of protein complexes. *J Prot* 72, 539–544.

7. Ndimba B.K., Chivasa S., Simon W.J., and Slabas A.R. (2005) Identification of *Arabidopsis* salt and osmotic stress responsive proteins using two-dimensional difference gel electrophoresis and mass spectrometry. *Proteomics* 5, 4185–4196.

8. Kieffer P., Schroder P., Dommes J., Hoffmann L., Renaut J., and Hausman J.F. (2009) Proteomic and enzymatic response of poplar to cadmium stress. *J Proteom* 72, 379–396.

9. Chivasa S., Hamilton J.M., Pringle R.S., Ndimba B.K., Simon W.J., Lindsey K., and Slabas A.R. (2006) Proteomic analysis of differentially expressed proteins in fungal elicitor-treated Arabidopsis cell cultures. *J Exp Bot* 57, 1553–1562.

10. Van Noorden G.E., Kerim T., Goffard N., Wiblin R., Pellerone F.I., Rolfe B.G., and Mathesius U. (2007) Overlap of proteome changes in *Medicago truncatula* in response to auxin and *Sinorhizobium meliloti*. *Plant Physiol* 144, 1115–1131.

11. Amey R.C., Schleicher T., Slinn J., Lewis M., Macdonald H., Neill S.J., and Spencer-Phillips PTN. (2008) Proteomic analysis of a compatible interaction between *Pisum sativum* (pea) and the downy mildew pathogen *Peronospora viciae*. *Eur J Plant Pathol* 122, 41–55.

12. Schenkluhn L., Hohnjec N., Niehaus K., Schmitz U.K., and Colditz F. (2010) Differential gel electrophoresis (DIGE) to quantitatively monitor early symbiosis- and pathogenesis-induced changes of the *Medicago truncatula* root proteome. *J Proteom* 73, 753–768.

13. Dornez E., Croes E., Gebruers K., Carpentier S., Swennen R., Laukens K., Witters E., Urban M., Delcour J.A., and Courtin C.M. (2010) 2-D DIGE reveals changes in wheat xylanase inhibitor protein families due to *Fusarium graminearum* Delta Tri5 infection and grain development. *Proteomics* 10, 2303–2319.

14. Tang W.Q., Deng Z.P., Oses-Prieto J.A., Suzuki N., Zhu S.W., Zhang X., Burlingame A.L., and Wang Z.Y. (2008) Proteomics studies of brassinosteroid signal transduction using prefractionation and two-dimensional DIGE. *Mol Cell Proteom* 7, 728–738.

15. Gallardo K., Job C., Groot S.P.C., Puype M., Demol H., Vandekerckhove J., and Job, D. (2002) Proteomics of *Arabidopsis* seed germination: a comparative study of wildtype and gibberelline deficient seeds. *Plant Physiol* 129, 823–837.

16. Damerval C., De Vienne D., Zivy M., and Thiellement H. (1986) Technical improvements in two-dimensional electrophoresis increase the level of genetic variation detected in wheat-seedling proteins. *Electrophoresis* 7, 52–54.

17. Corke F.M.K., and Roberts K. (1996) Large changes in the population of cell wall proteins accompany the shift to cell elongation. *J Exp Bot* 48, 971–977.

18. Laemmli U.K. (1970) Cleavage of structural proteins during the assembly of the head of bacteriophage T4. *Nature* 227, 680–685.

19. Hurkman W.J., and Tanaka C.K. (1986) Solubilization of plant membrane proteins for analysis by two-dimensional gel electrophoresis. *Plant Physiol* 81, 802–806.

20. Colditz F., Nyamsuren O., Niehaus K., Eubel H., Braun H.-P., and Krajinski F. (2004) Proteomic approach: Identification of *Medicago truncatula* proteins induced in roots after infection with the pathogenic oomycete *Aphanomyces euteiches*. *Plant Mol Biol* 55, 109–120.

21. Schägger H., and von Jagow G. (1987) Tricine-sodium dodecyl sulfate-polyacrylamide gel electrophoresis for the separation of proteins in the range from 1 to 100 kDa. *Anal Biochem* 166, 368–379.

22. Führs H., Hartwig M., Buitrago Molina L.E., Heintz D., Van Dorsselaer A., Braun H.-P., and Horst W.J. (2008) Early manganese-toxicity response in *Vigna unguiculata* L. – a proteomic and transcriptomic study. *Proteomics* 8, 149–159.

Chapter 24

Native DIGE of Fluorescent Plant Protein Complexes

Veronika Reisinger and Lutz Andreas Eichacker

Abstract

CyDye labeling and DIGE have not only been proven to work for soluble proteins but also at the level of whole membrane protein complexes. After complex solubilization and CyDye labeling, proteins can be separated by native PAGE which is often combined with SDS PAGE in a subsequent step. By this combination, sizes of complexes as well as their subunit composition can be compared after mixing samples from different physiological states. Plants interact specifically with light via protein-bound pigments. This can be used in combination with CyDye technology to extend the "classical" approach in plant research. As an example, chlorophyll can be excited for fluorescent scanning at the Cy5 excitation wavelength. This property can be used to identify pigment-binding plant complexes and complex subunits isolated from plastid membranes. In this protocol, we present a combination of the conventional CyDye labeling technique with 2D native/SDS PAGE and parallel scanning for CyDyes and fluorescence from endogenous bound chlorophyll for identification of pigment-binding complexes and complex subunits.

Key words: Native PAGE, DIGE, Membrane protein complexes, Chlorophyll fluorescence, Chlorophyll-binding proteins, Plant, Membrane, Protein complexes, Chlorophyll, CyDye

1. Introduction

Two-dimensional (2D) difference gel electrophoresis (DIGE) is a powerful tool to quantify proteomes isolated from different physiological states. Originally, this technique was invented for the overall hydrophilic water-soluble part of a proteome in combination with separation of proteins by 2D IEF/SDS PAGE (1). But different publications have proven the suitability of CyDye labeling for the analysis of membrane protein complexes as well (2–5). As IEF has been shown to work suboptimal with membrane proteins (6), CyDye labeling of membrane proteins is often combined with native PAGE. There are different protocols for native PAGE separations available (7–10), among which blue native (BN) PAGE represents the most popular technique nowadays. Although BN PAGE is up to now the technique which can be most easily applied to different sample types (11), the use of Coomassie 250 G as charge carrier is limited. As fluorescence is quenched by the Coomassie dye, the combination of BN PAGE with fluorescence detection is impossible. Therefore, BN PAGE is not compatible with CyDye detection. The following protocol describes a modified BN PAGE protocol which allows the separation of plastid membrane protein complexes and detection of chlorophyll and CyDyes by fluorescence detection. Fluorescent scanners used for visualization of fluorescent dyes in DIGE, perfectly match for the detection of chlorophylls and their derivatives as well. Hence, the classical DIGE technology can be extended in plant research.

2. Materials

2.1. Sample Preparation

1. Sample buffer: 10 mM Tris–HCl, pH 8.5, 10 mM $MgCl_2$, 20 mM KCl. Store at 4°C.
2. CyDyes: dilute and store according to instruction manual.
3. Lysine solution: 10 mM L-lysine. Store at 4°C.
4. Dodecyl maltoside solution: 10% (w/v) n-dodecyl-β-D-maltoside. Store at −20°C.
5. Digitonin solution: 10% (w/v) digitonin (see Note 1). Store at −20°C.
6. Lithium dodecyl sulfate buffer: 5% (w/v) lithium dodecyl sulfate. Store at −20°C.

2.2. Casting of Native Gradient PAGE Gels

1. Gel buffer (6×): 3 M ε-amino caproic acid, 0.3 M Bis-Tris–HCl pH 7.0. Store at 4°C (see Note 2). Adjust the pH of the Bis-Tris solution at room temperature.
2. Acrylamide solution: 30% (w/v) acrylamide/bisacrylamide solution (37.5:1, 2.6% C), acts in unpolymerized state as a neurotoxin. Store as stated by the manufacturer.
3. Glycerol (100%). Store at room temperature.
4. TEMED: N,N,N,N'-tetramethyl-ethylenediamine. Store at room temperature.
5. Ammonium persulfate (APS): 10% (w/v) solution. Stable at 4°C for up to 2 weeks.
6. Water-saturated isobutanol: 50% (v/v) isobutanol. Store at room temperature.

2.3. Casting of SDS PAGE Gels

1. Separating gel buffer (8×): 3 M Tris–HCl (pH 8.8) Store at 4°C (see Note 2).
2. Stacking gel buffer (2×): 0.25 M Tris–HCl (pH 6.8). Store at 4°C (see Note 2). Add 60 μL of a 2% (w/v) bromphenol blue solution to 100 mL of stacking gel buffer.
3. Acrylamide solution: 30% (w/v) acrylamide/bisacrylamide solution (37.5:1, 2.6% C), acts in unpolymerized state as a neurotoxin. Store as stated by the manufacturer.
4. TEMED. Store at room temperature.
5. APS: 10% (w/v) solution. Stable at 4°C for up to 2 weeks.
6. Water-saturated isobutanol: 50% (v/v) isobutanol. Store at room temperature.

2.4. Transfer of Native Gel Stripes to SDS PAGE Gels

1. Solubilization buffer: 2% (w/v) sodium dodecyl sulfate, 66 mM Na_2CO_3, 2% (v/v) β-mercaptoethanol. Store at 4°C for up to 1 month or prepare freshly (see Note 3).
2. Overlay solution: 0.5% (w/v) low melting agarose in 1× SDS running buffer. Store at room temperature.

2.5. Electrophoresis

1. Native PAGE (first dimension)
   (a) Running buffer cathode (5×): 400 mM tricine, 75 mM Bis-Tris–HCl (pH 7.0), 0.01% (w/v) lithium dodecyl sulfate. Store at −20°C for long time storage or at 4°C for some days. Adjust pH of buffer at room temperature.
   (b) Running buffer anode (10×): 500 mM Bis-Tris–HCl pH 7.0. Store at 4°C.
2. SDS PAGE (second dimension) running buffer cathode/anode (10×): 250 mM Tris, 1.92 M glycine, 1% (w/v) sodium dodecyl sulfate. Store at room temperature.

3. Methods

The protocol described below combines CyDye with chlorophyll fluorescence detection (see Note 4). In general, native PAGE (first dimension) is intended to separate protein complexes in a native state. Therefore, it is advisable to carry out all steps of sample preparation on ice to avoid protein complex degradation. This native PAGE system separates protein complexes depending on the volume/size of the single protein complexes. Therefore, the acrylamide concentration of the gel is responsible for the separation range. Instructions stated below correspond to a 6–12% linear gradient separating gel and a 4% stacking gel for the first native dimension which allows the separation of protein complexes in the molecular mass range from ~50 to 700 kDa. The second-dimension SDS gel consists of a homogeneous 12.5% separating gel and a 5% stacking gel. This setup allows the separation of proteins in the molecular mass range from ~10 to 200 kDa.

The entire experiment was carried out in a Protean II xi system (Bio-Rad Laboratories GmbH) which has a gel size of $20 \times 20 \times 0.075$ cm for the first dimension and $20 \times 20 \times 0.1$ cm for the second dimension.

3.1. Sample Preparation for Native Electrophoresis

1. Lyse 1×10^8 plastids (see Note 5) in 200 µL of sample buffer for 10 min. Pellet thylakoid membranes in a microfuge at $7,500 \times g$ at 4°C and remove the supernatant containing all soluble and peripheral proteins. To be sure that all peripheral proteins are removed, resuspend the pellet in 200 µL of sample buffer and centrifuge again. Remove the supernatant.
2. Resuspend the pellet in 70 µL of sample buffer.
3. Transfer 35 µL of the sample to a new tube (sample 1). CyDye detection requires a lower protein concentration than pigment fluorescence detection.
4. Add Cy2 or Cy3 (1 µL of a 400 pmol working solution) to sample 1, mix, and incubate the sample for 30 min on ice (see Note 6).
5. Add 1 µL of lysine to sample 1, mix, and incubate the sample for 10 min on ice.
6. Reunify the labeled sample with the unlabeled one.
7. Mix 30 µL of dodecyl maltoside solution with 60 µL of digitonin solution and 2 µL of lithium dodecyl sulfate buffer (see Note 7). Add 15 µL of mixed detergent solution to the sample and mix gently.
8. Incubate the sample on ice for 10 min to solubilize the protein complexes.
9. Centrifuge for 10 min at $16,000 \times g$ at 4°C in a microcentrifuge to pellet the unsolubilized material. Unsolubilized material can affect the subsequent electrophoretic separation of the protein complexes in a negative way.
10. Transfer the supernatant to a new tube and discard the pellet.

3.2. Casting of Native Gradient PAGE Gels (First Dimension)

3.2.1. Separating Gel

1. Clean glass plates and spacers with denatured ethanol (100%), assemble the glass plate sandwich, fix it on the casting stand, and adjust the assembled casting stand by a spirit level.
2. Place magnetic stirring rods in both chambers of the gradient mixer and ensure that all ports are closed.
3. Prepare the gel solutions stated in Table 1 (see Note 8) directly in the gradient mixer. Chamber 1 corresponds to the chamber next to the outlet tube (see Note 9).

Table 1
Solutions for the preparation of a native gradient PAGE gel

	12%	6%
Acrylamide (37.5/1)	4.20 mL	2.10 mL
6× Gel buffer	1.75 mL	1.75 mL
Glycerol	1.95 g	–
ddH$_2$O	3.00 mL	6.65 mL
Σ	10.50 mL	10.50 mL

4. Mix both solutions well by stirring them on a magnetic stirrer.

5. Place the gradient mixer on the magnetic stirrer and connect the pipette tip with the casting stand. The gradient mixer has to show a moderate incline towards the casting stand to achieve a directed flow of the gel solutions.

6. The tube which connects the gradient mixer with the glass plates is fixed in the middle of the assembled glass plates by tape. The tube ends in a cut yellow pipette tip.

7. Start the magnetic stirrer. Chamber 1 has to be stirred properly, whereas stirring of chamber 2 is not necessary. The stirring bar in chamber 2 acts as balancer for the solution levels in both chambers.

8. Add 5 μL of TEMED and 20 μL of the APS solution to both chambers and mix the solutions gently.

9. First, open the valve between the two chambers to remove the air in the valve between both chambers. Afterward, open the front valve. Let the solutions run between the glass plates.

10. Overlay the cast gel with water-saturated isobutanol. The gel should polymerize within 90 min.

11. Rinse the gradient mixer immediately with distilled water to prevent the polymerization of residual gel solution in the tubes.

3.2.2. Stacking Gel

1. After polymerization of the separating gel, remove the isobutanol by washing with distilled water (see Note 10).

2. Dry the area above the separating gel completely with Whatman paper.

3. Clean a 10-well comb with denatured ethanol (100%) and insert the dried comb in the glass plate sandwich.

4. Prepare the gel solution as stated in Table 2 in a beaker.

Table 2
Solutions for the preparation of a native stacking PAGE gel

	4% (mL)
Acrylamide (37.5/1)	1.33
6× Gel buffer	1.67
ddH$_2$O	7.00
Σ	10.00

5. Add 10 µL of TEMED and 100 µL of the APS solution to the solution and pipette the gel solution quickly between the glass plates up to the top of the glass plate. The gel should polymerize in about 15 min (see Note 11).

3.3. Native Electrophoresis (First Dimension)

1. After the stacking gel is polymerized, label the position of the wells on the outer glass plate and remove the comb carefully from the native gel. Assemble the electrophoresis apparatus.

2. Prepare the cathode running buffer by diluting 60 mL of the 5× cathode buffer with 240 mL of distilled water. Pour the 1× cathode buffer into the upper chamber.

3. Rinse the wells with buffer using a 100-µL microsyringe to remove the residues of acrylamide and entrapped air.

4. Underlay the samples into the wells using the microsyringe. Rinse the microsyringe with buffer before applying a new sample (see Notes 12 and 13).

5. Add 3 µL of a 2% (w/v) bromphenol blue solution to an empty well (see Note 14).

6. Dilute 100 mL of 10× anode buffer with 900 mL of distilled water and pour the 1× anode buffer in the lower buffer chamber.

7. Connect the tubes of the thermostatic circulator to the buffer chamber. Start the thermostatic circulator. The electrophoretic run is carried out at 4°C (see Note 15).

8. Assemble the electrophoresis unit completely and connect to a power supply.

9. Set the power supply to 12 mA, 1,000 V and 24 W and start the electrophoresis.

10. Stop electrophoresis when the bromphenol blue front has reached the bottom of the separating gel (about 3.5 h).

11. Stop the thermostatic cooler and disassemble the buffer chamber assembly. The glass plate sandwich can be directly used for detection of fluorescence complexes and is scanned in a fluorescence scanner for CyDye and pigment fluorescence. The use of low fluorescent glass plates is recommended. Adjustments for detection of Cy5 are used to scan for pigment fluorescence.

3.4. Casting of SDS PAGE Gels (Second Dimension)

3.4.1. Separating Gel

1. Clean glass plates and spacers with denatured ethanol (100%), assemble the glass plate sandwich, fix it on the casting stand and adjust the assembled casting stand by a spirit level (see Note 16).
2. Prepare the gel solution as stated in Table 3 in a beaker and dissolve the urea by stirring (see Note 17).
3. After dissolving the urea add 15 µL of TEMED and 50 µL of the APS solution to the solution and pipette the gel solution between the glass plates. The level of the gel should be approximately 5 cm below the top of the smaller glass plate.
4. Overlay the cast gel with water-saturated isobutanol. The gel should polymerize within 60 min.

3.4.2. Stacking Gel

1. After polymerization of the separating gel, remove the isobutanol by washing with distilled water (see Note 10).
2. Dry the area above the separating gel completely with Whatman paper.
3. Prepare the gel solution stated in Table 4 in a beaker, add 5 µL of TEMED and 50 µL of the APS solution, and pipette the gel solution between the glass plates. The stacking gel should be approximately 2 cm high (see Note 14).
4. Overlay the cast gel with water-saturated isobutanol. The gel should polymerize within 15 min.

Table 3
Solutions for the preparation of a SDS PAGE separating gel

	12.5%
Urea	7.21 g
Acrylamide (37.5/1)	12.5 mL
8× Separating gel buffer	3.75 mL
ddH$_2$O	9.00 mL
Σ	30.00 mL

Table 4
Solutions for the preparation of a SDS PAGE stacking gel

	5%
Acrylamide (37.5/1)	0.8 mL
2× Stacking gel buffer	2.50 mL
ddH$_2$O	1.70 mL
Σ	5.00 mL

3.5. Sample Preparation for SDS PAGE Electrophoresis

1. Disassemble the glass plate sandwich of the native gel and separate the strips of the first-dimension gel by cutting them with the spacer (see Note 18).
2. Remove the stacking gel.
3. Fill 30 mL of solubilization buffer in a 50-mL centrifuge tube.
4. Transfer the strip in the tube containing the solubilization buffer by grabbing the native strip at the 12% end.
5. Shake the strip at room temperature for 20 min (see Note 19).
6. After polymerization of the SDS PAGE stacking gel, rinse the gel with distilled water to remove the isobutanol. Remove the water on top of the SDS PAGE stacking gel with Whatman paper.
7. Position the strip between the glass plates of the SDS PAGE gel touching the surface of the SDS PAGE stacking gel. Take care that the native and the SDS PAGE stacking gel touch well. Avoid air bubbles between the first and second-dimension gel.
8. Overlay the first-dimension strip with overlay solution and let the agarose thicken.

3.6. SDS PAGE Electrophoresis (Second Dimension)

1. Prepare the SDS PAGE running buffer by diluting 200 mL of the 10× SDS running buffer with 1,800 mL of distilled water.
2. As soon as the agarose is set, assemble the electrophoresis apparatus and pour the 1× buffer into the upper and the lower buffer chamber.
3. Assemble the electrophoresis unit completely and connect to a power supply.
4. Set the power supply to 5 mA, 600 V and 24 W and start the electrophoresis for an overnight run (see Note 20).
5. Stop electrophoresis when the bromphenol blue front has reached the bottom of the separating gel.

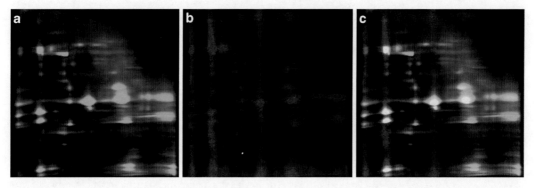

Fig. 1. Detection of pigment-binding thylakoid membrane complex subunits after native/SDS PAGE. Thylakoid membrane complexes were labeled by Cy2 and separated by 2D native/SDS PAGE. After electrophoresis, fluorescent proteins were detected by a Thyphoon trio scanner scanning for Cy2 ((**a**) *green*) and Cy5 ((**b**) *red*) Scanned images of fluorescent proteins were overlayed for difference analysis ((**c**) *green/red/yellow*).

6. Disassemble the buffer chamber assembly. Clean the glass plate sandwich with distilled water and scan with a fluorescence scanner for CyDye and pigment fluorescence. Scanning for Cy5 fluorescence is used to image pigment fluorescence (see Fig. 1).

7. After scanning, gels can be used for Coomassie/silver staining or blotting.

4. Notes

1. Digitonin is dissolved by heating up the solution to 100°C. After heating, digitonin can immediately be mixed with the other detergents.

2. If larger amounts of buffers are prepared, it is recommended to store just a small amount at 4°C and the rest at −20°C.

3. Before use, warm the solution up in warm water, but do not boil it. Sodium dodecyl sulfate precipitates at 4°C.

4. In general, this protocol can also be used to compare thylakoid membrane complexes from, e.g., mutant and wild-type plants. If different thylakoid samples are compared, the amount of proteins must at least be halved. For the comparison of two samples, samples are mixed together after labeling and solubilization and are subsequently separated together by native PAGE. As the composition of the gel and the anode buffer correspond to the gel and buffer used in blue native PAGE, this protocol can also be used to perform blue native PAGE if the cathode buffer is replaced by the blue native cathode buffer.

5. Plastids can be counted in an Abbe-Zeiss counting cell chamber. Instead of counting, the protein amount can be calculated via

chlorophyll or protein determination; 1×10^8 chloroplasts correspond approximately to 50 μg of chlorophyll or 400 μg of membrane proteins.

6. If pigment fluorescence detection is used via the Cy5 channel, only Cy2 or Cy3 can be used for protein labeling.

7. The rest of the detergent mix solution can be stored by −20°C and reused. The detergent mix is soluble at room temperature.

8. The simplest way to transfer the glycerol in the gradient mixer is to pipette it with a cut 1-mL pipette tip.

9. It is also possible to cast the gradient gel from the bottom of the gel. In that case, a peristaltic pump is needed and the risk of air bubbles in the gel during the casting process is higher.

10. Polymerization of the separating gel is completed when a layer of water is formed between the upper edge of the gel and the isobutanol layer.

11. After wrapping the gel sandwich in wet tissues and a plastic bag, it can be stored up to 1 week at 4°C. In case of SDS gels, the gel should be stored without the stacking gel because of the different pH of the separating and stacking gel.

12. You can add 2 μL of 80% glycerol to facilitate the application of the samples.

13. Beside the use of microsyringes, it is also possible to use gel loader tips to underlay the samples into the wells. As gel loader tips are much more flexible than a microsyringe and end in a capillary tip, it is easier to apply the samples between the glass plates.

14. The addition of bromphenol blue facilitates tracing of the running front during electrophoresis.

15. The electrophoretic run can also be performed in a cold room at 4°C if no thermostatic cooler is available. It is advisable to cool the gel during electrophoresis to keep the complexes intact, maintain their structure, and protect them against proteolysis.

16. The second-dimension gel must be thicker than the first-dimension gel for inserting the first-dimension gel strip between the glass plates. Therefore, 0.75-mm spacers are used for the first dimension, and 1-mm spacers are used for the second dimension in this protocol.

17. The addition of urea to the gel aids to keep the secondary structure of the membrane proteins denatured during SDS PAGE electrophoresis.

18. The experiment can be interrupted after cutting the first-dimension gel lanes. Lanes can be stored at −20°C in a plastic folder for some weeks before processed further.

19. The incubation time should not be extended as proteins could be eluted from the first-dimension gel in the solubilization buffer in the course of time.

20. The electrophoresis can be also performed in approximately 6 h at 30 mA. In this case, it is advisable to cool the gel during electrophoresis by a thermostatic cooler at 15°C.

References

1. Marouga R, David S, Hawkins E (2005) The development of the DIGE system: 2D fluorescence difference gel analysis technology. Analytical and Bioanalytical Chemistry 382: 669–678.
2. Heinemeyer J, Scheibe B, Schmitz UK, Braun HP (2009) Blue native DIGE as a tool for comparative analyses of protein complexes. J Proteomics 72: 539–544.
3. Reisinger V, Eichacker LA (2007) How to analyze protein complexes by 2D blue native SDS-PAGE. Proteomics 7 Suppl 1: 6–16.
4. Gillardon F, Rist W, Kussmaul L, Vogel J, Berg M, Danzer K, Kraut N, Hengerer B (2007) Proteomic and functional alterations in brain mitochondria from Tg2576 mice occur before amyloid plaque deposition. Proteomics 7: 605–616.
5. Reisinger V, Hertle AP, Plöscher M, Eichacker LA (2008) Cytochrome b_6f is a dimeric protochlorophyll a binding complex in etioplasts. FEBS Journal 275: 1018–1024.
6. Rabilloud T (2009) Membrane proteins and proteomics: Love is possible, but so difficult. Electrophoresis 30: 174–80.
7. Dreyfuss BW, Thornber JP (1994) Organization of the Light-Harvesting Complex of Photosystem I and Its Assembly during Plastid Development. Plant Physiol. 106: 841–848.
8. Wittig I, Braun HP, Schagger H (2006) Blue native PAGE. Nat Protoc 1: 418–428.
9. Wittig I, Karas M, Schagger H (2007) High resolution clear native electrophoresis for in-gel functional assays and fluorescence studies of membrane protein complexes. Mol Cell Proteomics 6: 1215–1225.
10. Wittig I, Schagger H (2005) Advantages and limitations of clear-native PAGE. Proteomics 5: 4338–4346.
11. Krause F (2006) Detection and analysis of protein-protein interactions in organellar and prokaryotic proteomes by native gel electrophoresis: (Membrane) protein complexes and supercomplexes. Electrophoresis 27: 2759–2781.

Chapter 25

An Overview of 2D DIGE Analysis of Marine (Environmental) Bacteria

Ralf Rabus

Abstract

Microbes are the "unseen majority" of living organisms on Earth and main drivers of the biogeochemical cycles in marine and most other environments. Their significance for an intact biosphere is bringing environmental bacteria increasingly into the focus of genome-based science. Proteomics is playing a prominent role for providing a molecular understanding of how these microbes work and for identifying the key biocatalysts involved in the major biogeochemical processes. This overview describes the major insights obtained from two-dimensional difference gel electrophoresis (2D DIGE) analyses of specific degradation pathways, complex metabolic networks, cellular processes, and regulatory patterns in the marine aerobic heterotrophs *Rhodopirellula baltica* SH1 (Planctomycetes) and *Phaeobacter gallaeciensis* DSM 17395 (*Roseobacter* clade) and the anaerobic aromatic compound degrader *Aromatoleum aromaticum* EbN1 (*Betaproteobacteria*).

Key words: 2D DIGE, Anaerobic degradation, Aromatic compounds, Carbohydrates, Catabolism, Environmental bacteria, Marine bacteria, Metabolic networks, Regulation

1. Introduction

1.1. Microbes Drive Marine Biogeochemical Cycles

The marine system covers a large part of the Earth's biosphere and harbors a tremendous abundance and diversity of microbes. Activity and dynamics of the latter are major drivers of biogeochemical matter turnover and fate and therefore pivotal for the functioning of the ocean ecosystem.

1.1.1. Marine Environments

The oceans cover about 70% of the world's surface, comprising a wide range of different environments. The water column (pelagial) extends from shallow shelf areas (e.g., the North Sea) to the deep sea ranging from around 0.5 to 11 km water depth. While the top

200 m of the ocean waters receives sunlight (euphotic zone), the deep sea is characterized by absence of light, high pressure, and low temperatures (−1 to 4°C) and covers about 95% of the seabed (benthos). High organic carbon input and mainly anoxic conditions prevail in sediments at upwelling areas, continental margins, hydrothermal vents, and cold seeps. In contrast, remote ocean areas, e.g., in the South Pacific, may have oligotrophic oxic sediments (1).

1.1.2. Abundance and Diversity of Prokaryotes in the Oceans

Microbes thrive in all ocean environments, from the sunlit surface waters across the deep sea into the seafloor sediments and the deep subseafloor biosphere, as well as from tropic waters to the arctic regions. Marine microbes represent the "unseen majority." It is estimated that 3.6×10^{28} prokaryotic cells occur in the euphotic zone, 6.5×10^{28} cells in the ocean water below 200 m, 1.7×10^{28} cells in the top 10 cm of the seafloor sediment, and 2.1×10^{30} below 10 cm sediment surface. In total, the marine microbial biomass corresponds to about one-third of the estimated total carbon in plants (2).

In seawater, less than 0.1% of the total cell counts can be covered by cultured microbes, i.e., for more than 99% of the microbial diversity in the oceans, only molecular fingerprints are available (3). In general, half of the more than 50 presently known bacterial phyla do not contain a single cultured representative (4). For the deep subseafloor biosphere, investigations into abundance, diversity, and activity of prokaryotes are at their early beginnings (5). The pioneering large-scale metagenomic analyses of near-surface seawater samples have revealed a hitherto unknown large number of novel phylotypes and protein families (6, 7), further underscoring the vastness of unexplored microbial diversity in the oceans and opening a new avenue for microbial ecology. For instance, comparison of 45 distinct microbiomes revealed that metagenomic differences actually correlated with the biogeochemical conditions prevailing in the studied environments (8). Moreover, the diversity of marine microbial populations is being cataloged by massive DNA sequencing (9), and large-scale metagenomic/transcriptomic studies are being conducted across spatial and temporal gradients of marine habitats (10). This trend will be further propelled by emerging ultrafast next-generation sequencing technologies (11). Besides resolving biogeographical patterns, the functional challenge will be to extract from the increasing (meta)genome data experimentally validated key redox biocatalysts of biogeochemical processes and to assess their distribution within microbial communities and diversity (Fig. 1) (12).

1.1.3. Fate of Organic Carbon

The top 200 m of the water column receive enough sun light for the phytoplankton to generate about one-half of the global primary production of organic carbon (13). About half of the fixed carbon

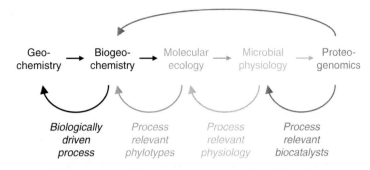

Fig. 1. Modern marine microbiology pursues identification of key redox biocatalysts underlying biogeochemical processes. In the 1970s–1980s, many geochemical processes were discovered to be biologically driven, coining the term biogeochemistry. In the 1990s, cultivation-independent (16S rRNA–based) diversity analysis allowed correlating in situ presence and abundance of microorganisms (phylotypes) with in situ biogeochemical processes, thereby detecting potential key players in the environment. Enrichment and isolation of such key players or representative model organisms form the basis for explaining biogeochemical processes with physiological capacities. Finally, since about 2000, genomics and proteomics are increasingly applied to identify key biocatalysts in process-relevant model organisms and to determine their regulation in response to changing environmental conditions.

is rapidly degraded by heterotrophic bacteria in the near-surface waters. A minor part of the remaining organic carbon is further transformed to refractory dissolved organic matter (DOM) persisting in deeper ocean waters, while the rest sinks as particulate organic matter (POM) to the seafloor (14, 15). In the upper sediment layers, organic matter (OM) is largely oxidized by sulfate-reducing bacteria (16). However, over geological time scales, the remaining small fraction of OM has accumulated in the deep sediments to the largest global reservoir of organic carbon or has thermally transformed into oil and gas. Hydrocarbons can enter the highly active upper sediment layers following tectonic activities, by recent formation in hydrothermal systems or due to anthropogenic impact (Fig. 2).

1.2. Anaerobic Degradation of Aromatic Compounds

The aromatic ring belongs, next to the glycosyl ring, to the most abundant organic chemical structures in the biosphere. Aromatic compounds are major constituents of proteins, lignin, flavonoids, tannins, and crude oil and are widely used as solvents or starting compounds in industrial chemical synthesis. Thus, aromatic compounds represent microbial substrates that are abundant in nature and structurally diverse. The aromatic system conveys high chemical stability, posing biochemical hurdles to the biodegradability of these compounds. Aerobic microorganisms employ O_2-derived highly reactive oxygen species in oxygenase-catalyzed reactions for activation and cleavage of the aromatic ring (17, 18). Due to rapid oxygen consumption by aerobic heterotrophs, anoxic conditions

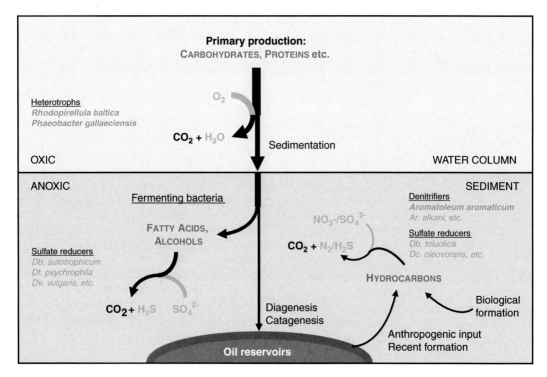

Fig. 2. Microbial mineralization of organic carbon in the marine/aquatic systems. Indicated organisms are studied in our group (*bold*, analyzed to date by 2D DIGE).

are established and prevail in many environments, such as marine and freshwater sediments, ground water aquifers, soils, deep biosphere, and oil reservoirs. Due to the absence of molecular oxygen under anoxic conditions, a variety of alternative, oxygenase-independent reactions are employed by diverse anaerobic bacteria that have both (reactions and bacteria) been discovered only during the last 10–20 years (19–22). Considering that aromatic compounds were present in the biosphere long before the evolution of an oxic atmosphere, their anaerobic biodegradation is probably an evolutionary rather old metabolic trait of bacteria.

1.3. Importance of Pure Culture-Based Proteogenomics

The sheer size of emerging (meta)genomic data is impressive, providing unprecedented insights into the genetic blueprints of microbes, as well as their diversity, biogeography, and in situ gene pools. However, the "true biology" unfolds at the proteome level, as the functional output of a (meta)genome in response to prevailing and/or changing environmental conditions. To gain causal understanding of an organism's functions and dynamics in ecosystems, pure culture studies mimicking nutritional or other environmental conditions are essential. Model organisms should be selected according to their ecosystem relevance: cultivated member of a highly abundant group (23), process relevant (24), and/or intriguing metabolism (25).

Here, the term proteogenomics refers to differential proteomic analysis based on the complete genome sequence of the study organism in contrast to the general definition of proteomics-enabled improvement of genome annotation. At present, about 1,000 prokaryotic complete genome sequences have been determined, and more than 3,400 genome projects are ongoing (26). In addition, the Genomic Encyclopedia of Bacteria and Archaea (GEBA) aims at a "phylogenetically balanced genomic representation of the microbial tree of life" (27). Considering these recent genomic developments as well as the well-established proteomic technologies (28), it can be expected that any prokaryotic isolate representing a promising model system will be genome-sequenced and thus applicable for subsequent functional proteomic studies.

2. Establishing 2D DIGE for Studying Environmental Bacteria

Proteomic analyses of genome-sequenced environmental bacteria provide unprecedented insights at the molecular level. However, the overall picture of an organism's physiology cannot be derived by a single comprehensive proteomic measurement. The actual challenge is to precisely capture distinct subproteomes (proteome signatures), each reflecting the organism's response to a specific carbon source, redox state, nutrient limitation, host interaction, or other environmental condition. Thus, quantitative differential protein profiling as it is possible with two-dimensional difference gel electrophoresis (2D DIGE) plays the central role for physiological proteomics of environmental bacteria.

The cyanine dye-based 2D DIGE was originally described by Ünlü et al. (29). 2D DIGE, based on three cyanine dyes, the 3-mode Typhoon 9400 scanner, and the DeCyder software, was evaluated for the first time using a marine bacterium (30). 2D DIGE-based studies from our group with environmental bacteria are summarized in Table 1 and described in more detail in the following sections.

2.1. 2D DIGE Method

The 2D DIGE procedures (general experimental setup, labeling reaction, 2DE separation, image acquisition, and image analysis) optimized in our laboratory over the last decade for environmental model organisms have recently been summarized (31). Design and workflow of a typical 2D DIGE experiment is divided into four major steps: fluorescence labeling, preparation of mixtures of labeled protein extracts, protein separation by 2DE, image acquisition, and image analysis (Fig. 3).

2.1.1. Fluorescence Labeling

Each protein extract is derived from an individual culture to account for biological variation. In the following, the typical 2D DIGE experiment is based on 12 parallel gels for reason of

Table 1
Overview of 2D DIGE-based studies with marine (environmental) bacteria from the author's laboratory

Organism	Research topic	Reference
Marine aerobic heterotrophs		
Rhodopirellula baltica SH1	Evaluation of 2D DIGE	(30)
	Peripheral/central pathways of carbohydrate metabolism	(40)
	Growth rate-dependent regulation of protein composition	(41)
Phaeobacter gallaeciensis DSM 17395	Central metabolism	(47)
Anaerobic degradation specialist		
Aromatoleum aromaticum EbN1	Anaerobic degradation of ethylbenzene and toluene	(54)
	Catabolic network across >20 substrate conditions	(56)
	Anaerobic degradation of *p*-ethylphenol	(55)
	Solvent stress response	(57)

simplicity—composition of the internal standard ultimately limits the number of test states that can be combined in a single working package. Protein extracts are generated, for example, from 16 distinct cell pellets, representing four biological replicates for each of the four different growth conditions: one reference state (R1–R4) and three test states (A1–A4, B1–B4, and C1–C4). For each labeling reaction, 50 μg of protein extract is labeled with 200 pmol cyanine dye. The type of cyanine dye applied depends on the experimental designation of the protein extract(s): reference state (R; Cy5 dye), test states (A, B, C; Cy3 dye), or pooled internal standard (IS; Cy2 dye). Each biological replicate of the reference state is labeled at threefold amount. Each biological replicate of the test states is labeled at onefold amount. The internal standard contains all samples (4× reference state, 12× test states) included in the working package and is labeled at 12-fold amount.

2.1.2. Mixtures of Labeling Preparation

The labeled protein extracts are mixed according to the experimental design to yield 12 separate mixtures, each containing equal amounts of the reference state, one of the three test states, and the internal standard.

2.1.3. Two-Dimensional Gel Electrophoresis

Separation of protein mixtures is performed by 2DE using immobilized pH gradients essentially according to the procedures developed by Görg and coworkers (32).

25 An Overview of 2D DIGE Analysis of Marine (Environmental) Bacteria 361

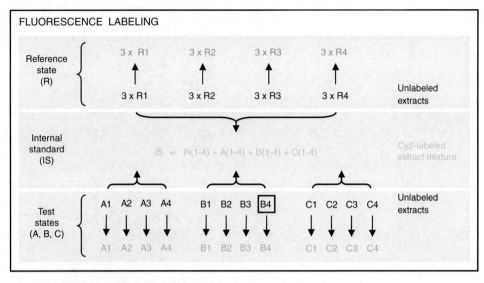

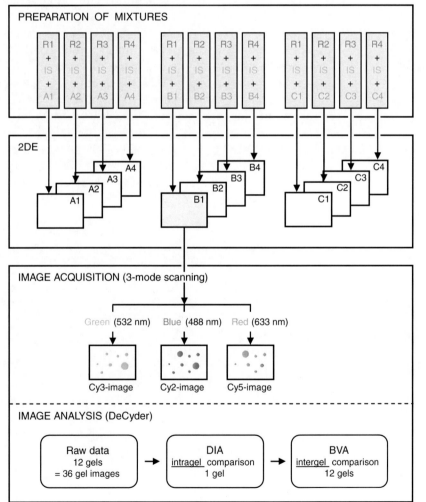

Fig. 3. Scheme of a typical 2D DIGE experiment consisting of fluorescence labeling, preparation of mixtures of labeled protein extracts, 2DE-based protein separation, image acquisition, and analysis (modified from (31)).

2.1.4. Image Acquisition

Following 2DE, gels are imaged with a three-mode fluorescence scanner (Typhoon 9400), yielding three independent CyDye-specific images from one gel. The scanner is operated at a resolution of 100 μm, and gel images are generated with optimal maximum pixel intensities of 50,000–80,000.

2.1.5. Image Analysis

Prior to analyses with the DeCyder software, gel images are cropped to exclude nonessential parts. In our laboratory, a spot detection limit of 3,000 spots is used. The spot detection parameters are set to slope >1, area <200, height <190, and volume <6,000. The gel with the best resolution and highest number of spots is selected as master gel. For intergel comparison in the biological variation analysis (BVA) module, 10–20 common landmarks are set on the master gel and subsequently across all other gels. Proteins with changed abundances are manually controlled and have to fulfill the following criteria: average ratio of <|2–2.5|, ANOVA P-value of <0.05, t-test value of <10^{-4}, and matched in at least 75% of all analyzed gels.

2.2. Ongoing Developments

Current methodological work in our group on 2D DIGE is concerned with evaluating the influence of biological vs. technical variation on fold-changes of protein abundance (average ratio) (33). Moreover, we are interested in defining the lowest, while still robust thresholds of significance, i.e., what is the minimal protein abundance fold-change that represents a biological meaningful response. In our experience, catabolic proteins display strong abundance fold-changes in response to the respective substrate(s). In such cases, a conservative threshold (>|2.5|) is reasonable since it allows avoiding too many false positives. However, in case of cellular or global regulatory processes, lower fold-changes may actually be of biological significance. Finally, the threshold of significance may have to be determined separately for each model organism.

3. Rhodopirellula baltica SH1: Heterotrophic Marine Planctomycete

3.1. Background

Planctomycetes are widespread in nature (freshwater, marine, soils); display unusual morphologies, such as rosettes, stalks, and intracellular compartmentalization (34, 35); and belong to a distinct phylogenetic superphylum (36). The marine, aerobic, and heterotrophic *Rhodopirellula baltica* SH1 (37) was the first planctomycete to be genome-sequenced (38) and investigated on the proteome level, including establishment of the first 2D DIGE master gel for planctomycetes (39, 40). The presumptive lifestyle of *R. baltica* SH1 in the marine water column involves energy generation from aerobic carbohydrate oxidation and transition between planktonic (free-living swarmer cells) and particle-associated (attached via holdfast substances at the cell pole to nutrient-rich marine snow) states.

3.2. Carbohydrate Metabolism

Considering the nutritional specialization of R. baltica SH1, architecture and regulation of peripheral and central pathways for carbohydrate utilization were investigated (Fig. 4a) (41). Almost all enzymes (25 out of 32 predicted) of glycolysis, tricarboxylic acid (TCA) cycle, and oxidative branch of the pentose phosphate cycle were identified. They displayed rather constant fold-changes of protein abundances (<|1.5|) across all eight applied substrate conditions (ribose, xylose, glucose, lactose, N-acetylglucosamine, melibiose, maltose, and raffinose), agreeing with constant in vitro activities of six selected enzymes. Most of the 22 upregulated protein species were either dehydrogenases/oxidoreductases of unknown substrate specificities or proteins of unknown function, which are unique for R. baltica SH1. This pointed to thus far unknown routes for peripheral carbohydrate catabolism (Fig. 4b, c). The coding genes of many substrate-specifically co-regulated proteins are not organized in clusters on the chromosome, demanding special regulatory processes.

3.3. Growth Phase-Dependent Regulation

Microscopic analyses indicated motile single swarmer cells of R. baltica SH1 to dominate during linear growth phase, while rosette-like cell aggregates were more abundant during stationary growth phase. Under standard conditions, entry into stationary growth phase resulted from carbon source (glucose) depletion. Based on these observations, 2D DIGE was applied at different time points distributed from early linear to late stationary growth phase (41–432 h of incubation). The goal was to define protein signatures of cells representing different morphotypes and stages in cell cycle, or reflecting adaptation to nutrient limitation (42). The number of regulated protein spots (fold-changes in protein abundance >|2|) increased from early (10 spots) to late stationary growth phase (179 spots), with fold-changes reaching maximal values of 40. The regulated protein spots represented 98 different protein species. TCA cycle and oxidative pentose phosphate cycle were oppositely regulated, while several enzymes of amino acid biosynthesis were downregulated and the alternative sigma factor σ^H upregulated in stationary growth phase. Interestingly, 26 and 10 proteins of unknown function were specifically regulated in the stationary growth phase and during growth on solid surface (agar plates), respectively. This group of proteins could harbor promising candidates for the development of the different R. baltica SH1 morphotypes, i.e., swarmer cells and rosettes.

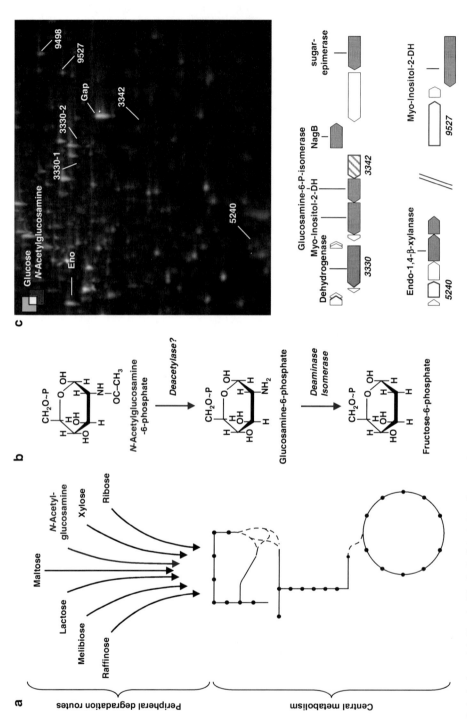

Fig. 4. Carbohydrate metabolism in *Rhodopirellula baltica* SH1. (**a**) Carbohydrates are channeled via peripheral pathways into central metabolism; (**b**) the reaction sequence for *N*-acetylglucosamine degradation is unclear at present, since a known deacetylase for initial acetyl-group removal is not encoded in the genome. (**c**) However, 2D DIGE analysis revealed the specific formation of proteins of unknown function that could possibly be involved in the peripheral degradation of *N*-acetylglucosamine. Eno (enolase) and Gap (glyceraldehyde-3-phosphate dehydrogenase) of central glycolysis are not regulated. Upregulated proteins are marked in red. Coloring of genes: *gray*, predicted function; hatched, conserved hypothetical; *white*, unknown function.

4. Phaeobacter gallaeciensis DSM 17395: Heterotrophic Marine Roseobacter

4.1. Background

The *Roseobacter* clade belongs to the *Alphaproteobacteria* and covers all major marine habitats: coastal and open ocean waters, phytoplankton and algal blooms, marine snow, sediments, biofilms, and surfaces of marine plants and invertebrates. In open ocean waters, roseobacters can account for up to 25% of all detected bacterial phylotypes. The *Roseobacter* clade harbors a wide range of metabolic capacities, including heterotrophy, aromatic compound degradation, CO oxidation, aerobic anoxygenic photosynthesis, and secondary metabolite production. Thus, the *Roseobacter* clade represents a major lineage of marine bacteria (43–45). Recent comparison of genomes of 32 roseobacter isolates provides a first comprehensive perspective of patterns in genome features and distributions of genes and pathways (46). *Phaeobacter gallaeciensis* DSM 17395 is an aerobic versatile heterotroph, utilizing a wide range of carbohydrates and amino acids, and producing the antibiotic tropodithietic acid (47).

4.2. Growth Phase-Dependent Regulation of Central Metabolism

Global protein (2D DIGE) and metabolite (GC-MS) profiles were determined at five different time points during growth of *P. gallaeciensis* DSM 17395 with glucose. Among the identified 215 proteins and 101 metabolites, 60 proteins and 87 metabolites displayed changed abundances upon entry into stationary growth phase. Glucose oxidation apparently proceeds via the Entner-Doudoroff (ED) pathway, since (1) the key enzyme of the Embden-Meyerhoff-Parnass pathway (phosphofructokinase, PFK) is not encoded in the genome and (2) 2-keto-3-desoxygluconate as key metabolite of the ED pathway and the respective enzymes were detected. Comparative genomics verified this metabolic trait also in the majority of other genome-sequenced roseobacters. Upon entry into stationary growth phase due to glucose depletion, sulfur assimilation (incl. cysteine biosynthesis) and parts of cell envelope synthesis (e.g., 1-monooleoylglycerol) were downregulated, while cadaverine formation was upregulated. In contrast, proteins and metabolites of the ED pathway, pyruvate oxidation, and TCA cycle remained essentially unchanged, pointing to a metabolic "standby" modus as an ecophysiological adaptation strategy. This "standby" and the absence of an *rpoS* gene demonstrated that stationary phase response of *P. gallaeciensis* DSM 17395 differs fundamentally from that of standard organisms such as *Escherichia coli* (48). Glucose breakdown via the ED pathway in the roseobacters *P. gallaeciensis* DSM 17395 and *Dinoroseobacter shibae* were confirmed by flux analysis (49).

5. Aromatoleum aromaticum EbN1: Anaerobic Degradation Specialist

5.1. Background

The denitrifying aquatic bacterium *Aromatoleum aromaticum* EbN1 was the first pure culture demonstrated to anaerobically degrade ethylbenzene (50). Its versatility is reflected by the wide range of utilized aromatic growth substrates. *A. aromaticum* EbN1 belongs to a betaproteobacterial cluster of anaerobic degradation specialists. This *Aromatoleum-Azoarcus-Thauera* cluster consists of more than 20 different strains degrading aromatic compounds, alkanes, and monoterpenes under nitrate-reducing conditions and dominated enrichment cultures with crude oil (51). The genome of *A. aromaticum* EbN1 was the first to be determined for a member of this phylogenetic cluster and of anaerobic hydrocarbon degraders in general (25). The genome sequence together with the below described differential proteomic studies allowed reconstructing the complex network of *A. aromaticum* EbN1 for anaerobic degradation of aromatic compounds (Fig. 5).

5.2. Anaerobic Toluene and Ethylbenzene Degradation

A. aromaticum EbN1 is unique in anaerobically degrading ethylbenzene and toluene via two fundamentally different pathways, converging at the central intermediate benzoyl-CoA. Ethylbenzene is initially hydroxylated at the methylene carbon, then dehydrogenated, and carboxylated, until acetyl-CoA is thiolytically removed. In contrast, the methyl group of toluene is radically added to fumarate, followed by modified β-oxidation and thiolytic removal of succinyl-CoA. Directed by Edman sequences of 2DE-resolved substrate-specific proteins (52), the complete gene clusters for both pathways were assembled from whole-genome shotgun sequencing data (53, 54).

The substrate-dependent regulation of both pathways was studied at the RNA (real-time RT-PCR and DNA microarray) and protein (2D DIGE) level. The toluene-related *bss* and *bbs* operons were specifically expressed in toluene-adapted cells and suggested to be coordinately regulated by a single two-component system. In contrast, ethylbenzene-related genes (*ebd/ped* and *apc* operons) were expressed in ethylbenzene- and toluene-adapted cells, suggesting gratuitous induction by toluene or cross talk between the sensor/regulator systems of both pathways. Differential gene expression and protein profiles of ethylbenzene- and acetophenone-adapted cells suggested a sequential regulation by two distinct two-component systems responsive to ethylbenzene and acetophenone, respectively. Moreover, the profiles indicated involvement of several thus far unknown proteins in the degradation of both alkylbenzenes (55).

25 An Overview of 2D DIGE Analysis of Marine (Environmental) Bacteria 367

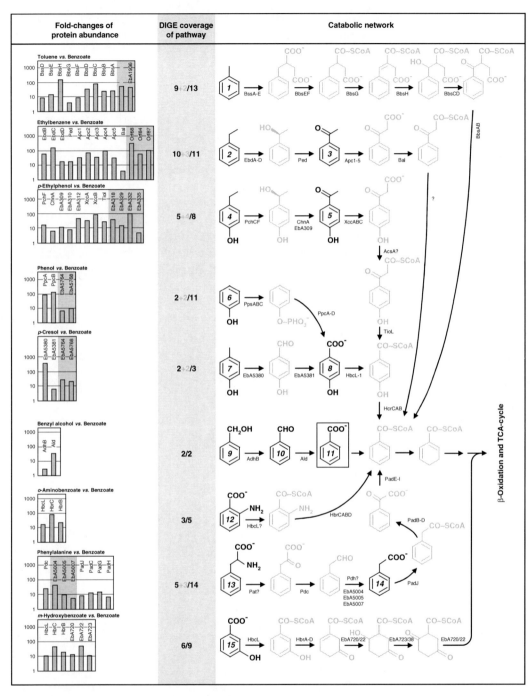

Fig. 5. Metabolic network of *Aromatoleum aromaticum* EbN1 for anaerobic degradation of aromatic growth substrates. Numbered compounds represent adaptation substrates underlying these comprehensive 2D DIGE analyses (57). Compound numbering: *1*, toluene; *2*, ethylbenzene; *3*, acetophenone; *4*, *p*-ethylphenol; *5*, *p*-hydroxyacetophenone; *6*, phenol; *7*, *p*-cresol; *8*, *p*-hydroxybenzoate; *9*, benzyl alcohol; *10*, benzaldehyde; *11*, benzoate; *12*, *o*-aminobenzoate; *13*, phenylalanine; *14*, phenylacetate; *15*, *m*-hydroxybenzoate. Enzymes are assigned to reactions as previously predicted (25, 55–57). Benzoate-adapted cells (*boxed*) served as references state. Fold-changes of protein abundance are displayed for most analyzed test states on the *left* panel, shaded proteins have not been predicted by original genome annotation to be involved in substrate metabolism. DIGE coverage of pathways (X + Y/Z) compares DIGE-identified proteins (X) vs. predicted proteins (Z), considering also additional co-regulated proteins (Y).

5.3. Anaerobic p-Ethylphenol Degradation

Inspired by metabolic network proteomics (see below) p-ethylphenol and related aromatic compounds were recently discovered as new anaerobic growth substrates of *A. aromaticum* EbN1. Integrating identification of p-ethylphenol-specific proteins (2D DIGE-directed) and metabolites (GC-MS) allowed proposing a degradation pathway analogous to that for ethylbenzene: hydroxylation and dehydrogenation of the methylene carbon, terminal carboxylation, thiolytic removal of acetyl-CoA, and additionally reductive dehydroxylation to benzoyl-CoA. The coding genes for p-ethylphenol degradation are organized in a single 15-kb operon-like structure, which is probably regulated by a single σ^{54}-dependent sensor/regulator. The sharply demarcated protein profiles of p-ethylphenol- vs. ethylbenzene-adapted cells reflect a strict sensory distinction between these two structurally similar compounds. Notably, previous genome annotation could not assign this gene cluster to the catabolic functions subsequently inferred by proteomics. In addition, co-regulated proteins of currently unknown function (e.g., EbA329) were suggested to represent a novel efflux system possibly involved in p-ethylphenol-specific solvent stress response and related to other aromatic solvent-induced proteins of *A. aromaticum* EbN1 (56).

5.4. Anaerobic Phenylpropanoid Degradation

Cinnamate and hydrocinnamate (*trans*-3-phenylacrylate and 3-phenylpropionate) as well as their hydroxylated derivatives p-coumarate (4-hydroxycinnamate) and 3-(4-hydroxyphenyl) propionate represent building blocks of lignin and are recently discovered substrates for anaerobic growth of *A. aromaticum* EbN1. 2D DIGE analysis of benzoate- *vs.* cinnamate-, hydrocinnamate-, p-coumarate-, and 3-(4-hydroxyphenyl)propionate-grown cells revealed the specific upregulation of the same set of protein spots (up to 44-fold), most likely involved in the β-oxidation of all four phenylpropanoids to benzoyl-CoA, as supported by metabolite analysis. The coding genes are organized in an operon-like structure (*ebA5316* through *ebA5320*) and could previously not be assigned to defined substrate specificities (Trautwein and Rabus, unpublished).

5.5. Metabolic Network

The genome sequence of *A. aromaticum* EbN1 predicted multiple interconnected pathways for anaerobic aromatic compound degradation controlled by a fine-tuned regulatory network (25). To assess the genome-wide metabolic and regulatory predictions, a comprehensive 2D DIGE-based analysis (285 digital gel images containing in each case 1,047–1,548 detected protein spots) across 22 different substrate and redox conditions was conducted (57). In total, 354 different proteins were identified. The identified members of the "constitutive" subproteome (core proteome) comprised 155 proteins displaying low average abundance fold-changes of ~|1.3|. The "regulated" subproteomes (abundance fold-changes >|2.5|) collectively contained 199 proteins, which mainly represented enzymes of the various degradation pathways. Noteworthy, proteins with other predicted or unknown functions were co-regulated and could therefore play hitherto not conceived roles in the respective pathways or reflect a more general stress response to toxic aromatic substrates (e.g., phenols). In several cases, the genome-predicted pathway involvement of proteins could not be confirmed, e.g., phenylacetaldehyde dehydrogenase (PDH) in anaerobic phenylalanine degradation. Instead, co-regulation of previously not considered proteins shed new light on the respective pathways. For instance, a predicted aldehyde:ferredoxin oxidoreductase together with a ferredoxin-regenerating system were upregulated in phenylalanine-adapted cells and could substitute PDH. Moreover, strong evidence was obtained for thus far unpredicted degradation pathways of three hitherto unknown substrates (e.g., o-aminobenzoate, anoxic).

5.6. Solvent Stress Response

A. aromaticum EbN1 grows anaerobically with the aromatic solvents ethylbenzene, toluene, p-cresol, and phenol, the hydrophobicity of which determines their toxic properties. Anaerobic cultivation of *A. aromaticum* EbN1 at semi-inhibitory (about 50% growth inhibition) concentrations of the four substrates resulted in impaired growth, which was paralleled by decelerated nitrate-nitrite consumption. In addition, ethylbenzene- and toluene-utilizing cultures accumulated poly(3-hydroxybutyrate) (PHB) up to 10% of the cell dry weight. These physiological responses were also reflected by 2D DIGE-resolved protein profiles, e.g., upregulation of PHB granule-associated phasins, cytochrome cd_1 nitrite reductase of denitrification, and several proteins involved in oxidative (e.g., SodB) and general (e.g., ClpB) stress responses. One may speculate that alkylbenzene-derived acetyl-CoA is rerouted from the TCA cycle to PHB synthesis, decreasing the NAD(P)H pool under conditions of impaired denitrification (Fig. 6). PHB, functioning as sink for reducing equivalents, could thus ensure continuous alkylbenzene consumption (58).

6. Conclusion

2D DIGE-based studies have provided new insights into the metabolic capacities of *A. aromaticum* EbN1 that have not been recognized during initial genome annotation, i.e., new substrates, new degradation pathways, and novel efflux system. To date, 2D DIGE-derived protein signatures were generated for more than 50 different substrate, redox, and growth conditions. They provide in-depth insights into regulatory hierarchies and overlaps of individual degradation pathways and general metabolic responses, as well as into what can be considered as the constitutive core proteome of *A. aromaticum* EbN1.

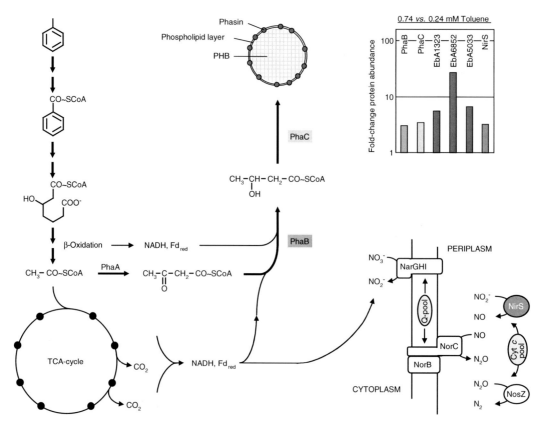

Fig. 6. Solvent stress response of *A. aromaticum* EbN1 during anaerobic growth with semi-inhibitory concentrations of toluene (0.74 mM). Rerouting of acetyl-CoA from TCA cycle to PHB synthesis allows keeping alkylbenzene degradation operative under conditions of impaired denitrification.

References

1. Jørgensen BB, Boetius A (2007) Feast and famine – microbial life in the deep sea bed. Nature Rev Microbiol 5:770–781
2. Whitman WB, Coleman DC, Wiebe WJ (1998) Prokaryotes: the unseen majority. Proc Natl Acad Sci USA 95:6578–6583
3. Amann R, Ludwig W, Schleifer K-H (1995) Phylogenetic identification and in situ detection of individual microbial cells without cultivation. Microbiol Rev 59:143–169
4. Schloss PD, Handelsmann J (2004) Status of the microbial census. Microbiol Mol Biol Rev 68:686–691
5. Fry JC, Parkes RJ, Cragg BA et al (2008) Prokaryotic biodiversity and activity in the deep subseafloor biosphere. FEMS Microbiol Ecol 66:181–196
6. Venter JC, Remington K, Heidelber JF et al (2004) Environmental genome shotgun sequencing of the Sargasso Sea. Science 304:66–74
7. Yooseph S, Sutton G, Rusch DB et al (2007) The Sorcerer II Global Ocean Sampling expedition: expanding the universe of protein families. PLOS Biology 5:0432–0466
8. Dinsdale EA, Edwards RA, Hall D et al (2008) Functional metagenomic profiling of nine biomes. Nature 452:629–632
9. Huse SM, Dethlefsen L, Huber JA et al (2008) Exploring microbial diversity and taxonomy using SSU rRNA hypervariable tag sequencing. PLOS Genetics 4:e1000255
10. Hazen TC, Dubinsky EA, DeSantis TZ et al (2010) Deep-sea oil plume enriches indigenous oil-degrading bacteria. Scienceexpress 10.1126/science.1195979
11. Metzker ML (2010) Sequencing technologies – the next generation. Nature Rev Genet 11:31–46
12. Falkowski PG, Fenchel T, DeLong EF (2008) The microbial engines that drive Earth´s biogeochemical cycles. Science 320:1034–1039

13. Field CB, Behrenfeld MJ, Randerson JT et al (1998) Primary production of the biosphere: integrating terrestrial and oceanic components. Science 281:237–240
14. Dittmar T, Paeng J (2009) A heat-dissolved molecular signature in marine dissolved organic matter. Nature Geosci 2:175–179
15. Jiao N, Herndl GJ, Hansell DA et al (2010) Microbial production of recalcitrant dissolved organic matter: long term carbon storage in the global ocean. Nature Rev Microbiol 8:593–599
16. Jørgensen BB (1982) Mineralization of organic matter in the sea bed – the role of sulphate reduction. Nature 296:643–645
17. Parales RE, Resnick SM (2006) Aromatic ring hydroxylating dioxygenases, p. 287–340. In: Ramos JL Levesque RC (eds.) Pseudomonas, vol. 4., Molecular biology of emerging issues. Springer, New York
18. Vaillancourt FH, Bolin JT, Eltis LD (2006) The ins and outs of ring-cleaving dioxygenases. Crit Rev Biochem Mol Biol 41:241–267
19. Heider J, Fuchs G (1997) Anaerobic metabolism of aromatic compound. Eur J Biochem 243:577–596
20. Widdel F, Rabus R (2001) Anaerobic biodegradation of saturated and aromatic hydrocarbons. Curr Opin Biotechnol 12:259–276
21. Gibson J, Harwood CS (2002) Metabolic diversity in aromatic compound utilization by anaerobic microbes. Annu Rev Microbiol 56:345–369
22. Fuchs G (2008) Anaerobic metabolism of aromatic compounds. Ann NY Acad Sci 1125:82–99
23. Giovannoni S, Stingl U (2007) The importance of culturing bacterioplankton in the "omics" age. Nature Rev Microbiol 5:820–826
24. Strittmatter AW, Liesegang H, Rabus R et al (2009) Genome sequence of *Desulfobacterium autotrophicum* HRM2, a marine sulfate reducer oxidizing organic carbon completely to carbon dioxide. Environ Microbiol 11:1038–1055
25. Rabus R, Kube M, Heider J et al. (2005) The genome sequence of an anaerobic aromatic-degrading denitrifying bacterium, strain EbN1. Arch Microbiol 183:27–36
26. Liolios K, Chen IMA, Mavromatis K et al (2010) The Genomes On Line Database (GOLD) in 2009: status of genomic and metagenomic projects and their associated metadata. Nucleic Acids Res 38:D346–D354
27. Wu D, Hugenholtz P, Mavromatis K et al (2009) A phylogeny-driven genomic encyclopedia of Bacteria and Archaea. Nature 462:1056–1060
28. Hufnagel P, Rabus R (2006) Mass spectrometric identification of proteins in complex post-genomic projects. Soluble proteins of the metabolically versatile, denitrifying '*Aromatoleum*' sp. strain EbN1. J Mol Microbiol Biotechnol 11:53–81
29. Ünlü M, Morgan M, Minden JS (1997) Difference gel electrophoresis: a single gel method for detecting changes in protein extracts. Electrophoresis 18:2071–2077
30. Gade D, Thiermann J, Markowsky D et al (2003) Evaluation of two-dimensional difference gel electrophoresis for protein profiling. Soluble proteins of the marine bacterium *Pirellula* sp. strain 1. J Mol Microbiol Biotechnol 5:240–251
31. Rabus R, Trautwein K (2010) Proteogenomics to study the anaerobic degradation of aromatic compounds and hydrocarbons. In: KN Timmis (ed.), Handbook of Hydrocarbon and Lipid Metabolism. Springer-Verlag Berlin, DOI 10.1007/978-3-540-77587-4_344
32. Görg A, Drews O, Lück C et al (2009) 2DE with IPGs. Electrophoresis 30:122–132
33. Zech H, Echtermeyer C, Wöhlbrand L, Blasius B, Rabus R (2011) Biological versus technical variability in 2D-DIGE experiments with environmental bacteria. Proteomics 11: 3380–3389
34. Fuerst JA (1995) The planctomycetes: emerging models for microbial ecology, evolution and cell biology. Microbiology 141:1493–1506
35. Fuerst JA (2005) Intracellular compartmentation in Planctomycetes. Annu Rev Microbiol 59:299–328
36. Wagner M, Horn M (2006) The *Planctomycetes*, *Verrucomicrobia*, *Chlamydiae* and sister phyla comprise a superphylum with biotechnological and medical relevance. Curr Opin Biotechnol 17:241–249
37. Schlesner H, Rensmann C, Tindall BJ et al (2004) Taxonomic heterogeneity within the *Planctomycetales* as derived by DNA-DNA hybridization, description of *Rhodopirellula baltica* gen. nov., sp. nov., transfer of *Pirellula marina* to the genus *Blastopirellula* gen. nov. as *Blastopirellula marina* comb. nov. and emended description of the genus *Pirellula*. Int J Syst Evol Microbiol 54:1567–1580
38. Glöckner FO, Kube M, Bauer M et al (2004) Complete genome sequence of the marine planctomycete *Pirellula* sp. strain 1. Proc Natl Acad Sci USA 100:8293–8303
39. Rabus R, Gade D, Helbig R et al (2002) Analysis of *N*-acetylglucosamine metabolism in the marine bacterium *Pirellula* sp. strain 1 by a proteomic approach. Proteomics 2:649–655
40. Gade D, Theiss D, Lange D et al (2005) Towards the proteome of the marine bacterium *Rhodopirellula baltica*: mapping the soluble proteins. Proteomics 5:3654–3671
41. Gade D, Gobom J, Rabus R (2005) Proteomic analysis of carbohydrate catabolism and regulation

in the marine bacterium *Rhodopirellula baltica*. Proteomics 5:3672–3683

42. Gade D, Stührmann T, Reinhardt R et al (2005) Proteomic analysis of growth stages and morphotypes in the marine bacterium *Rhodopirellula baltica*. Environ Microbiol 7: 1074–1084

43. Buchan A, González JM, Moran MA (2005) Overview of the marine roseobacter lineage. Appl Environ Microbiol 71:5665–5677

44. Wagner-Döbler I, Biebl H (2006) Environmental biology of the marine Roseobacter lineage. Annu Rev Microbiol 60:255–280

45. Brinkhoff T, Giebel HA, Simon M (2008) Diversity, ecology, and genomics of the *Roseobacter* clade: a short overview. Arch Microbiol 189:531–539

46. Newton RJ, Griffin LE, Bowles KM et al (2010) Genome characteristics of a generalist marine bacterial lineage. ISME J 4:784–798

47. Martens T, Heidorn T, Pukall R et al (2006) Reclassification of *Roseobacter gallaeciensis* Ruiz-Ponte et al. 1998 as *Phaeobacter gallaeciensis* gen. nov., comb. nov., description of *Phaeobacter inhibens* sp. nov., reclassification of *Ruegeria algicola* (Lafay et al. 1995) Uchino et al. 1999 as *Marinovum algicola* gen. nov., comb. nov., and emended descriptions of the genera *Roseobacter*, *Ruegeria* and *Leisingera*. Int J Syst Evol Microbiol 56:1293–304

48. Zech H, Thole S, Schreiber K et al (2009) Growth phase-dependent global protein and metabolite profiles of *Phaeobacter gallaeciensis* strain DSM 17395, a member of the marine *Roseobacter*-clade. Proteomics 9:3677–3697

49. Fürch T, Preusse M, Tomasch J et al (2009) Metabolic fluxes in the central carbon metabolism of *Dinoroseobacter shibae* and *Phaeobacter gallaeciensis*, two members of the marine *Roseobacter* clade. BMC Microbiol 9:209

50. Rabus R, Widdel F (1995) Anaerobic degradation of ethylbenzene and other aromatic hydrocarbons by new denitrifying bacteria. Arch Microbiol 163:96–103

51. Rabus R, Wilkes H, Schramm A et al (1999) Anaerobic utilization of alkylbenzenes and *n*-alkanes from crude oil in an enrichment culture of denitrifying bacteria affiliating with the β-subclass of *Proteobacteria*. Environ Microbiol 1:145–157

52. Champion KM, Zengler K, Rabus R (1999) Anaerobic degradation of ethylbenzene and toluene in denitrifying strain EbN1 proceeds via independent substrate-induced pathways. J Mol Microbiol Biotechnol 1:157–164

53. Rabus R, Kube M, Beck A et al (2002) Genes involved in the anaerobic degradation of ethylbenzene in a denitrifying bacterium, strain EbN1. Arch Microbiol 178:506–516

54. Kube M, Heider J, Hufnagel P et al (2004) Genes involved in the anaerobic degradation of toluene in a denitrifying bacterium, strain EbN1. Arch Microbiol 181:182–184

55. Kühner S, Wöhlbrand L, Hufnagel P et al (2005) Substrate-dependent regulation of anaerobic ethylbenzene and toluene metabolism in a denitrifying bacterium, strain EbN1. J Bacteriol 187:1493–1503

56. Wöhlbrand L, Wilkes H, Halder T et al (2008) Anaerobic degradation of *p*-ethylphenol by "*Aromatoleum aromaticum*" strain EbN1: pathway, involved proteins and regulation. J Bacteriol 190:5699–5709

57. Wöhlbrand L, Kallerhoff B, Lange D et al (2007) Functional proteomic view of metabolic regulation in "*Aromatoleum aromaticum*" strain EbN1. Proteomics 7:2222–2239

58. Trautwein K, Kühner S, Halder T et al (2008) Solvent stress response of the denitrifying strain EbN1. Appl Environ Microbiol 74:2267–2274

Chapter 26

Application of 2D DIGE in Animal Proteomics

Ingrid Miller

Abstract

Two-dimensional electrophoresis (2 DE) is one of the most important proteomic tools and allows studying the complexity of proteomes of different origin. This chapter describes a setup for 2D DIGE with minimal labeling for qualitative and quantitative applications. It relies on homemade gels of medium size and in our hands has been found useful for a wide variety of separation problems involving complex protein mixtures of animal or human origin. The basic method is given for serum proteins of different species, but with minor modifications the method may be easily adapted to other sample materials (other body fluids, cells, tissues), conditions, or size. Examples are given for simple pattern comparisons (e.g., quality control, fast comparison of just two samples) as well as for quantitative applications to larger sample sets.

Key words: 2D DIGE, Animal body fluids, Animal sera, Animal tissues, Homemade IPGs

1. Introduction

Two-dimensional electrophoresis (2 DE) is one of the key methods to study the proteomes of complex biological fluids and tissues. It relies on isoelectric focusing (IEF) under reducing and denaturing conditions in the first dimension and reducing SDS-PAGE in the second. This combination allows seeing single proteins and/or protein subunits, separated according to their isoelectric points and molecular masses and with a very high resolution (1). Especially in its early days, 2 DE has suffered from the fact that each sample is separated on a single gel, and gel-to-gel variation often makes pattern comparison between gels difficult. The new variety of 2D DIGE helps to minimize this problem and thus allows detecting more reliably small concentration changes between samples. Protein detection relies on pre-electrophoretic minimal labeling of the proteins via (some of their) lysine residues by fluorescent CyDyes, which permits scanning the gels right after the electrophoretic run without any additional staining step. Furthermore, in this way up to three different samples may be compared on one gel (2, 3).

In its basic methods, animal proteomics is not different from other proteomic varieties, relying on the same principles and applying the same methods. In most cases, the 2 DE protocol has to be adapted to the sample type rather than to the species. "Species-specific" problems may derive from sample collection or pretreatment (most commercial kits are designed for human sample material), properties of homologous or species-specific proteins, and/or the incompleteness of animal protein databases (for subsequent identification by mass spectrometric analysis).

The way to perform 2 DE described here is—apart from applying modern DIGE technology—a bit "old-fashioned", meaning that it does not rely on commercial precast gels. It is thus more time-consuming, includes more manual steps, and therefore requires some experience. But, on the other hand, it is more flexible, as every detail, particularly pH gradient and size of the IPG strip, but also additives and second dimensional running conditions, can be adapted to one's own needs and thus may include specifications or varieties not commercially offered. The system has served quite well over years and is rather robust, allowing application to all kinds of samples, among them different body fluids, cells, and tissue preparations, mainly of animal origin (e.g., (4–6)). The main sections describe 2D DIGE of serum, with an optional step for MS-compatible silver staining. Tips and comments on other sample material, gel size, or separation systems are also provided.

2. Materials

All solutions are prepared with ultrapure water (corresponding to Milli Q quality) and chemicals of p.a. quality.

2.1. Sample Preparation/DIGE Labeling

1. DIGE labeling buffer: 8 M urea, 4% CHAPS, 30 mM Tris–HCl, pH 8.5; the pH is checked with pH indicator paper (see Note 1). The buffer may be stored in appropriate aliquots at −20°C.
2. Dyes: CyDye DIGE fluor Cy2, Cy3, Cy5 minimal dyes (GE Healthcare, Munich, Germany); stock solutions for the dyes are prepared in anhydrous dimethylformamide (DMF; see Note 2). The tube with the solid dye is briefly spun, and DMF is added to achieve a 1-mM dye solution (e.g., 5 nmol dye/5 μL), mixed vigorously, and centrifuged to collect the solution at the bottom of the tube. As a working solution, a 1:2.5-dilution either with DMF or with DIGE labeling buffer may be used.
3. Stop solution: 10 mM lysine, freshly prepared.

4. IPG sample buffer: Dissolve 100 mg dithiothreitol (DTT), 400 mg CHAPS, and 5.4 g urea in 6 mL of water and add 0.5 mL of alkaline carrier ampholytes (e.g., Pharmalyte pH 8–10.5, GE Healthcare; see Note 3). The buffer may be stored in appropriate aliquots at −20°C.

2.2. IPG Preparation

1. Gels are cast on GEL-FIX for PAGE (260×125 mm; Serva, Heidelberg, Germany) in a casting cassette (consisting of two glass plates of 125×260×0.5 mm, to one of them a rubber U-frame open at the longer side is attached; GE Healthcare). The inner side of the U-frame glass plate is made hydrophobic with Repel Silane (GE Healthcare); the procedure has to be repeated at intervals to avoid any sticking of polyacrylamide.

2. pH gradients are established by Acrylamido buffer pKa 1 (Fluka, Buchs, Switzerland), Immobilines II pK 4.6, 6.2, 7.0, 8.5, 9.3 (GE Healthcare), 1 M acetic acid.

3. T30C4 solution: 28.8% (w/v) acrylamide and 1.2% (w/v) N,N'-methylenebisacrylamide (see Note 4). The T30C4 solution is best stored in aliquots at −20°C.

4. 70% (v/v) glycerol.

5. TEMED.

6. 10% (w/v) ammonium persulfate, freshly made.

7. Gradient maker (15 or 30 mL; Hoefer Scientific Instruments, San Francisco, CA, USA).

8. Magnetic stirrer.

9. Peristaltic pump.

10. Dryer/oven (for heating up to 50°C).

11. Glass tray to hold 1 L of solution.

12. Laboratory shaker.

13. Washing solution: 10 mL of glycerol in 1 L of water.

14. Ventilator or fan (cold air) to dry the gel.

15. Freezer (−20°C).

16. Plastic cover (transparent envelope) for storage.

2.3. First Dimension/IEF

1. Re-swelling solution: 12 g urea, 0.5 g CHAPS, 37.5 mg DTT dissolved in 16 mL of water, and 0.25 mL of ampholytes (see Notes 5 and 6). The solution is stored in aliquots at −20°C.

2. DryStrip Aligner (GE Healthcare).

3. Glass plate.

4. Cover for the strips while swelling (glass tray/container) and a quiet place for the strips to swell at room temperature.

5. n-Decan (Merck) as contact fluid.

6. Electrode strips and sample application pieces (GE Healthcare).
7. Filter paper (Whatman No. 1).
8. Multiphor II (GE Healthcare).
9. Refrigerated cooling bath.
10. Power supply for up to 2,000–3,000 V.
11. Transparent envelope.
12. Stapler.

2.4. Preparation of SDS-PAGE Gels

1. Separation gel buffer: 1.5 M Tris–HCl, pH 8.8 (see Note 7).
2. Stacking gel buffer: 0.5 M Tris–HCl, pH 6.8, and 0.4% SDS (w/v).
3. T30C2.7 solution: 29.2% (w/v) acrylamide and 0.8% (w/v) N,N'-methylenebisacrylamide, stored at 4–6°C.
4. 70% (v/v) glycerol.
5. TEMED, stored at 4–6°C.
6. 10% (w/v) ammonium persulfate (not older than 1 week), stored at 4–6°C.
7. Glycerol-bromophenol blue solution: Mix 10 mL of water with 20 mL of glycerol and 25–50 µL of saturated bromophenol blue solution. This solution is just for filling the void volumes in the casting system and the bottom of the gel cassette; the glycerol used here does not need to be of highest purity.
8. Low-fluorescence glass plates for the SE600 vertical electrophoresis chamber (18 × 16 cm; Hoefer Scientific Instruments).
9. Spacers (1.5 mm thick).
10. 4-gel caster (including additional glass plates, filler sheets; Hoefer Scientific Instruments).
11. Gradient maker (500 mL; Hoefer Scientific Instruments).
12. Magnetic stirrer.
13. Peristaltic pump.
14. Dual gel casting stand (Hoefer Scientific Instruments).
15. Preparative comb (for one strip and two single samples; Hoefer Scientific Instruments).

2.5. Equilibration

1. Equilibration solution (stock solution, for two strips): 2.5 mL of stacking gel buffer (see Subheading 2.4), 9 g urea, 10 mL of 70% (v/v) glycerol, 0.5 g SDS, 5 mL of water; each strip is first equilibrated in 5 mL of the stock solution supplemented with 100 mg DTT then incubated in 5 mL of the stock solution with 125 mg iodoacetamide.
2. Glass tubes with screw caps (20 mL; Schott, Jena, Germany).
3. Laboratory shaker.

2.6. Second Dimension/SDS-PAGE

1. Running buffer: 25 mM Tris, 192 mM glycine, 0.1% SDS.
2. Sample buffer (for additional standards or controls): Stacking gel buffer diluted to 1:4 and made up to 3% (w/v) SDS; include a trace of bromophenol blue.
3. SE600 vertical electrophoresis chamber (Hoefer Scientific Instruments).
4. Refrigerated cooling bath.
5. Power supply for up to 600 V.
6. Agarose: 1% in running buffer (e.g., Indubiose A37). Agarose stock solution is prepared by dissolving solid agarose in buffer on a magnetic stirrer at 100–150°C; then a trace of bromophenol blue is added and the solution stored in aliquots at 4–6°C. For 2 DE, the appropriate number of aliquots is heated at 95°C on a Thermomixer 5436 (Eppendorf, Hamburg, Germany). After complete melting, it is cooled down to 65°C for overlaying the IPG strip after transfer (see Note 8).
7. Standards: molecular weight markers; in the second pocket, either another marker or a reference sample may be used (see Note 9). For application of these samples through the running buffer, use a long form of tips (e.g., Gel-Load; Greiner Bio-One, Kremsmünster, Austria, see Note 10).

2.7. Scanning and Post-detection

1. Typhoon 9400 Variable Mode Imager (GE Healthcare).
2. Software DeCyder Version 5.02 and ImageQuant Version 5.2 (both GE Healthcare), or higher.
3. Optional: MS-compatible silver stain, needing the following solutions (see Note 11):
   (a) Fixing solution: 30% ethanol, 10% acetic acid.
   (b) Sensitizer: Dissolve 0.5 g sodium thiosulfate ($Na_2S_2O_3 \cdot 5H_2O$) and 17 g sodium acetate ($CH_3COONa \cdot 3H_2O$) in 175 mL of water, add 75 mL of ethanol.
   (c) Silver nitrate: 0.4 g silver nitrate/200 mL.
   (d) Developer: Dissolve 18.75 g anhydrous sodium carbonate (see Note 12) in 750 mL of water and add 75 µL of 37% formalin.
   (e) Stop solution: 1% glycine.

3. Methods

3.1. Sample Preparation/DIGE Labeling

1. Setup of the experiment: For qualitative DIGE, any three samples can be labeled with the three different dyes. For quantitative applications, a pool of all samples of the respective sample set

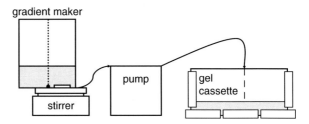

Fig. 1. Setup to cast gradient gels for IPG gel plates.

is made and used as an internal standard throughout the whole study (labeled with Cy2). In the latter case, samples are only labeled with Cy3 and Cy5. Dye swapping (reverse labeling, i.e., labeling aliquots of the same specimen with different dyes) is recommended to avoid bias from preferential labeling.

2. Serum is diluted 1:6 with DIGE labeling buffer and precooled in an ice bath (see Notes 13 and 14); 2.5 µL of this pre-diluted sample is mixed with 0.5 µL of the CyDye working solution (see Note 15) and incubated in the ice bath for 30 min in the dark. The reaction is stopped by addition of 0.5 µL of the lysine solution and vigorous mixing. After further 10 min in the ice bath under light protection, 7 µL of the IPG sample solution is added. The sample is mixed and after 10 min cooling (in the ice bath or fridge) ready for immediate use. Alternatively, it can then be frozen and used at a later time-point (see Notes 16–19).

3.2. IPG Preparation

1. The supporting film is applied to the glass plate of the casting cassette with the help of a roller (see Note 20). The carefully cleaned glass plate with the rubber U-frame is wiped with a paper towel moistened with some milliliters of Repel Silane, left to dry, and afterward rinsed with plenty of water. After drying it, it is put on the glass plate with the film and the casting cassette is fully assembled with the clamps (see Note 21).

2. The casting setup is arranged as shown in Fig. 1: The gradient maker with a small magnetic bar in the front chamber is put on the magnet stirrer and its outlet connected to the peristaltic pump (see Note 22). The tubing outlet of the pump is put into the middle notch of the assembled casting cassette (e.g., via a butterfly needle or pipette tip).

3. The two casting solutions are prepared in glass vials or tubes according to the composition given in Table 1. After addition of the polymerizing agents, they are immediately filled into the respective compartments of the gradient maker. The alkaline solution is put into the back compartment. The valve between back and front chamber is carefully filled with this solution avoiding trapping air bubbles that would hamper liquid flow

Table 1
Composition of IPG gel for a nonlinear pH 4–10 gradient ($T = 4.0\%$, $C = 4.0\%$; see Notes 25 and 26)

Solution	Acidic (µL)	Basic (µL)
Acrylamido buffer pKa 1	193	24
Immobiline II pK 4.6	193	–
Immobiline II pK 6.2	228	69
Immobiline II pK 7.0	–	91
Immobiline II pK 8.5	–	34
Immobiline II pK 9.3	–	64
1 M Acetic acid	–	21
T30C4	1,015	1,015
70% Glycerol	1,500	–
Water	4,471	6,282
TEMED	4.6	4.0
10% Ammonium persulfate	40	40

later on; it is closed again immediately afterward. Then, the acidic solution is added in the front (mixing) compartment. The connecting valve is opened, and at the same time, the pump (at low speed) and the magnet stirrer are started. The cassette is thus slowly filled with gel solution (see Note 23). Take care to adjust/lower mixing speed when the solution volumes decrease. At the very end of the gradient when the solutions run out, incline the gradient maker to get all remaining liquid droplets into the pump to achieve the complete desired pH range (see Note 24). When the entire gel solution is filled into the cassette, disconnect the cassette and wash the gradient maker, pump, and tubing immediately with plenty of water. The cassette with the gelling solution is put into the—still cold—dryer/oven and the gradient is allowed to settle for 5–10 min. Overlaying the completed gelling solution with water is optional. Although it helps forming a good meniscus and keeping off oxygen, there is the risk of mixing the two layers, resulting in a "diluted" and softer gel at the cathodic end.

4. After the short settling period, the oven is turned on to 50°C and the gel allowed to polymerize for 1 h. Thereafter, the gel cassette is removed from the oven and allowed to cool to room temperature. Optionally, this can be followed by shortly putting

the gel assembly in a cold room to facilitate disassembling. In either case, the clamps are then removed, and a long thin blade is introduced carefully below the supporting sheet alongside the edges of the rubber frame. If the assembly does not split immediately, the blade may be turned slightly. The gel on the supporting film should then separate from the glass plate. It is put immediately into a large glass tray and washed with dilute glycerol, under constant and very slow shaking.

5. After 1 h the IPG gel is removed from the washing solution, put on a dry glass plate (backing downwards), fixed at the sides with some bulldog clamps, turned to a vertical position and dried under a cold stream of air with a ventilator or fan. This usually takes 1–1.5 h (see Notes 27 and 28).

6. Dried gels are stored in a transparent envelope and a plastic bag at −20°C (see Note 29).

3.3. First Dimension/IEF

1. The IPG plate is taken out of the freezer and out of the plastic bag. At the anodal side, there is a 5-mm margin of plastic support; on the cathodal side, the margin is wider. Strips are cut over the complete width of the plastic backing. Use an overhead marker to mark 5-mm-wide strips on the back of the plastic cover and then cut along the line, as straight as possible (see Note 30). Use normal, good scissors and cut the gel when it is still in the transparent envelope. This will protect the sensitive gel side from damage and fingerprints. Cut as many strips as you plan to include in your experiment. If this is a new gel, discard the first strip (the quality is not good enough: Usually, edges are not nicely polymerized and the iso-pH gradient is not straight).

2. The appropriate volume of re-swelling solution is thawed at room temperature (315 μL per strip). Put a plastic DryStrip Aligner on a glass plate of equal size (see Note 31). Distribute 315 μL of re-swelling solution in one lane, making the line of liquid approximately as long as the IPG strip. Put the strip gel side down onto the liquid, trying to avoid trapping of air bubbles. Air bubbles will mean "holes" in the re-swollen strip and disturb the electric field. Proceed in a similar way with all your strips, taking care not to spill the solution between the lanes (you may leave every second lane empty). Having this completed, cover everything with a glass tray or another container to prevent excessive drying of the incubating strips. Do not move the container for the first 15 min (this is the time needed for the strips to take up most of the liquid). Afterward, you can shift the whole assembly to a quiet place where gel strips can swell undisturbed for the next 6–7 h.

3. Preparing the samples: Thaw the labeled samples, spin them briefly, and mix the ones you want to run on each gel strip. If necessary, spin again.

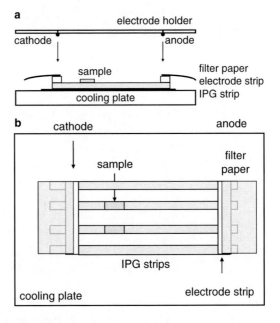

Fig. 2. Schematic drawing showing how IPG strips are arranged on the cooling plate of a horizontal electrophoresis chamber (e.g., a Multiphor) for running the first dimension separation step. Sectional view (**a**) and top view (**b**).

4. Preparing the run: Dismantle the IPG strip assembly and put the rehydrated strips with the gel side up onto a wet piece of kitchen roll. Work with gloves and touch the strips only at the cathodal end of the support. Apply a thin line of contact fluid on the cooling plate of the Multiphor and place a strip on it. When picking up the strips from the wet paper, gently wipe the back of the strips. Proceed with the other strips and align them side by side, about 7–10 mm apart. Wet a piece of filter paper large enough to cover all strips, blot it between two pieces of kitchen paper and then carefully place it on the strips. Peel it off slowly—it has taken up any excess of liquid and this will prevent formation of urea crystals during the run. Cut two pieces of electrode strips, wet them with water (carefully blot excess water with kitchen paper), and place them at both ends of the gel strips, i.e., the anodic and cathodic sides. Put the sample application pieces at the appropriate positions on your strips and apply your samples. Each piece will hold 20 μL of liquid; you may stack 2–3 pieces. Cut two pieces of filter paper as wide as the electrode strips and about 3–4 cm long; put them on the electrode strips (pointing outwards); they will collect excessive water during the run. The completed assembly is shown in Fig. 2. Adjust the position of the electrodes on the electrode holder, put them on the gel, and close the chamber. Turn on the cooling circulation (15°C), connect the cables, and start the run (see Note 32).

5. Running conditions: Usually, strips are run with the following program consisting of a series of voltage steps: 1 h at 100 V, 1 h at 200 V, 9.8 h at 500 V, 5 h at 1,000 V, and 1.5 h at 2,000 V (up to 13.2 kVh). Then the run is terminated, the chamber opened, and the sample application pieces removed. The strips are inspected: They should be more or less dry, with slight "wrinkles" (corresponding to the iso-pH-lines generated by the ampholytes). In serum, CyDye labeled albumin will be visible as colored multiple bands. The electrodes are replaced and the run is continued (1 h at 200 V, 1 h at 500 V, 0.5 h at 1,000 V, and 0.25 h 2,000 V, i.e., to 14.9 kVh in total). This focusing program is slow, but has the advantage that it can cope with higher amounts of salts or considerable protein load (see Notes 33 and 34).

6. Immediately at the end of the run, the strips are transferred into a transparent cover (ends fixed with staples), labeled, and frozen at –20°C (see Note 35). They may be stored for days/weeks. Alternatively, they may be directly prepared for the second dimension (see Subheading 3.5) if the SDS-PAGE gels have already been prepared (see Subheading 3.4).

3.4. Preparation of SDS-PAGE Gels

1. Gels are cast between two low-fluorescence glass plates, with 1.5-mm spacers. Up to three of these assemblies fit into the gel-casting cassette (the name 4-gel caster given by the manufacturer is misleading for this gel thickness; with 1.5-mm spacers, three gels are the limit); the rest is filled with glass plates, filler sheets, or acrylic spacer plates (see Note 36).

2. The separation gel is cast as a gradient gel with a similar setup as described in Subheading 3.2 but with a larger gradient maker and a higher pumping speed. In this case, the casting cassette is better filled from the bottom. First, the tubing is filled with water, without trapping air bubbles, and additional 6–8 mL of water are put into the cassette, to achieve an even and flat gel surface. Then the gradient is started between the light and the dense gel solution. The composition of the two gel solutions for casting 10–15%T-gradient gels is listed in Table 2 (see Note 37). Be careful to exclude any air bubbles between the two reservoirs of the gradient maker and to adjust the stirring speed during casting. After the last drops of the gel solutions are pumped into the cassette (slightly tilt the gradient former for this, but avoid introducing air), immediately continue with the glycerol-bromophenol blue solution. When the appropriate height of the gels is reached (about 5 mm below the upper edge of the glass plate), the pump is stopped, and the gel left overnight at room temperature for polymerization (see Note 38). Next morning, the cassette can be opened and the individual gel assemblies either processed immediately or stored at 4–6°C (see Note 39).

Table 2
Gel solutions to cast three gradient gels ($T=10–15\%$, $C=2.7\%$) with the equipment described in Subheading 2.4

Gel composition

	10%	15%
T30C2.7	18.711 mL	28.083 mL
Separation gel buffer	14.355 mL	14.355 mL
Water	23.1 mL	–
70% Glycerol	–	13.728 mL
10% Ammonium persulfate	122 µL	122 µL
TEMED	16.5 µL	16.5 µL

Table 3
Composition of stacking gel ($T=5\%$, $C=2.7\%$)

Per gel

T30C2.7	0.683 mL
Stacking gel buffer	1.035 mL
Water	2.427 mL
10% Ammonium persulfate	40 µL
TEMED	5.5 µL

3. Fix the separation gel assembly with the long clamps and put it upright in the dual gel casting stand. The polymerized gel should be about 2–3 cm below the upper edge of the glass plates. Decant any water on top and fill the residual space almost completely with stacking gel solution (for composition see Table 3). Insert the comb and leave to polymerize for about 30 min. The gel is then ready for inserting the equilibrated IPG strip (see Subheadings 3.5 and 3.6).

3.5. Equilibration

Equilibration solutions are freshly prepared and aliquots transferred into the glass tubes. The IPG strips are taken out of the freezer, the protruding anodal part of the support film is cut off, and the cathodal one is only shortened so that it can still easily be held with forceps. When half-thawed, the protecting transparent covers are peeled off, each strip put into one glass tube with equilibration solution 1 and put on a horizontal shaker at 100 rpm for 10 min (see Note 40). Afterward, strips are transferred to a fresh tube with equilibration solution 2. After 5 min horizontal shaking, the strips are recovered and immediately transferred to the stacking gel of the second dimensional gel.

3.6. Second Dimension/SDS-PAGE

Take the equilibrated IPG strip out of the glass tube, thereby holding the cathodal plastic with forceps (see Note 41). Place the strip into the trough of the stacking gel with the anode to the left and the plastic backing of the strip toward the back glass plate of the gel assembly. Make sure that the backing is sticking to this glass plate and that there is some room between the gel side of the strip and the front glass plate. This space is now filled with warm agarose (about 700 µL per gel), which helps to keep the strip in place. Carefully avoid bubbles as they would prevent an even current. Fill the rest of the gel cassette with running buffer, fix the upper cathode chamber onto the two gels (or one gel and the place holder), and transfer the completed assembly into the anodal buffer tank. Fill the cathodic compartment with running buffer, making sure that the chamber does not leak (see Note 42). Start the cooling circulation (set to 15°C). Apply standards and references in the left and right pockets of the stacking gel, respectively, and start the run (25 mA constant current per gel). It will take about 4–5 h until the bromophenol blue reaches the bottom of the separation gel.

3.7. Scanning and Post-detection

1. Switch on the scanner about 20 min before the run is finished. When the bromophenol blue has reached the end of the gel, stop electrophoresis, remove the gels (gels and glass plates) from the tank, and rinse the assembly carefully with warm tap water. Then dry them and put the gels (still between the glass plates) in the scanner. The scanner is operated according to the manufacturer's protocol. Scanner settings for a first measurement should be around 450 V PMT for each channel and 100-µm pixel size (for a first pre-scan you may select a lower resolution setting, i.e., larger pixel size). Make sure to select the appropriate focal plane ("+3 mm"). The albumin spot should not be saturated, otherwise re-scan (see Notes 43 and 44).

2. Further processing depends on the experiment:

    (a) Qualitative analysis: The three channels of the false-color representation of the original scans (red, green, blue) can be overlaid and—depending on the protein distribution—result in mixed colors (for a complete overlap of spots and equal amounts of proteins, this will be white). The images may be looked at singly or in combination (in false colors), adjusted in intensity, and also converted to gray scales.

    (b) Semiquantitative analysis: This may be done with the software module DIA (Differential In-gel Analysis) in DeCyder, which allows pairwise comparison of two channels. Spots can be detected, filtered (according to size, intensity, etc.), and then compared by using pre-fixed factors (e.g., 2.0; 1.5; 1.2-fold). This gives a good first overview for comparison of just a few samples. Comparability of gels/sample patterns is limited. Even when used in only one gel, it is no true and reliable quantification, as it does not take into account the sometimes occurring preferential labeling (7) or quenching of components in the gel.

(c) Quantitative analysis: This can only be achieved with including an internal standard labeled with Cy2, and using appropriate software (e.g., DeCyder, module BVA, standing for "Biological Variation Analysis"). Details of operation can be found in the manual. This type of analysis results in lists of differentially regulated spots with low variation within the respective group. Statistics features of DeCyder may be used or data exported for use in other software.

3. Any of the previous steps (2a–c) may raise the wish to find out more about one or several of the separated spots. One way to do this is by staining the protein pattern in the gel with a visible, but MS-compatible dye, followed by cutting out the respective spot/gel plug, digesting the proteins to peptides and subjecting this mixture to mass spectrometric analysis for identification (see Note 45). We use an MS-compatible silver-staining protocol to visualize the protein pattern (see next step).

4. Optional MS-compatible silver staining: Gel cassettes are opened after the scanning, and gels are transferred to the fixing solution. After initial shaking for about 1 h, the gels are left at room temperature until the next day (see Note 46). The further incubations are performed on a laboratory shaker: 1 h in the sensitizer solution, 3×20 min in water, and 30 min in silver nitrate solution. After a 1-min wash with water, the gel is shortly rinsed with developer and the protein pattern is developed in two baths of developer (approximately 5 min each, depending on the protein load and the intensity of the pattern). The staining is stopped by a glycine solution (20 min). After three 10-min washes in water, the gel is ready for spot cutting and further processing for MS analysis (see Notes 47–49).

3.8. Modifications of the Method and Applications

3.8.1. Modifications

1. Besides serum, the described method is applicable to various types of samples:

    – Body fluids with lower protein content than serum/plasma may be lyophilized, followed by dissolving in DIGE labeling buffer. For high salt-to-protein ratios desalting by gel filtration, dialysis or TCA precipitation is recommended (8).

    – Cells from cell culture should be serum-free, i.e., washed with isotonic solutions and then centrifuged (see Note 50).

The pellet may be either directly dissolved in DIGE labeling buffer or lyophilized first (see Notes 51 and 52). Try to work out the best cell count-to-buffer volume ratio for your type of specimen.

- Tissues/organs: Depending on its rigidity, the minced tissues may either be dissolved in DIGE labeling buffer, or homogenized in it (e.g., using a Potter-Elvehjem homogenizer or manually, using a sample grinding kit; GE Healthcare; see Notes 53 and 54). Ultrasonication as well as freezing/thawing support solubilization.

- For all samples with variable protein content, it is necessary to determine the protein concentration. The easiest way to do this is a Coomassie G-based photometric protein determination according to Bradford (9), e.g., optimized for microplate readers. Samples may be diluted with water to avoid quenching through additives such as urea and detergents in the assay (see Note 55).

2. Sample application: Sample application as described here, i.e., cup loading, allows applying the sample at a chosen position—away from the bulk of proteins in your sample or at a pH most beneficial (or least degrading) for your proteins. However, for DIGE application, this means that a sample with a high protein and a considerably low salt concentration is needed. Body fluids other than serum may be lyophilized and dissolved in a small volume of DIGE labeling buffer. Cellular proteins, especially with hydrophobic properties, are often soluble only to a small extent and in low concentrations and, thus, in-gel rehydration (active or passive, i.e., with or without current) seems advisable (see Note 56).

3. Alkaline proteins need special care (and equipment): running under oil to protect from oxidation, additional reducing agent, and usually shorter runs and anodic sample application (for details see (10)). When using alkaline pH gradients in the IPG strips, lower dye-to-protein ratios may be chosen to avoid staining artifacts (over-labeling).

4. There are several modifications of 2DE, e.g., without reduction or using different additives (for a review see (10)). In addition, even when using a standard 2DE system, pH range and gel pore size (in SDS-PAGE) may be adapted to be able to either get a good general overview or to zoom into different parts of the proteome (11–13). For homemade and hand-cut IPG strips, the strip width may be varied in order to increase sample volume.

5. A larger gel system (Ettan-DALT) is described in detail in (8) including commercial IPG strips of 24 cm-length (pH 4–7 and 6–9) with 7 M urea and 2 M thiourea and large size homogenous SDS-PAGE in the second dimension. As for sample preparation, TCA precipitation is described for dilute protein samples.

3.8.2. Applications

The concept of 2D DIGE offers new possibilities for 2DE applications, especially regarding sensitivity and reproducibility. It needs only small amounts of samples, is sensitive, and measured signals are linear over several orders of magnitude. For detection and quantification, no additional staining step is necessary. Minute differences in pI and molecular mass between samples may be measured, and with inclusion of an internal standard, also smaller variations in spot intensities (correlating to protein concentration) may reliably be determined compared to conventional stains. Besides all these major advantages, a few points should be considered: As with all other methods, protocols need to be carefully followed (e.g., controlling the pH in the labeling step), and as the "stain" is based on a different principle (fluorescent dye labeling of lysines), results are not always comparable with other detection methods. Labeling properties of the three different fluorophores and possibly interfering compounds present in the sample need to be considered. Specific equipment and fluorophores are necessary, making the method expensive. For post-electrophoretic protein identification, in most cases additional colorimetric staining or additional equipment (e.g., robotic spot picker) is required.

In the following, some examples illustrate diverse applications of 2D DIGE in the field of animal proteomics:

(a) Qualitative DIGE: This may be used for simple and rapid comparison of two or three samples, for instance for detection of contaminations. All samples are separated on one gel, and differences are seen at a glance without the need for spot matching. Figure 3 shows such an example, a check of sample preparation: Lysates from primary murine bone marrow macrophages that have been kept for a few days in cell culture medium containing fetal calf serum (FCS) were compared to the pattern of (dilute) FCS and of (dilute) mouse serum. Although the cells had been carefully washed before lysis, traces of highly abundant FCS proteins could be detected (14). Their protein spots are seen in violet in the image, resulting from the color mix of red macrophage and blue FCS protein spots. No mouse serum proteins were found in traceable amounts (green spots). Detection of residual proteinaceous medium supplements in specimens derived from cell culture at an early stage of the experiment may help to optimize sample preparation and/or to prevent putting these spots on the list of candidates for further investigation.

(b) Semiquantitative DIGE: Two to three samples may be compared in a rapid and simple way, even semiquantitatively (if dye-bias has been excluded, see also Subheading 3.7, step 2). An example for isoform detection is given in Fig. 4, showing a close-up of a gel with different horse serum samples. Spots in the encircled area belong to α_1-antitrypsin (also called Pi). This protein is

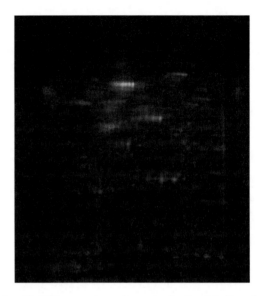

Fig. 3. Qualitative 2D DIGE comparison of samples for the detection of contaminations using IPG pH 4-10NL for the first dimension and SDS-PAGE (10–15% T) for the second. *Red* spots show the protein pattern of a lysate from peritoneal murine macrophages (25 μg protein, 4 nmol Cy2 per mg protein), *green* ones belong to mouse serum (protein amount corresponding to 0.1 μL of serum; Cy5). *Pure blue* spots derive from FCS (protein amount corresponding to 0.1 μL of serum; Cy3), but appear *violet* due to the overlap with *red* spots from the macrophage lysate.

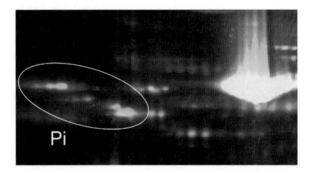

Fig. 4. Qualitative or semiquantitative 2D DIGE comparison of samples for the mapping of genetic polymorphism. Two different horse serum samples were labeled with Cy3 and Cy5, respectively, using IPG pH 4-10NL for the first dimension and SDS-PAGE (10–15% T) for second. A "zoom-in" of the gel image is shown with the protein "Pi" (α_1-antitrypsin) marked with an ellipse.

known to exist in numerous phenotypes, giving complex electrophoretic patterns. In "pre-genomics time" this feature was used in different electrophoretic setups for parentage testing and breeding control. When separating samples with different isoform patterns on conventional and separate 2DE gels, exact assignment of spot positions can be difficult.

(c) Quantitative DIGE: These applications require an internal standard, dye swapping, and a sufficient number of biological replicates. We have used quantitative 2D DIGE in two more extensive studies involving rodent animal models. Macrophages from wild-type and Tyk2-deficient mice were compared in their protein patterns before and after treatment with lipopolysaccharide (LPS). Five of the differentially regulated proteins with interesting protein expression patterns in the genotypes were further investigated at the mRNA as well as the protein level (immunoblots). Data showed that the lack of a single protein brought about changes in proteins from different functional categories, suggesting involvement of this tyrosine kinase in various cellular processes (8, 15). In a rat model of endotoxic shock, the influence of LPS on the proteome of different organelles was studied and changes of the protein levels correlated to functional, genomic, and histological findings. Organelles are affected to a different degree, but first results suggest major impairment of endoplasmic reticulum, based also on changed expression levels of proteins with functions in protein folding, transport, and detoxification (6, 16). Conventional 2DE would not have been sufficiently sensitive to detect all the DIGE-detected candidates and their regulations, in both studies.

4. Notes

1. In practice, 200 µL of separation gel buffer (1.5 M Tris–HCl pH 8.8) per 10 mL of final solution volume is used. This gives almost the correct pH, but should be fine-adjusted if necessary.

2. DMF can be kept in a desiccator to prevent water uptake and extend shelf-life.

3. Carrier ampholytes are highly viscous solutions. Take care to pipette slowly and to use pipette cones, which are not too small; you may also cut tips to the right size with sharp scissors.

4. Acrylamide monomers are neurotoxic. They should be handled with care (gloves; use fume cabinet if you prepare solutions from solid chemicals). Polyacrylamide is only toxic if polymerization is incomplete and the gel still contains monomers.

5. Carrier ampholytes from a different source may be used, either broad-range (pH 4–10) alone or spiked/supported with narrow-range ampholytes (e.g., pH 4–6). This largely depends on the brand used and on the sample.

6. A trace of bromophenol blue may be added to the re-swelling solution. The migration of this dye toward the anode can then be checked during the IPG run.

7. It is not necessary to have SDS in the separation gel if you have an SDS-containing stacking gel.

8. Temperatures given are best to handle Indubiose A37. Other types of agarose may need different settings. The lowest possible temperature should be selected which still ensures good handling without solidifying the agarose too early.

9. It is good to have the standards and reference samples ready for use in aliquots in the freezers so that you only need to thaw and centrifuge the tubes. For the described silver staining protocol (see Subheadings 2.7 and 3.7), the following amounts are recommended per gel: Take 5 µL of 1:20-diluted molecular weight markers (e.g., LMW, GE Healthcare) or 5 µL of 1:100-diluted serum, respectively, add 5 µL of sample buffer (with freshly added DTT to 20 mg/mL), heat to 95°C for 5 min, cool by centrifugation for 5 min at $16,000 \times g$ (e.g., on a Centrifuge 5415C, Eppendorf), add 5 µL of 70% (v/v) glycerol and mix.

10. Alternatively, standards may be applied on filter application pieces which are put directly on the flat surface of the stacking gel, side by side to the IPG strip. Take care that no proteins leak out of the filter and into the space reserved for the IPG strip, thus contaminating the sample.

11. Staining is best performed in glass vessels. Except for the developer, which is needed for more than one bath, 200–250 mL of each solution are enough for this gel size.

12. Good quality of the sodium carbonate is essential to keep background staining low.

13. Styrofoam boxes with lids serve well for ice baths and keeping samples in the dark.

14. The pH of the protein sample dissolved or diluted in labeling buffer should be checked, at least for each sample type and/or condition and, if necessary, readjusted. This is of special importance if samples have been TCA-precipitated.

15. Before starting the incubation, make sure that the components are thoroughly mixed and that all liquid is at the bottom of the tubes. Small droplets of sample or dye elsewhere on the walls of the tubes are brought down by spinning the samples briefly at high speed. Pipetting of such small volumes is not easy. Wiping the tips on the wall of the dye tube helps to have more uniform dye amounts in the samples. This ensures comparable sample labeling and facilitates the selection of optimal scanning conditions for each gel. There are special pipettes which allow measuring such small volumes, e.g., Research 0.1–2.5 µL (Eppendorf).

16. Labeled single samples are kept apart and are combined only shortly before sample application, to avoid cross-reactions.

17. We have made good experiences with the stability of the DMF-dissolved CyDyes, even beyond the recommended shelf-life, provided that the dyes are always properly stored and only removed from the freezer for short intervals. Take special care to always use fresh pipette tips in order to avoid contaminations. When using larger packages, aliquot the dissolved dyes and use them one by one.

18. For larger sample volumes, the volumes of the stop solution and the IPG sample buffer are increased accordingly. However, make sure that overall sample volumes are not getting too large (after mixing up to three single samples), as this protocol describes point application via sample application pieces on the IPG strips.

19. The ratio for dye-to-protein labeling recommended by the manufacturer is 200 pmol dye/25 μg protein. However, this is minimal labeling, and works also with lower amounts of dyes. If DIGE serves only for analytical purposes, both dye and protein amounts may be cut down, as seen for instance in Fig. 3.

20. The supporting film should stick firmly to the back of the casting cassette. Too much water, large air bubbles, or small particles trapped below the film make it camber, thus affecting gel thickness.

21. If smaller gels (fewer IPG strips) are needed, a casting cassette of the same dimensions but with a U-frame open at the shorter side ("portrait format") is commercially available (e.g., from GE Healthcare).

22. Gradients can also be poured without a peristaltic pump, just by gravity. Liquid flow is then controlled by differences in height (putting gradient maker and magnetic stirrer well above the casting cassette), as shown in Fig. 4 of (17). The once established setup has to be used consistently to ensure reproducibility in casting.

23. Avoid getting bubbles into your gel cassettes, as they will form holes in the gel. However, if they happen, small bubbles may be removed by gentle tapping against the front glass plate, by tapping the whole cassette onto the table or by carefully tilting the cassette. But mind that all these measures will also disturb your gradient.

24. To ensure having the complete pH range without disturbance of the gradient especially in the alkaline range, you may include pH plateaus in the front and at the end: For instance, keep 0.5 mL of the acidic and the alkaline gel solution and make the gradient only from the rest; the acidic solution is pipetted into the cassette before, the alkaline after the gradient.

25. The selection of acrylamido buffers or Immobilines II depends on the pH gradient used and may vary. For detailed protocols, see (17).

26. Always use freshly prepared ammonium persulfate solution to ensure good and reproducible polymerization. Depending on the brand and the age of your solid chemical you might need to adjust the amount to achieve good polymerization in appropriate time.

27. Do not prolong the washing time of the IPG gel. It tends to collect even more water then and the gel layer may peel off the support. When the gel is not well polymerized, it will also tend to detach from the film, even during a 1 h-wash. Those gels are hardly ever to be rescued (except if only the very edge is affected); then it is better to cast a new gel. Also avoid any mechanical damage of the gel and take care that gel drying takes place in a dust-free location.

28. Drying the gels should quickly take place in a stream of cold air. Drying at room temperature without the help of an air stream takes too long and affects the performance of the gel.

29. Shelf-life for homemade IPG plates is several months. Upon aging water exudation ("sweating") of the strips may start, getting worse with prolonged storage. Although some water collecting at the surface of IPG strips is not uncommon in highly loaded gels, larger amounts cannot be tolerated as proteins collect there, too, and are lost for the analysis. In addition, if the water is not removed, it tends to blur the focused zones of the strip.

30. Do not worry if the hand-cut strips are not completely straight or slightly "crooked". The important thing is that the gradient in the gels is well-cast. However, not so nicely cut strips have to be embedded in agarose with some more care, especially avoiding air bubbles (see Subheading 3.6).

31. Alternatively, strips may be swollen in an upright standing re-swelling cassette in an excess of re-swelling solution. The assembly is similar to the described gel casting in Subheading 3.2, step 1 (see also Note 21). In addition to the rubber U-frame attached to the glass plate, a U-frame of the same dimensions is cut from a double-layer of Parafilm. This compensates for the thickness of the gel support film and allows swelling of the strips to original thickness. Rehydration in the cassette needs much more re-swelling solution, but is recommended when strips are rehydrated overnight, as there is less risk of drying out and formation of urea crystals.

32. Depending on the geometry of the electrophoresis chamber a drying-out phenomenon may occur, even urea crystals may form on the edges or the surface of the strips. This can partly be overcome by placing a wet piece of filter paper (size approximately 3–4 × 2 cm) on either side of the strips and/or by including "blank" strips left and right to the sample-loaded strips (Fig. 2b).

33. Running conditions may be adjusted to the separation problem. The run can be further slowed down by prolonging the 100- and 200 V-steps (e.g., if the sample is known to contain high amounts of salt). It may also be speeded up by shortening or omitting these steps, especially in the second run. The idea of the second run is to refocus the bands below the sample applications pieces. Alternatively, the 2,000 V-step of the first run may be prolonged to achieve the recommended 14.9 kV and the second run canceled. Including an additional low-voltage step at the end of the run may allow a more flexible timetable while keeping bands well focused.

34. In case of samples with high salt content or when focusing in narrow pH gradients it may be advisable to change electrode strips and/or filter paper. This removes components that may disturb the run, especially when switching to higher voltages.

35. Sometimes IPG strips tend to collect water on their surface (either upon aging of the strips or due to high protein load). Partly, this phenomenon can be handled by shortly blotting the strip surface with filter paper (careful—do not contaminate the strips—the soft gel is quite sticky!) or by turning the strips over-edge before putting them into the transparent cover for storage.

36. The gel casting cassette has to be filled the same way every time because a more loose or tight packing will influence the length of the separation gel.

37. Pore size of the gels has to be selected depending on the protein composition of the sample. Besides gradient gels, homogeneous gels may be used. They can be cast directly in the casting cassette (closing the bottom outlet and using the rubber wedge) or singly in the dual casting stand. For these it is recommended to overlay the gel surface with water (each gel separately with about 1 mL, carefully overlaying the more dense gel solution).

38. Meanwhile fill the empty gradient maker with water, opening the connection between the reservoirs. Cover the casting cassette with Parafilm to avoid evaporation of the water layer, but not earlier than 1 h after casting. The gel will have settled then and begun to polymerize. If you are short of time, leave it uncovered until the next day.

39. Separation gels can be stored in plastic bags in a cold room for several days. Pour the stacking gel right before use and use the gel immediately. Otherwise, due to diffusion, the two buffer systems will mix and affect the quality of your separation.

40. 20-mL glass tubes with screw caps are usually tight enough to be horizontally shaken on a laboratory shaker. Take care to insert the strips in a standardized way, e.g., always with the anodic side first. They can then be easily taken with forceps at

the protruding film of the cathodic side. The strips should float in the solution and not stick to the glass tube.

41. If the plastic support protrudes less than 5 mm, the strip should fit nicely into the trough in the stacking gel. Otherwise you need to shorten it.

42. The anodic running buffer may be used up to 5–10 times, the cathodic buffer only once.

43. Saturation of spots is easily noticed during scanning, as these spots appear red. When the run is completed and the software ImageQuant opens immediately, the different channels of the scan may be split (click on the appropriate buttons) and the cursor moved slowly over the albumin spot. If "100004.25 Counts" appears, the spot is saturated, and for the next scan less volts PMT should be chosen. If on the other hand the value is below 40,000–50,000, PMT settings should be increased.

44. If gels cannot be scanned immediately or if you want to re-scan them at a later time, storage in fixing solution is possible. Although without the support of the glass plates the shape of the gels will change slightly (due to their swelling in the solution), fluorescence is preserved for several days/weeks at least for qualitative purposes and when gels are kept in a (dark) cold room.

45. If you intend to include mass spectrometric identification of protein spots in your workflow, you have to avoid introducing contaminant proteins (e.g., keratins). This means special emphasis on purity of all labware and solutions, wearing gloves, etc.

46. Fixing solution is also a good choice for storing 2DE gels, for instance while waiting for the results of your quantification. In this solution, gels may be kept for days, even weeks. However, there may be changes in the patterns, e.g., loss of small molecules for which the fixing solution is not "strong" enough to denature and immobilize them in the gel. This will need to be checked from case to case.

47. Solutions containing silver ions should be collected for separate disposal. This can best be done by combining them with the first wash of developer. In this alkaline solution, silver precipitates as hydroxide and may be collected in a concentrated form.

48. It is not always possible to find DIGE-detected spots in silver-stained gels, especially spots of smaller proteins. This is due to the different staining properties of the dyes/stains used (7).

49. Silver-stained gels can be kept at 4–6°C in water for days/weeks before cutting spots for MS analysis. However, usually best results are obtained with "fresh" spots.

50. If your cells are adherent, do not "trypsinize" them but scrape them off the culture plate/flask with a cell scraper.

51. Any supernatant will dilute your labeling buffer and may influence the unfolding of the proteins and the salt composition in your sample.

52. Instead of DIGE labeling buffer, cells may be disrupted by detergent-containing or low-salt buffer.

53. Samples should be put in an ice bath, either during homogenization or between different (short) homogenization cycles.

54. Tissues may also be homogenized in other solutions, especially for fractionation purposes (e.g., differential solubilization (18) and subcellular fractionation (19)). In addition, DIGE labeling buffer does not work equally well for all types of organ samples. For kidney, for instance, it is better to use Tris buffer with Triton X-100.

55. Optimize your assay to achieve high sensitivity, so that you will be able to use high sample dilutions. In any case, for the blank and the calibration curve use a similarly diluted sample buffer. This will reduce the interference of additives.

56. In-gel rehydration may harm sensitive proteins as they are exposed for hours to room temperature (during the swelling period) and to different pH regimes (in the IPG strip). In addition, not all of them may enter the gel only by passive re-swelling.

References

1. Westermeier R, Naven T, Höpker H-R (2008) Proteomics in Practice. A Guide to Successful Experimental Design, 2nd edn. Wiley-VCH Verlagsgesellschaft GmbH, Weinheim, Germany.

2. Uenlue M, Morgan ME, Minden JS (1997) Difference gel electrophoresis: a single gel method for detecting changes in protein extracts. *Electrophoresis* 18, 2071–2077.

3. Timms JF, Cramer R (2008) Difference gel electrophoresis. *Proteomics* 8, 4886–4897.

4. Miller I, Friedlein A, Tsangaris G, Maris A, Fountoulakis M, Gemeiner M (2004) The serum proteome of *Equus caballus*. *Proteomics* 4, 3227–3234.

5. Miller I, Teinfalt M, Leschnik M, Wait R, Gemeiner M (2004) Nonreducing two-dimensional gel electrophoresis for the detection of Bence Jones proteins in serum and urine. *Proteomics* 4, 257–260.

6. Kozlov AV, Duvigneau JC, Miller I, Nürnberger S, Gesslbauer B, Kungl A, Öhlinger W, Hartl RT, Gille L, Staniek K, Gregor W, Haindl S, Redl H (2009) Endotoxin causes functional endoplasmic reticulum failure, possibly mediated by mitochondria. *Biochim Biophys Acta - Mol Basis Dis* 1792, 521–530.

7. Miller I, Crawford J, Gianazza E (2006) Protein stains for proteomic applications: Which, when, why? *Proteomics* 6, 5385–5408.

8. Radwan M, Miller I, Grunert T, Marchetti M, Vogl C, O'Donoghue N, Dunn MJ, Kolbe T, Allmaier G, Gemeiner M, Müller M, Strobl B (2008) The impact of Tyrosine kinase 2 (Tyk2) on the proteome of murine macrophages and their response to lipopolysaccharide (LPS). *Proteomics* 8, 3469–3485.

9. Bradford MM (1976) A rapid and sensitive method for the quantitation of microgram quantities of protein utilizing the principle of protein-dye binding. *Anal Biochem* 72, 248–254.

10. Miller I, Eberini I, Gianazza E (2010) Other than IPG-DALT: two-dimensional electrophoresis variants. *Proteomics* 10, 586–610.

11. Gianazza E, Celentano F, Magenes S, Ettori C, Righetti PG (1989) Formulations for immobilized pH gradients including pH extremes. *Electrophoresis* 10, 806–808.
12. Hoving S, Voshol H, van Oostrum J (2000) Towards high performance two-dimensional gel electrophoresis using ultrazoom gels. *Electrophoresis* 21, 2617–2621.
13. Tastet C, Lescuyer P, Diemer H, Luche S, van Dorsselaer A, Rabilloud T (2003) A versatile electrophoresis system for the analysis of high- and low-molecular-weight proteins. *Electrophoresis* 24, 1787–1794.
14. Miller I, Radwan M, Strobl B, Müller M, Gemeiner M (2006) Contribution of cell culture additives to the two-dimensional protein patterns of mouse macrophages. *Electrophoresis* 27, 1626–1629.
15. Radwan M, Stiefvater R, Grunert T, Sharif O, Miller I, Marchetti-Deschmann M, Allmaier G, Gemeiner M, Knapp S, Kovarik P, Müller M, Strobl B (2010) Tyrosine kinase 2 controls IL-1beta production at the translational level. *J Immunol* 185, 3544–3553.
16. Miller I, Gemeiner M, Gesslbauer B, Kungl A, Piskernik C, Haindl S, Nürnberger S, Bahrami S, Redl H, Kozlov AV (2006) Proteome analysis of rat liver mitochondria reveals a possible compensatory response to endotoxic shock. *FEBS Lett* 580, 1257–1262.
17. Westermeier R (2004) Method 10: IEF in immobilized pH gradients. In: Westermeier R (ed) Electrophoresis in Practice. A Guide to Methods and Applications of DNA and Protein Separations, 4th edn. VCH Verlagsgesellschaft GmbH, Weinheim, Germany.
18. Molloy MP, Herbert BR, Walsh BJ, Tyler MI, Traini M, Sanchez J-C, Hochstrasser DF, Williams KL, Gooley AA (1998) Extraction of membrane proteins by differential solubilization for separation using two-dimensional gel electrophoresis. *Electrophoresis* 19, 837–844.
19. Pasquali C, Fialka I, Huber LA (1997) Preparative gel electrophoresis of membrane proteins. *Electrophoresis* 18, 2573–2581.

INDEX

A

Acrylamide 13, 22, 56, 60, 117, 121, 139, 140, 143, 146–148, 150–152, 158, 162, 165, 183, 184, 189, 191, 226, 235, 242, 247, 251, 255, 256, 266, 275, 276, 280, 284, 290, 344–350, 375, 376, 389
Adenocarcinoma 37, 253
Amino acid substitution 37
Ampholyte 9–10, 13, 14, 108, 116, 119–120, 213, 226, 242, 246, 247, 250, 255, 262, 263, 275, 382, 389
Analysis of variance (ANOVA) 32, 35, 39–42, 44, 45, 103, 104, 110, 163–164, 282, 312, 362

B

Bacteria
    anaerobic degradation 357–358, 360, 366–369
    carbohydrates 360, 362–365
    catabolism 360, 362, 363, 368
    environmental 355–370
    marine 355–370
    metabolic networks 367–369
    regulation 357, 363–364
Biological
    fluids 386
    replicates 10–12, 14, 31–32, 34, 35, 37, 38, 43, 100, 109, 213, 245, 300, 304, 360, 389
    variation analysis 35, 102, 120, 163, 202, 217, 230–231, 240, 263–264, 282, 362, 385
Biomarker
    discovery 176, 181, 195–219, 223–236
    research 199, 202, 203
Blue native PAGE 343–344, 351
Body fluids
    animal 386
    serum 385, 386
    tissue 92, 374, 386

C

Cancer proteomics 253–267
Carbamylation 84, 126, 251
Cardiovascular tissue 287–297
Cell lysis 4, 93, 97, 183, 189, 283
Cell surface proteins 319
Chlorophyll fluorescence 344, 345, 352
Combinatorial ligand libraries (CLL) 170, 182
Coomassie brilliant blue 122, 226, 229
Cup loading 104, 166, 176, 190, 213, 242, 246, 247, 386
Cyanine dyes (CyDyes)
    benzoxazolium Cy2 .... 68–71, 82
    ethyl Cy5 .... 68, 82
    fluorescence 4, 18, 248, 258–259, 359–360
    propyl Cy3 .... 68, 72–75, 82
    synthesis 67–84, 116, 126, 357, 365, 369, 370
    thiol-reactive 120, 123
Cysteine labeling 5, 93, 111, 113–127
Cytoplasmic fractions 37, 326–327

D

Data analysis 48, 50, 53, 108, 125, 140, 163, 170, 202
Decellularization 288, 292, 293
Depletion 60, 108, 182, 207–219, 224, 229, 230, 294, 320, 363, 365
Destreak™ 209, 242, 246, 250, 275, 278, 302, 305, 306, 317, 323
Detergent
    based fractionation 146, 152, 395
    nonionic 146, 152, 292
Difference/differential analysis 10, 27–29, 32–33, 37, 39, 47, 67, 79–81, 93, 99, 184, 240, 248, 265
Difference gel electrophoresis (DIGE)
    experimental design 34, 40, 43, 99, 102, 130–131, 184–185
    historical perspective 3
    labeling 156, 178, 183, 209, 213, 219, 374–375, 377–378, 385, 386, 395
    minimal labeling 12, 90, 92, 157, 172
    narrow pH range 7, 109, 224, 230, 231
    saturation labeling 6, 10, 14, 89–111, 253–267
    study design 170, 195–206, 254
    two-dimensional 114, 208, 240, 339, 359

DIGE. *See* Difference gel electrophoresis (DIGE)
Dimethylformamide (DMF) 13, 14, 17, 68, 75, 77, 82, 83, 93, 116, 133, 147, 152, 157, 159–160, 165, 172, 175, 177–179, 183, 189, 193, 209, 218, 241, 244, 245, 250, 255, 275, 277, 283, 302, 305, 314, 317, 322, 332, 336–337, 341, 374, 389, 391
Dye
    bias 17, 41, 45, 120, 387
    cyanine 6, 10, 115, 116, 175, 224, 359, 360
    fluorescent 167, 314
    maleimide 133, 136, 256
    swap 11, 14, 17, 38, 40, 165, 175, 179, 184–185, 378, 389

E

ECM. *See* Extracellular matrix (ECM)
EDA. *See* Extended data analysis (EDA)
Eigenvector 34
Electrophoresis
    denaturing 9, 49, 146, 164, 292, 293, 296, 352, 373, 394
    gel 146, 150–152, 158–159, 183
    native 346, 348–349
    one-dimensional 54, 146
    polyacrylamide gel electrophoresis (PAGE) 14–15, 255–256, 262–263, 308–309, 323–324, 327–329, 350–352
    SDS PAGE 183, 350–352
    two-dimensional 9, 155, 208–210, 213–216, 224, 256, 262–263, 323–324, 327–329, 373
Epithelial-mesenchymal transition (EMT) 269–284
Equalizer beads 182
Expression pattern 33, 44, 164, 204, 389
Extended data analysis (EDA) 29, 35, 104, 163, 202, 282
Extracellular matrix (ECM) 270, 287, 288, 291–294, 296

F

False color representation 384
False discovery rate (FDR) 103–104, 110, 197–206
Family-wise error rate (FWER) 197, 198, 203, 205–206

G

Gel
    electrophoresis 146, 150, 151, 158–159
    imaging 15, 20–30
    polymerization 290
    scanning 24, 140, 276, 281, 301, 303, 309–312
    -to-gel matching 241

Gel spot
    dehydration 48, 52
    destaining 48, 52, 59, 227, 231
    detection 10, 11, 15, 21–23, 27, 28, 79, 102, 163, 217, 229, 240, 263, 362, 387
    digestion 16, 48–49, 52, 58, 60, 63, 64, 118, 124, 257, 265, 302, 313–314
    picking 59, 312, 313
Glycosylation 7, 287, 296

H

HAP. *See* High abundance protein (HAP)
HCC. *See* Hepatocellular carcinoma (HCC)
Heart 7, 294
Hepatocellular carcinoma (HCC) 224, 225, 227, 228, 230–235
Hexapeptide library beads 169–180
Hierarchic clustering 33, 35, 36, 44, 104, 164, 369
High abundance protein (HAP) 173, 182, 208, 209, 211, 229, 230, 236, 240
Home-made IPGs 386, 392

I

Image
    acquisition 21–27, 35, 159, 162, 191–192, 194, 276, 281, 330–331, 359, 362
    analysis 6–7, 12, 15, 21–30, 91, 93, 100–104, 117, 120–122, 163–164, 172–174, 176, 184, 192, 217, 230–236, 248–249, 263–264, 281–282, 301, 303, 309–312, 362
    overlay 10, 22, 28, 98, 99, 102, 121–123, 151, 256, 259
    warping 28, 44, 102
Immobiline dry strip 117, 120, 158, 162, 184, 190, 225, 226, 242, 246–248, 250, 275, 278–279, 290, 294
Immobilized pH gradient (IPG) 9, 13, 92, 158, 183, 289
    buffer 116, 134, 141, 158, 166, 172, 175, 177, 183, 190, 209, 213, 226, 229, 275, 278, 289, 290, 302, 306, 317, 323, 328, 339
    home-made 256, 386, 392
    strip 13–15, 60, 68, 78, 121, 134, 138, 158, 160–162, 166, 172, 175, 176, 190, 194, 209, 210, 213, 214, 301, 302, 305–309, 312, 317, 323, 327–329, 337, 374, 377, 395
Immunoaffinity 182, 208
Immunodepletion 182, 229–230
Internal standard 10–12, 14–16, 29, 34, 35, 37–39, 41–43, 45, 67, 68, 78, 79, 152, 157, 160, 165, 174, 175, 179, 184, 185, 190, 193, 202, 204, 213, 228, 245, 247, 257, 259–261, 265, 271, 272, 277–279, 282–284, 304–306, 310, 314, 320, 331, 332, 338, 339, 360, 385, 387

Index

Isoelectric
  focusing (IEF) 4–5, 7, 9, 68, 78, 94, 97, 98, 104, 106, 107, 121, 134, 137, 146, 158, 172, 175–176, 190–191, 214, 229, 242, 246–247, 255, 257, 275, 288, 301, 302, 306–308, 323, 328
  point ($_pI$) 7, 9, 10, 64, 98, 106, 129, 146, 232–233, 250, 315–316, 373

L

Labeling
  amino acid 5, 14, 90, 109, 129, 130, 141, 170
  cysteine 5, 93, 113–127
  live cell 319–332
  lysine 5, 115
  optimization 6, 92, 100, 314
  post-translational modification 129–143, 224
  preferential 165, 378, 385
  total protein 54, 90, 92, 94, 95, 104, 108, 109, 133, 136, 137, 190
Laser capture microdissection 256
Ligand 169–180, 182, 207
Lung 253–267

M

Madin Darby canine kidney (MDCK) 273, 274, 276, 277, 279
Mammalian oocytes 90
Mass spectrometry
  (nano)ESI 302
  MALDI-TOF/TOF 50, 51, 53, 115, 118, 124, 125, 127, 230
  MS/MS (tandem) 48–50, 53, 55–58, 61–65, 115, 124–125, 225, 227, 232–235, 272, 282, 288, 297, 300, 302, 313–314, 318
Metabolic syndrome 169
MFA. See Multifluorescence analysis (MFA)
Microdissection 12, 90, 110, 254, 256–261, 263, 264, 266
Minimal dye labeling 10, 13, 68, 157, 165, 167, 183, 225, 241, 244–246, 250, 275, 294, 320, 322, 374
Mitochondria 151, 315
Multifluorescence analysis (MFA) 23–25, 27–29
Multiple Affinity Removal System (MARS) 208, 209, 211, 216, 219, 225, 227, 229, 230, 235
Multiplexing 10, 12, 67, 123
Multivariate
  ANOVA 32, 35, 39–42, 44, 45, 103, 104
  statistics 44, 68
  tests 32, 33, 44, 104

N

Native PAGE 343, 345, 351
N-ethylmaleimide (NEM) 115, 127, 133
Neuroproteomics 129

N-hydroxysuccinimide (NHS) ester 67, 70, 133, 155, 271
Normalization 11, 15, 29, 35, 36, 68, 91, 101, 163, 217, 234, 240, 271, 272, 282

O

Oocytes 90
Organelles 142, 146, 149, 152, 179, 240, 245, 248, 249, 287, 336, 389
Oxidation 53, 62, 63, 114, 115, 122, 123, 125, 127, 130, 132, 135–137, 250, 363, 365, 366, 368, 369, 386

P

PAGE. See Polyacrylamide gel electrophoresis (PAGE)
Palmitoylation 130, 133, 136
Pattern analysis 28, 35, 164, 178
PCA. See Principal component analysis (PCA)
Peptide extraction 49, 60, 61, 63–64, 124
Peptides 16, 37, 48–50, 52–65, 89, 90, 108, 111, 115, 118, 124, 125, 127, 133, 182, 207, 208, 219, 227, 234, 235, 265, 282, 300, 312, 315, 316
Phosphorylation 129–131, 135, 140, 141, 230
Photomultiplier tube (PMT) 23, 25–26, 54, 79, 122, 140, 176, 191, 192, 194, 236, 248, 263, 281, 284, 310, 311, 327, 330–331, 384, 394
Plant 12, 151, 335–341, 343–353, 356, 365
Plasma 115, 169–182, 193, 204, 223–236, 288, 292, 294, 385
Polyacrylamide gel electrophoresis (PAGE) 90, 146, 183, 255–256, 262
Pooled standard 228, 229, 235
Post-translational modification (PTM) 16, 50, 129–143, 217, 224, 265, 271, 287, 336
Power calculation 180
Principal component analysis (PCA) 33, 34, 36–45, 50, 104, 164, 282
Protein
  carbamylation 84, 126, 251
  cell surface 12, 319, 320
  chlorophyll-binding 344, 345, 351–352
  complexes 144–153, 343–353
  digestion 48, 49, 52, 217–218, 234, 385
  high abundance (HAP) 170, 171, 173, 177, 181–182, 208, 209, 211, 224, 229, 230, 236, 294
  identification 5, 15, 27, 37, 48–50, 54, 55, 65, 92, 104–106, 127, 196–198, 210–211, 259, 265, 271–273, 300, 315
  labeling 6, 78, 82, 93, 130, 133, 137, 157, 159–160, 165, 172, 175, 193, 244, 245, 250, 255, 259, 263, 275, 277–278, 335, 352, 394l

Protein (*Continued*)
  low abundance 171, 177, 208, 211, 230, 239, 319
  membrane 7, 90, 142, 152, 319, 322, 326, 343, 352
  quantification 8, 31–45
  visualization 30, 32, 34, 45, 48, 54, 113, 151–152, 249
Proteogenomics 358–359
Proteome
  heart 7, 294
  lung 253–267
  membrane 7, 90, 142, 152, 239–240, 292, 319, 320, 322, 326
  muscle 155–167
  plasma 169–180, 223–236
  reference 227, 259–261, 263
  serum 108, 169–194, 207–219
  signatures 359, 363, 369
Proteomics
  cancer 253–267
  cardiovascular 287–297
  clinical 170, 195, 201
  plasma 169–180, 223–225, 232, 288, 292
  quantitative 31–45, 90, 169, 171, 172
  redox 113–127, 356
  serum 108, 207–219, 387
ProteoMiner™ 171, 174, 177, 178, 181–194
PTM. *See* Post-translational modification (PTM)

Q

Quantification 21, 29, 126, 159, 190, 203, 310, 327, 343

R

Ramos cells 239–251
Randomization 35, 79, 100, 245, 304
Ras 273, 274, 279
Reductant 10, 15
Reference proteome 257, 259–261, 263, 265
Rehydratation 13, 14, 48, 60, 78, 104–105, 120, 134, 137, 142, 158, 160, 166, 172, 175, 177, 183, 186, 188, 190, 193, 209, 213, 218, 229, 242, 246, 247, 250, 275, 278–279, 283, 289, 294, 301, 302, 305–308, 317, 323, 327, 328, 337, 339, 381, 386, 392, 395
Replicates
  biological 10–12, 14, 31–32, 34, 35, 37, 38, 43, 100, 109, 213, 245, 300, 304, 360, 389
  technical 10, 31, 43, 63, 109, 175, 179, 320
Resolubilization 188, 326

S

Sample
  pooling 10, 11, 14, 17, 32, 43, 67–68, 184, 185, 190, 235, 377–378

pre-fractionation 182, 208, 271
preparation 14, 34, 36, 43, 51, 53, 62, 78, 92, 110, 169–180, 183, 209, 211–212, 225, 227–229
size calculation 198–201, 203
Saturation dye labeling 10, 92, 93, 104, 255, 258
Scanning 18, 22–26, 30, 59, 78–79, 107, 117, 122, 127, 140, 151–152, 173, 174, 179, 243, 248–249, 265, 377, 384–385, 390, 394
SDS PAGE. *See* Sodium dodecyl sulfate polyacrylamide gel electrophoresis (SDS PAGE)
Secretome 273–274, 276–279, 282, 283
Serum
  depletion 208, 216
  profiling 169
Signal-to-noise 31–45, 53, 62, 98–100, 127
Sodium dodecyl sulfate polyacrylamide gel electrophoresis (SDS PAGE) 9, 15, 16, 54, 68, 78–79, 82, 92, 97, 117, 120, 121, 125, 134, 138–140, 147–148, 150–151, 174, 175, 183, 184, 189–191, 247–248, 275–276, 279–281, 351, 373, 376, 377, 382, 384, 386, 388
Solubilization 93–94, 108–109, 147, 149, 152, 164, 212, 218, 240, 241, 244, 288, 292, 293, 296, 297, 323, 326, 327, 345, 350–353, 386, 395
Somatic mutation 339
Spot
  detection 15, 21–23, 27, 28, 79, 82, 100, 102, 163, 217, 229, 240, 263, 267, 362, 387
  picking 15–16, 48, 51, 59, 107, 117, 121, 123, 124, 301, 310, 313, 387
Staining
  Coomassie brilliant blue (CCB) 122, 123, 226, 229
  eosin 254, 257, 258, 266
  fluorescent dye 167, 314
  hematoxylin 254, 257, 258
  immunohistology 265
  silver 184, 312, 351, 374, 385, 390
Statistical power 43, 100, 199–201, 203, 312
Student's t-test 32, 44, 45, 80–81, 103, 104, 110, 163, 202, 204, 205, 230, 264
Sub-cellular
  components 240, 271
  organelles 142, 146, 179, 240, 245
  proteome 243
Sub-fractionation 287–297
Sub-proteome 239–251, 271, 288, 296, 297, 359, 368

T

Type 2 diabetes mellitus (T2DM) 299–318

U

Ubiquitination 130, 132–133, 136, 140, 141
Univariate
    statistics 32, 40, 44, 104
    test. 32

V

Validation 67–84, 115, 118, 125–126, 170, 196, 200, 202–203, 205, 206, 265, 282, 356

W

Warping 28, 44, 102